Automated Guided Vehicle Systems

IFS

International Trends in Manufacturing Technology

AUTOMATED GUIDED VEHICLE SYSTEMS

Edited by
Professor R. H. Hollier

IFS (Publications) Ltd, UK

Springer-Verlag
Berlin   Heidelberg   New York
London   Paris   Tokyo

R. H. Hollier
University of Manchester Institute
   of Science and Technology (UMIST)
PO Box 88
Manchester M60 1QD
England

British Library Cataloguing in Publication Data

Automated guided vehicle systems. –
(International trends in manufacturing technology)
   1. Automated guided vehicle systems
   I. Hollier, R. H.   II. Series
   629.04'9   TS191

ISBN 0-948507-27-6   IFS (Publications) Ltd
ISBN 3-540-17432-X   Springer-Verlag Berlin
ISBN 0-387-17432-X   Springer-Verlag New York

© 1987 **IFS (Publications) Ltd,** 35–39 High Street, Kempston,
                    Bedford MK42 7BT, UK
        Springer-Verlag Berlin Heidelberg New York
                    London Paris Tokyo

Phototypeset by Fleetlines Typesetters, Southend-on-Sea
Printed and bound by Short Run Press Ltd, Exeter

International Trends in Manufacturing Technology

The advent of microprocessor controls and robotics is rapidly changing the face of manufacturing throughout the world. Large and small companies alike are adopting these new methods to improve the efficiency of their operations. Researchers are constantly probing to provide even more advanced technologies suitable for application to manufacturing. In response to these advances IFS (Publications) Ltd is publishing a series of books on topics that highlight the developments taking place in manufacturing technology. The series aims to be informative and educational.

Books already published in the series are:

Robot Vision, Programmable Assembly, Robot Safety, Robotic Assembly, Flexible Manufacturing Systems, Robot Sensors, Education and Training in Robotics, Robot Grippers, Human Factors, and Simulation.

Forthcoming titles include:

Robotic Welding, Artificial Intelligence, Electronics Assembly and Just-in-Time Manufacturing.

The series is intended for manufacturing managers, production engineers and those working on research into advance manufacturing methods. Each book is published in hard cover and is edited by a specialist in the particular field.

This, the eleventh in the series – Automated Guided Vehicle Systems – is under the editorship of Professor R. H. Hollier of the University of Manchester Institute of Science and Technology. The series editors are: Michael Innes and Brian Rooks.

Finally, the Publisher's gratitude is expressed to the authors whose works appear in this book.

Acknowledgements

IFS (Publications) Ltd wishes to express its acknowledgement and appreciation to the following publishers/organisations for granting permission to use some of the papers reprinted within this book, and for their cooperation and assistance throughout the production process.

IFS (Conferences) Ltd
35–39 High Street
Kempston
Bedford MK42 7BT
England

American Institute of Industrial Engineers, Inc.
25 Technology Park
Atlanta
Norcross, GA 30092
USA

International Journal of Production Research
Taylor & Francis Ltd
4 John Street
London WC1N 2ET
England

Japan Industrial Robot Association
c/o Kikaishinko Bldg
3–5–8 Shiba Koen
Minato-ku
Tokyo
Japan

National Science Foundation
Washington, DC 20550
USA

Istel Ltd
Highfield House
Headless Cross Drive
Headless Cross
Redditch
Worcs B97 5EU
England

Contents

Preface

Interest in Automated Guided Vehicle Systems (AGVS) has increased rapidly in the past five years. The first International Conference on the subject took place at Stratford-upon-Avon in June 1981 and three subsequent events in this series have been held in Stuttgart in June 1983, Stockholm in October 1985 and Chicago in June 1986. All these conferences have attracted good papers, from suppliers, consultants and users, and large audiences eager to hear about the development of the technology and the way in which it is being applied in manufacture and distribution. The first book in English devoted entirely to the subject was published in 1983, and it now seems opportune to publish an edited set of papers drawn from the above conferences and other sources to provide an authoritative view of the present state of knowledge.

On the supply side, the number of companies now offering AGVs has expanded enormously and, although not all of these will probably survive in the longer term, the rate of growth in the market continues to be healthy worldwide. There is a distinct separation in the marketplace between AGVs used in relatively small numbers in various material handling functions and those used in larger numbers as assembly work platforms.

So far as industrial users are concerned, there is constant pressure to reduce operating costs to remain internationally competitive. Material handling costs are a highly significant element of these operating costs, whether it is in component manufacture, assembly or warehousing. Although an AGV system may appear to be an expensive solution compared to more conventional material handling methods, it does offer several advantages which are not always easy to quantify in terms of potential savings. Examples of these intangible savings are increased control over material movement, greater routing flexibility, less product damage, reduction in space needed, ability to operate in hazardous environments and interface with other automated equipment, etc. In assembly, an AGV system can provide opportunities for better work design creating enhanced job satisfaction and product quality.

Depending on the definition of an AGV which is adopted, vehicles which are rail-guided or guided remotely by a human operator are usually excluded. Most current systems use wires in the floor, but free-ranging capability is now becoming available which may be valuable in certain cases. Improvements have taken place in the design of the vehicle itself, in communications between the vehicle and the central controller and in the control software needed for the complete system. Development has resulted in an increase in the variety of basic vehicle designs now available in terms of size, type and features such as lifting capacity. Attention has also been paid to load transfer for interfacing to other automated equipment, to battery performance and to questions of safety.

AGVs are not new – their early use can be traced back to the late 1950s in the USA and a little later in the UK. However, it was the pioneering installation at the Kalmar assembly plant of Volvo in 1974 and later at Fiat that started to excite interest. Many of the early applications were in the automotive industry, especially in assembly operations, and it is for this reason that the greatest user experience belongs to those concerned with these early developments. It is they who have met and overcome many of the initial problems and seen the technology mature.

Enthusiasm for AGV systems, however, is not confined to the automotive industry. Distribution warehousing was also an early user of such systems. More recently they have been adopted in many flexible manufacturing systems for moving work-in-progress and tools in the mechanical engineering industries generally. Specialist versions of AGV have been designed for the electronics industry, for example, where handling is required in clean rooms. This type of application then leads on to possible use in the food and pharmaceutical industries where similar considerations apply.

Although not dealt with explicitly here, the possible application of AGVs in many non-industrial areas is not difficult to imagine. They could be used in offices for document handling, in postal operations, in hospitals for transporting supplies, in agriculture for seed sowing, etc. Some of these applications are already a reality and there is no doubt that further uses will be found in the future.

This book comprises 30 carefully selected papers, many of them modified and edited to bring them right up-to-date, divided into five chapters. Chapter One is partly introductory but also deals with the current European and American markets for AGVs and some of the main management issues involved in their use. Design of AGV systems is covered in Chapter Two, not only of the vehicles themselves, but their guidance and control as well as the overall system. The emphasis in Chapter Three is on the operational aspects of AGV systems, although most of the topics discusssed need to be addressed at the design stage. Of particular importance is the evaluation of the route network by means of simulation, so that capacity and control logic can be validated.

Chapter Four is concerned with typical applications in manufacture and warehousing. It is sometimes difficult to persuade users to write about their experience with new developments, and thus it is often the equipment supplier who presents the applications made by his customers, some of whom may be within his own organisation. This is partly the case here, but care has been taken to ensure that a balanced view is given. In manufacturing, the AGV has played a major role in achieving flexibility, both in component production and in assembly. Although it is not a particularly fast means of material movement compared to a conveyor system, it can provide great flexibility of routing through production processes and service those processes with tools and fixtures, perhaps allowing unmanned operation. In assembly it can be designed around the needs of the human operator to remove the pacing effects of the traditional line or conveyor system and thus improve job satisfaction. With regard to warehousing, movement to and from storage locations and interfacing with automatic storage and retrieval systems are the principal applications.

Future developments is the theme of Chapter Five. Mention has already been made of free-ranging AGVs and a paper in Chapter Four discusses mobile robots. These topics are reiterated again in this chapter, which contains some interesting glimpses of work being carried out in Japan. AGVs are in wide use in Japan and it is significant that the next International Conference will be held in Tokyo in October 1987. Beyond, the future for AGVs looks assured, since there are certainly many industrial and non-industrial application areas yet to be developed with the existing technology, besides new markets generated by technical advance.

R. H. Hollier
January 1987

1

Introduction to AGVS – Definition and History

The first two papers discuss the historical development of AGV systems, the various applications and the American and European market trends. Next, vehicles for horizontal movement are described together with the more recent use of the same guidance principles for automating narrow-aisle trucks. The steps involved in purchasing an AGV system are then outlined, stressing the importance of careful planning and good project management. This is followed by a review of the experience of Volvo in using AGVs in assembly work over the past twelve years. The chapter finishes with a paper on safety considerations.

EVOLUTIONARY AGVS – FROM CONCEPT TO PRESENT REALITY

G. C. Hammond
GMI Engineering and Management Institute, USA

Although the birth of automated guided vehicles occurred in the USA over 30 years ago, the technology did not take hold there until recently. However, during this same period, AGVs experienced considerable acceptance and growth in Europe. This paper discusses many of the factors that caused this. It also analyses the phenomenal growth of the AGV industry in North America over the past seven years and the reasons for it.

North American industries are looking for methods to reduce their operating expenses, one of the greatest of which is material handling. This applies whether it is a warehousing operation, a manufacturing or an assembly operation, or even delivering mail in an office, or food and supplies in a hospital. AGVs are offering these industries a viable solution because of their flexibility and ability to offer material tracking and inventory, support asynchronous assembly, and readily integrate with other automation such as robots, automatic storage and retrieval systems, or CNC machines. It is also one of the few material handling systems that will allow unmanned manufacturing to become a reality.

Historical development

The first automated guided vehicles (AGVs) were developed in the USA by Barrett Electronics in the early 1950s. The first system was installed in 1954 at Mercury Motor Freight in Columbia, South Carolina. It was a tugger system, following a wire guidance path and with a controller based on vacuum tube technology. Barrett had the market to itself until 1961 when the Webb Company delivered a tugger system to the Ogden Army Depot in Utah. These early systems were all tugger systems used in distribution (warehousing) applications.

Many factors hindered the early growth of the AGV industry. The controllers were bulky and had very limited capabilities. Also, in North America the labour unions saw them as a direct threat and even resorted to sabotage of the AGV systems. Finally, the suppliers tended to offer only standard product lines and were reluctant to customise vehicles or controllers to meet customer's specific needs.

During the 1960s and early 1970s the controllers were first transistorised and then later replaced with integrated circuit (IC) technology. This permitted more compact controllers with more computing power. But even with this improvement in the on-board electronics, it was still too expensive for the complex manufacturing material-handling problems experienced in industry. Thus, during this period, a few tugger and pallet truck systems were installed in distribution applications, but the industry's growth was slow.

AGV technology, however, caught on much faster in Europe. Factors that contributed to this included:

- European companies have standardised the size and construction of pallets.
- The European workforce does not view automation as much of a threat to their jobs as they do in the USA.
- European companies accept longer term payback periods whereas US companies are orientated to short-term results.
- Strict regulations with regard to safety and the workplace environment often justify automation.
- A stable workforce, as found in European plants, tends to prefer well-trained operators and maintenance personnel.

Another important factor in the European growth of AGVs was their application in manufacturing and assembly use. One of the first applications of this type was the Volvo assembly plant in Kalmar, Sweden. A new plant was built in which the team approach for final assembly was a basic requirement. This required a highly flexible means of transporting the assembly to the team as well as supplying materials to them. Through the cooperative effort of Volvo and Schindler-Digitron, a large low-profit AGV was designed to carry the auto body on itself. This was a dramatic design change from the standard tugger and pallet truck AGVs. At this point the AGV became part of the overall material control and assembly process. Many other European companies followed suit and AGVs have become a popular conveyance method in assembly type operations.

Several factors have contributed to a new interest in the use of AGVs in the USA and include:

- Foreign competition has become a serious matter.
- The recession of the late 1970s and early 1980s was particularly damaging to the US automotive industry, forcing it to look for more efficient manufacturing and assembly techniques.

- The labour unions have taken a new attitude toward automation and are more willing to accept it if it means more job security.
- The European vehicle and system developments are now being transferred to the USA.
- The development of vehicles with on-board computers that can interface with other automation and support material resource planning is helping to promote industry modernisation.

Major milestones can be identified. One of the first in the USA was the introduction of European technology for direct computer control of the AGVs. This system was installed in 1978 at the Keebler Distribution Center in Chicago. Another milestone was when John Deere in Waterloo, Iowa, linked an AGV system to an automatic storage and retrieval system in 1981 to provide complete tracking of material being transported between manufacturing departments. A major milestone occurred in 1984 when General Motors commissioned its first flexible assembly system. Since then GM has become the largest user of AGVs.

Application areas

Today in the USA there is growth in the three general areas of distribution, assembly, and manufacturing. The distribution applications typically involve the movement of material to and from the production process or within warehousing operations. The major advantages of applying AGVs in these applications is increased efficiency and the ability to support material resources planning systems. While the growth of the number of AGVs in these applications has been steady, the total numbers are not that impressive. Many of these systems have one or two or at the most only a few vehicles. In terms of total vehicle numbers, the shift is away from distribution and towards manufacturing and assembly.

Manufacturing applications are limited to material movement between manufacturing cells. At present there are only a few FMS operations using AGVs in the USA. Some of the problems experienced is that FMS operations are hard to justify in high-volume operations such as experienced in the US automotive industries. Although FMS applications are still few in number, considerable interest is being given to them, with large companies establishing pilot systems. As a result there are very few AGVs used in this application even though they are the ideal material handling devices for FMS.

By far the most impressive application in terms of growth and vehicle numbers is assembly operations. Here the workpiece is transported through the assembly/build process. The advantages are tremendous and include:

- Ability to eliminate inefficiencies related to line balancing.

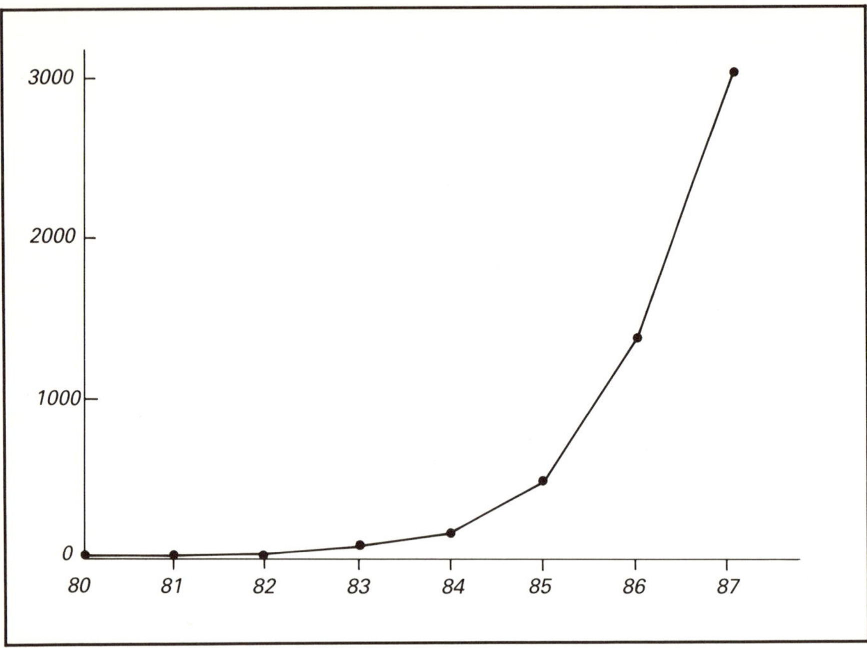

Fig. 1   Total number of AGVs in General Motors

- Asynchronous assembly increases equipment systems and flexibility.
- Improvement in product quality.
- Improvement in quality of work-life and ergonomics.

Automotive applications

Fig. 1 shows the total number of AGVs in General Motors in North America from 1980 through 1987. In 1980 there were 12 tugger vehicles, in 1986 there was a total 1407 tuggers, fork-type carriers and unit load carriers. For 1987 there are 1662 AGVs either purchased or planned to be purchased. This more than doubles the 1986 total AGV population.

Fig. 2 shows the growth of AGVs in assembly operations versus material handling operations. Of the 3069 vehicles in GM, 2739 are used for assembly while only 300 are used in material handling. Whereas over the past 10 years, the growth of AGVs in material handling has been steady, the growth of AGVs in assembly has gone from 0 in 1983 to 2739 vehicles projected for 1987. There are certain assembly applications where the number of AGVs purchased for just one facility far exceeds the total number of material handling AGVs in use. Orders of such magnitude have greatly changed the AGV industry.

Although the major advances in AGV technology were made in Europe in the 1960s and 1970s, in the 1980s the major advances are being made in the USA. This is as a result of industrial demands that include:

- More sophisticated computer controlled systems.
- Main line systems that are capable of handling 60 – 70 jobs/hour.
- Shorter index times in both manual and automated workstations (index times of approximately 10 seconds from the inbound waiting mode into the workstations).
- 24 hour (three shift) operations with in-line (opportunity) battery charging.
- Improved reliability of the vehicles and controllers.

Electronic applications

Other areas of AGV growth in North America include the electronic industry, heavy duty application, and in non-manufacturing operations. The electronics industry is extremely competitive requiring, wherever possible, methods to reduce production costs, work-in-process, and space requirements. Furthermore, the trend is towards moderate volume of many different models making conventional conveying methods difficult to implement. Thus, the electronics industry is turning more and more to using AGVs. They offer the flexible programmable transporting systems needed to meet these new industry demands.

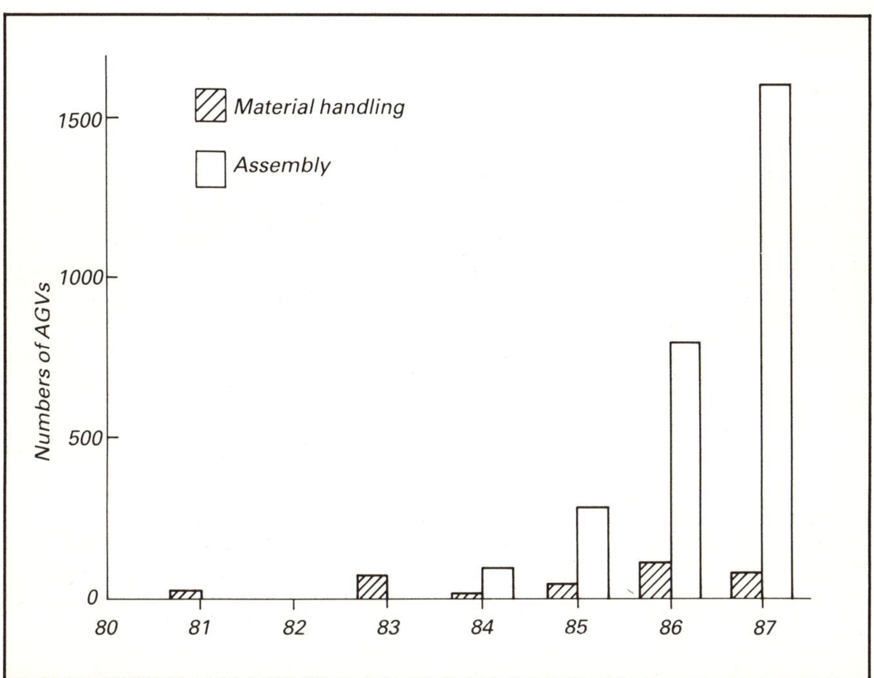

Fig. 2 AGVs purchased by GM (by year and application type)

In printed circuit board applications the AGVs can be readily programmed to route the boards through the insertion machines in any manner required. The benefits of using AGVs over other conveying systems include:

- Supports moderate and low-volume production.
- Readily integrated with other automation.
- Provides access space to machines.
- Provides floor-space savings.
- Is much simpler.

Some AGV suppliers are now concentrating on the electronics industries and, in particular, clean room applications. Vehicles are now available to meet 'Class 10' clean room operations and even offer on-board robots to perform tasks at the workstations. For these applications, inertial or optical guidance is offered rather than wire guidance techniques. Many suppliers are projecting significant growth in the use of AGVs in the electronics industries.

Heavy duty applications

Most carriers have maximum load capacities of 5000–10,000lb (2268–4536kg). However, there are some AGV suppliers who are specialising in producing AGVs with load capacities of up to 150,000lb (68,040kg). For press line applications, vehicles are being used to carry dies automatically in and out of the presses. This requires the use of a rolling bolster as part of the press and a mechanism on the AGV powerful enough to push and pull the dies either off or on.

In other heavy duty applications, AGVs are used to carry coil steel to blanking presses where it is sheered into rectangular blanks. The AGVs are then used to move the blanks to the press line or to the storage area for later use. In another application, a 25ft (7.6mm) tall six-axis robot, weighing 7.5tons (6800kg) was mounted on an AGV. The unit is used for painting aero frames. These heavy duty AGVs are expensive, but can often be justified for other than purely economic reasons, such as the elimination of dull, demanding, routine, and dangerous jobs.

Non-manufacturing applications

As in the manufacturing community, there likewise has been considerable interest in using AGVs in non-manufacturing applications. These applications range from carrying food, laundry, and other supplies in hospitals, to carrying mail, messages, and packages in offices. One company has even installed an AGV to carry customers through its various showrooms; the AGV also takes the customers through its factory area so that they can see how the product is manufactured. One of the benefits derived from doing this was a safer environment for the customers whilst in the manufacturing area.

Several hospitals have implemented AGV systems. Some are very simple systems with one or two vehicles carrying samples from the outpatient facilities to the central laboratories. Others are complex systems involving a multitude of vehicles transporting food, laundry, and medical and pharmaceutical supplies. These items are transported from a central material management centre to various floors of the hospital and to other hospitals within the complex. The AGV system even contains its own elevators which it controls to move the AGVs between floors.

The US Postal Service has installed AGV systems in several of its mail processing facilities. These facilities often operate 24 hours a day handling literally millions of items of mail. They range in size from 200,000ft^2 (18,000m^2) to 3.5 million ft^2 (325,000m^2). Tons of mail have to be moved hundreds of feet from inbound docks to the processing areas, between various processing areas, and then to outbound docks. Tugger systems capable of towing a multitude of mail carts and containers of various design are used. The guidepaths are complex in nature with lengths of many thousands of feet. The Pittsburgh General Mail Facility guidepath is 6500ft (2000m) long with 56 stops and served by 12 tuggers.

Many corporations and office buildings have incorporated AGVs to deliver mail, messages and packages to and from the central mail room. These same vehicles are used by employees to transport materials from one area to another. Some users have reported much better service, i.e. the vehicle making its rounds several times per day instead of one or two times per day as with clerks. This is accomplished at considerable labour savings, often producing relatively short pay-back periods.

Concluding remarks

Industry in North America is recognising the need for improved productivity and is committing itself to high-tech automation systems. Pressure to automate factories has made AGV systems the natural material-handling method for moving material to and from production areas and as carriers for assembly work-in-process, replacing synchronous assembly conveyor systems. The growing rate of the AGV industry is phenomenal with sales in the USA having tripled over the past five years.

AGVS IN EUROPE – CURRENT TECHNIQUES AND FUTURE TRENDS

T. Müller
Fachhochschule Hamburg, West Germany

In contrast to the previous paper which concentrated on North America, this paper briefly explains the development of AGVs in Europe and discusses the current techniques and future trends.

The development of AGV systems has made rapid progress in Europe, particularly since 1983. In 1985, approximately 450 plants with over 10,000 vehicles were produced and put into operation. Fig. 1 shows the distrubution of the market among the various manufacturers of AGVs with regard to the number of vehicles produced. The following is a brief survey of the key points of AGVs in the individual fields and various applications, concentrating on the European market.

57% of the AGVs market in Europe has found application in the automotive industry. The electronics industry is a long way behind at 9%, with mechanical engineering representing only 6%. In West Germany the automotive industry has an even higher share of the market at around 64%. With regard to the actual market opportunities, the electronics industry and mechanical engineering are strikingly under-represented (Figs. 2 and 3).

A glance at the main application areas indicates clearly that the field of FAS (flexible assembly systems) is predominant due to the particular significance of the automotive industry (Figs. 4 and 5). The quota falling upon other application areas is relatively low. In comparison with Europe, the situation in West Germany is even more distinctive, with FAS accounting for 60%. ASRS and FMS are practically without significance.

Due to the dynamic development witnessed during the past three years and the continuing high demand for automation of material flow, AGV systems will in future remain products with high growth rates.

The future of the automotive industry will see a particular increase

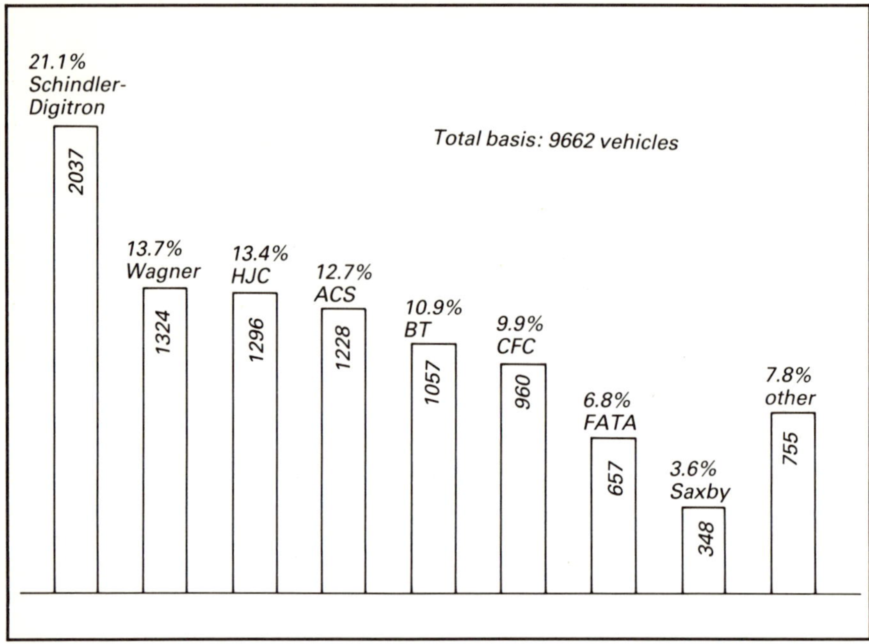

Fig. 1   Installation of AGVs by European producers up to 1985 (vehicles)

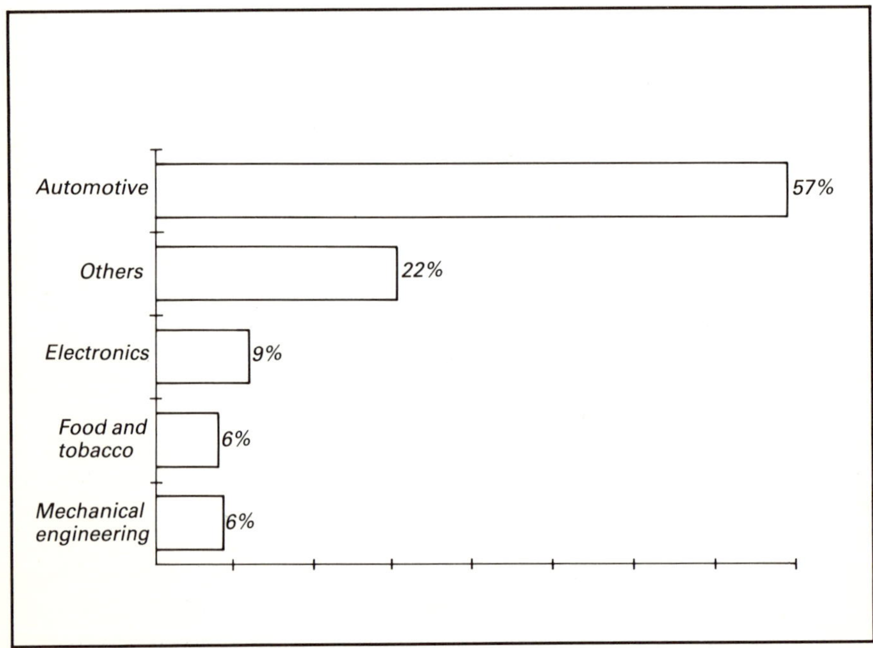

Fig. 2   AGVs (%) in European industries

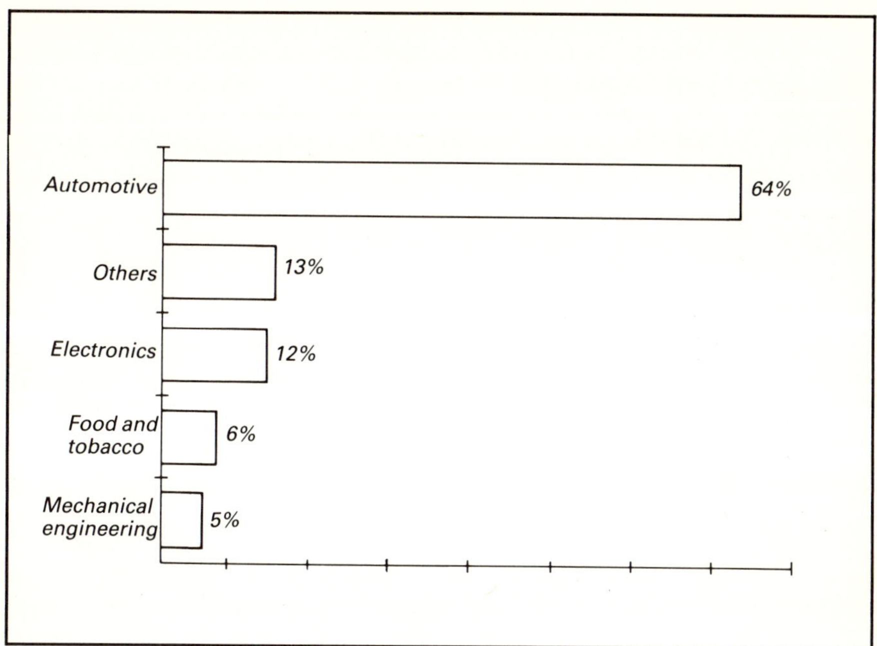

Fig. 3   AGVs (%) in West German industries

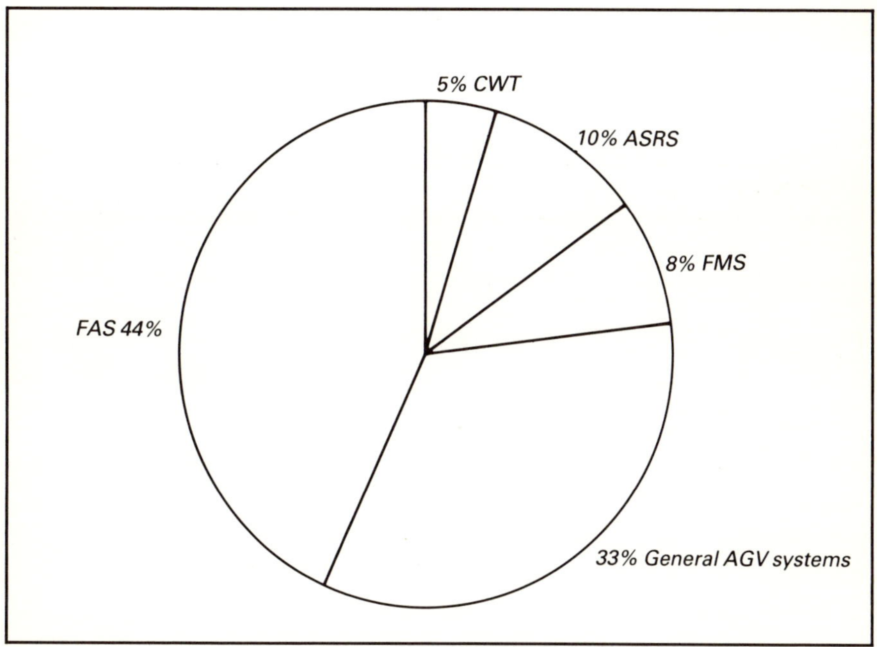

Fig. 4   AGVs (%) in different applications (Europe)

in such application areas as door assembly and lacquering, but the demand will continue in the classic areas; for example motor assembly, body assembly and welding stations.

It can be assumed that in other fields of industry too, the potential for AGVs is by no means exhausted. The experience gained in the automotive industry can be partially applied to the assembly systems of other industrial processes, mainly in the construction of other vehicles, but also in the electrical industry.

Improved and new techniques

The majority of attempts at improving present AGV system technology is aimed at partially or completely replacing inductive guidance. However, it can only be expedient to dispose of inductive guidance entirely if at the same time successful wireless communication of data is also achieved.

Considerable improvements can already be achieved by disposing of the guide wire in certain areas. If, for example, the vehicle is provided with the curve movement in programmed form, the complicated milling of curved parts will no longer be necessary, and switching points using the different frequencies principle can be dispensed with if the vehicle receives precise instructions, with the help of a reference point, as to whether it should follow the main route or the secondary route (Fig. 6).

In West Germany, the development of completely free-ranging vehicles continues to meet with difficulties. On the one hand, comparable systems such as inertial guidance cannot yet be applied economically, and on the other, wireless data communication is still posing problems. These problems basically consist of regulations imposed by the Federal Postal Administration, disturbances in the vicinity of the AGVs and, last but not least, insufficiently reliable data transfer techniques.

Aside from the gradually developing AGV technology, there are new aspects in the field of planning and application. One substantial area in which AGVs can contribute a furher step towards integration of material flow is the intersection between automated storage and retrieval systems (ASRS) and AGV systems. As a rule, conventional ASRS with increased throughput capacity utilise a continuous conveyor technique, as illustrated in Fig. 1 of the author's paper in Section Four of this book. The question which inevitably arises from this situation is: to what extent can one dispense with the continuous conveyor technique, by which the individual stacker crane P-D stations are directly accessed by the AGV? This results in four essential advantages:

• Considerable reduction of costs by disposing of the continuous conveyor technique.

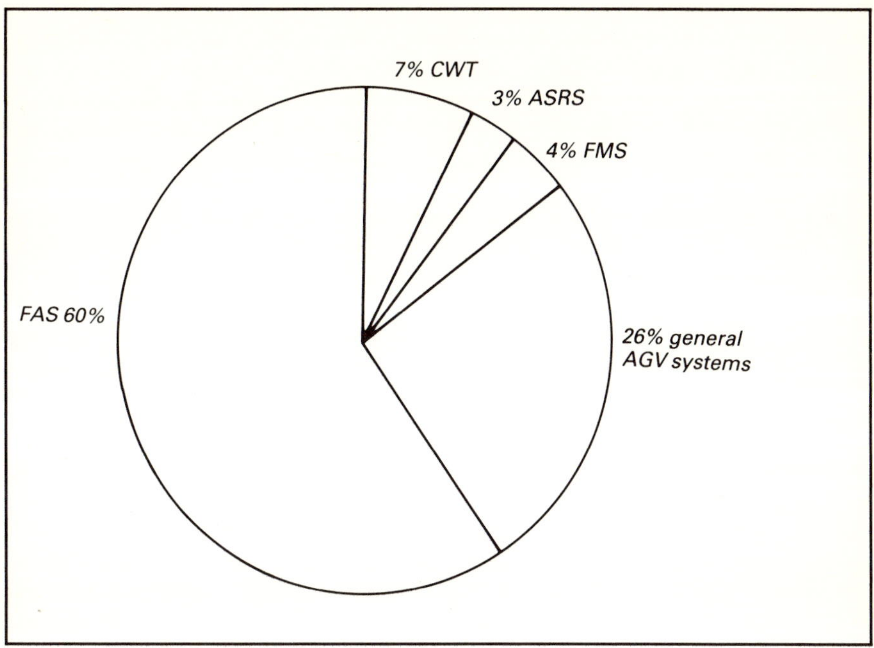

Fig. 5   AGVs (%) in different applications (West Germany)

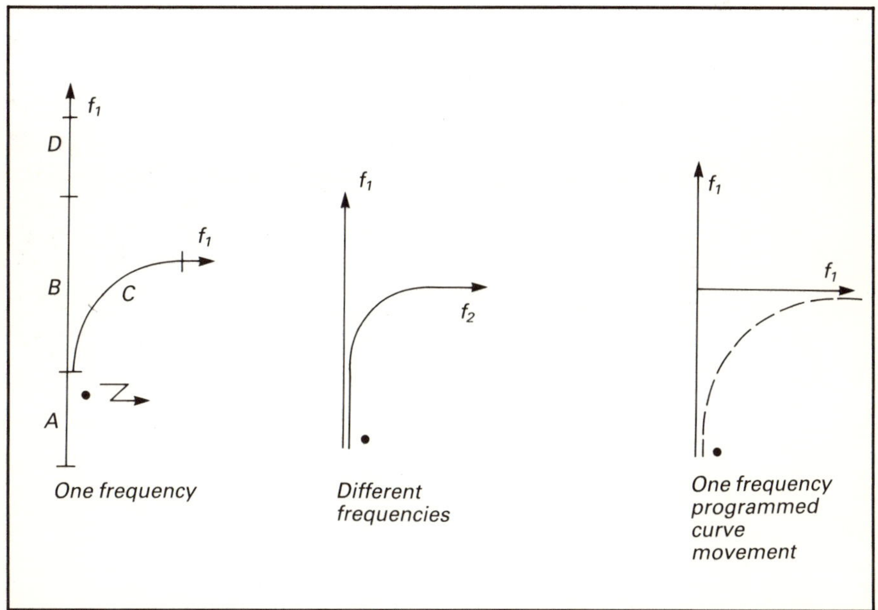

Fig. 6   Switch control

- Space can usually be saved.
- Higher availability by reducing or eliminating conventional conveyor techniques.
- Better access to ASRS or stacker cranes for maintenance or servicing.

The simplest solution to the problem of load transfer in the given situation is illustrated in Fig. 6 of the author's paper in Section Four. This solution can be applied equally to input as well as to output operation. In order to execute a double cycle, the vehicle must access an additional station. Maximum performance lies at around 15 double cycles (30 load units) per stacker crane per hour. A shorter cycle time is possible if the vehicle accesses a station which is equipped with a roller conveyor and lifting device as illustrated in Fig. 8 of the author's other paper, in which case the vehicle is also equipped with a roller conveyor and lifting device. A double cycle can then be carried out within one load transfer station, the maximum performance thus being increased to approx. 22 double cycles per stacker crane per hour.

Should the need for an intermediate run within the station be eliminated entirely as shown by the example in Fig. 11 of the author's other paper, maximum performance will be increased to 26 double cycles per stacker crane per hour. In this case the station would be equipped with chain conveyors and roller conveyors.

Fig. 7 (this paper) shows the incoming pallets area of an ASRS for the automotive industry. An AGV system replaces the conventional conveyor technique. A total of 2000 pallets is to be processed in two shifts, i.e. within 13 hours. The hourly rate would thus be somewhat in excess of 150 pallets. Given the layout illustrated and applying a

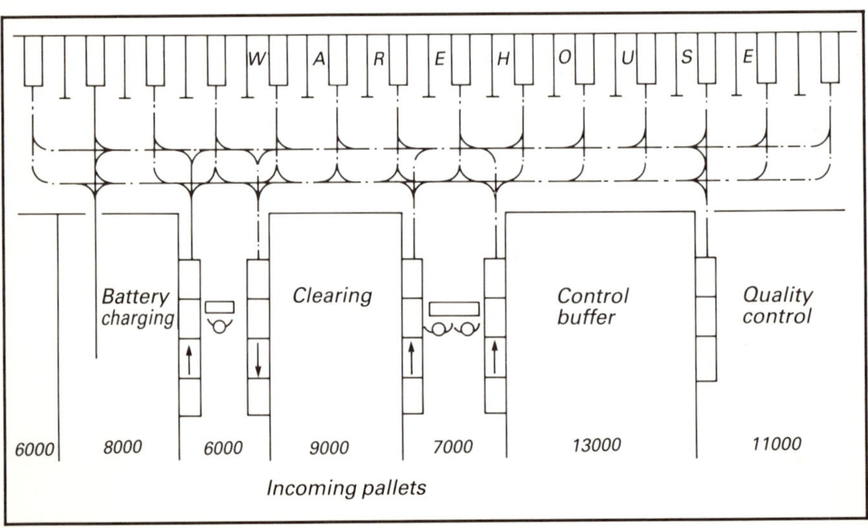

Fig. 7 *Direct intersection AGVS/ASRS*

special control strategy to the vehicles, it is possible at this relatively fast rate to make do with a total of eight vehicles. Every time a vehicle leaves one of the three identification stations upon transfer of the load, an empty vehicle is called out towards the identification station. Since the identification stations can accommodate no more than three vehicles, a further five vehicles are positioned at one of the unoccupied load transfer stations in the direct vicinity of a stacker crane. The result of this is that upon transfer of the load at the identification station, a vehicle will already be on its way from the waiting place to the identification station which is being vacated. The existence of two main travelling routes allows the vehicles to avoid each other, and one of the identification stations can be accessed at the frequency of the corresponsing empty runs from the ASRS to the identification station.

Future trends

The successful manufacturer of the future will be the one who can offer a high-performance control system in conjunction with a simple and reliable data transfer technique. This area produces the majority of the faults indicated by users of AGVs, whereby the purely mechanical components seldom give cause for complaint. Due to the increasing complexity of the systems, the interfaces within the computer hierarchy are of particular significance, so that in many cases turnkey responsibility is demanded.

In the foreseeable future, i.e. within the next five years, it can be expected that inductive guidance will gradually be replaced by other systems. The logical prerequisite for this development is that either a radio system or an infrared system be used for data communication.

With regard to the future market potential, the reader is referred to this paper's introduction. As for 1986, we can count on the production of a further 2000 installed vehicles, presumably to be produced by European AGV manufacturers.

AUTOMATED GUIDED VEHICLES

D. A. Clayton
Jungheinrich (GB) Ltd, UK

In recent years, with the increase in labour costs and the desire to cut
down workforces in order to make them, as the government tell us,
leaner and more efficient, more companies have been investigating
the desirability of automating the movement of goods within and
between factories and warehouses. This paper views this automation
in two distinct areas. Firstly, horizontal movement and, secondly, the
relatively new concept of automated narrow aisle stackers. With
automated guided vehicles the principal method of guidance is
inductive control from wire buried in the ground and for this reason,
a brief description of how this is achieved is given.

The optimum route for automated guided vehicles (AGVs) is selected
and a circular disc cutter is used to cut a groove approximately 2–3mm
wide and 15–20mm deep. A conventional plastic coated copper wire is
laid in the groove and grouted in. A point worth remembering is that
this is a relatively simple operation which causes minimal disruption
and can be caried out quite often in existing premises during normal
working hours. A high-frequency transmitter feeds the wire with a
10kHz alternating current which creates a magnetic field from which
the AGV's scanning head takes its route instructions. This scanning
head contains two antennae in which the magnetic field induces a
voltage. The extent of this voltage depends on the lateral distance
between the guide wire and the antennae. The AGV monitors these
voltages, interprets them and issues instructions to the steering motor
to control the direction of travel.

The trucks can be programmed through their own 'on board'
microprocessors which allow the preselection of routes and stopping
points. Instructions can be given direct to the AGV via its own ten
digit keyboard or remotely, either by manual input or computer
on-line instructions.

The AGV used to be controlled by installing patterns of two
magnets in the floor, either side of the wire. The AGV's scanning head
then 'read' the magnets and acted accordingly. The magnets were
buried in varying patterns which the AGV compared with the route

instructions in its microprocessor and reacted by slowing down, stopping, reversing, or any other function. Because there is a limit to the number of magnet patterns it is feasible to use, there was great difficulty in the larger installations. Programmable identification points are now used which have the advantage of needing only one unit, buried on one side of the wire only. Also, as their name implies they can be programmed for in excess of 4000 different signals which should cover even the most complex of installations.

Outside functions can also be triggered by the AGV such as automatic doors, traffic lights, lifts, etc.

Safety considerations are handled in two areas. Firstly, if more than one AGV is used there is a need to separate them to avoid accidents. This is done either by the introduction of the block system where a section of wire immediately behind an AGV has its current reduced, a condition which it recognises as a stop command, or by the location of ultrasonic detectors on the AGVs which give a 'no go' area of controllable distance. The other main safety consideration is to separate them from the people. Most installations of this type of equipment are in populated areas and the 'Health and Safety at Work Act' covers their use. The most obvious feature is the flexible safety guard at the front, this has to be of sufficient depth so that should it contact an object, the AGV will stop before its main body hits the object. The size or depth of this guard and the AGV's braking distance therefore, control the speed at which it is allowed to travel. Should the control wire or the ultrasonic detector for some reason go dead, it is essential that the truck immediately recognises this as an emergency stop situation and does not continue travelling in an uncontrolled condition.

In the service area the microprocessor fitted to Teletrak (Jungheinrich's registered trademark for its range of AGVs) systems can also be used to fault find on the AGV. The service engineer is able to interrogate the truck and identify the component to be repaired more efficiently than by using other types of test equipment giving considerable reduction in downtime.

Horizontal movers

This is by no means a new concept. There have been systems operating in this country since the early sixties with the early versions being mostly tow tractors pulling fairly simple trailers (Fig. 1). As experience grew, these trailers were equipped with conveyors which use the truck's own power pack so that they could be automatically loaded and this was the start of the fully automatic system.

A variation on the trailer theme which has proved fairly popular, is to use a simple hand pallet truck. These can be towed by a tractor in a train or 'conga' line and while they do have to be manually hooked-up and disconnected, they have two advantages. The first is a very low

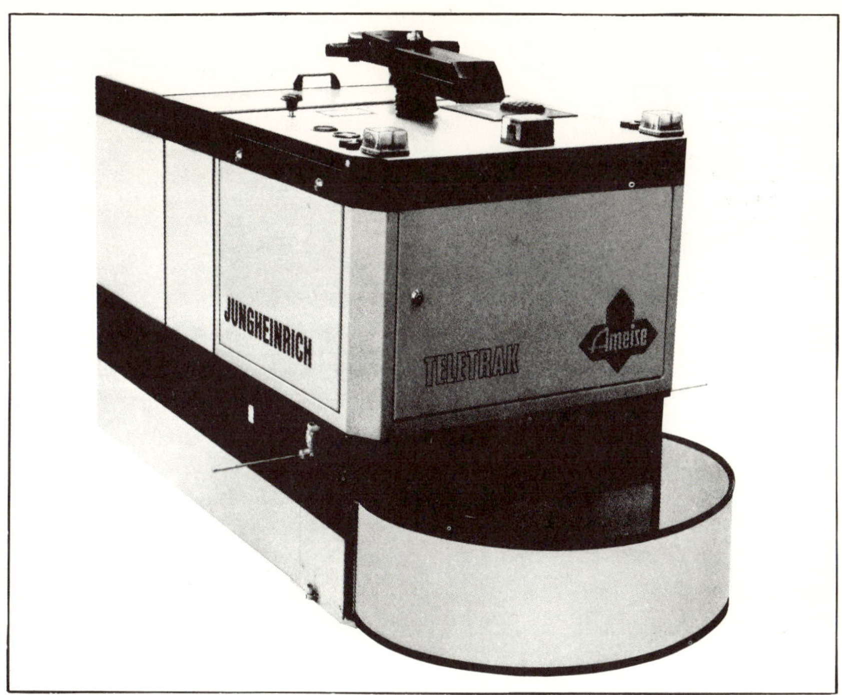

Fig. 1   Standard tractor unit

Fig. 2   Pallet truck type

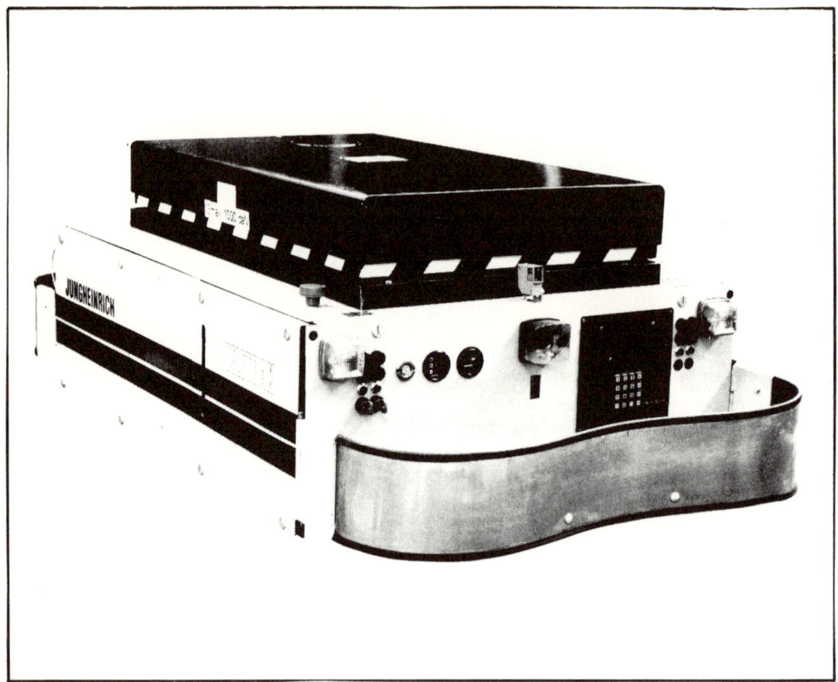

Fig. 3   *Unit load carrier*

unit cost, and the second that they retain local mobility within a delivery area.

In recent years, accurate methods of reversing the equipment automatically have been produced and an AGV similar to a conventional pedestrian electric power pallet truck has been introduced (Fig. 2). The truck reverses its forks under the load, lifts automatically and drives away with the load. Other developments have seen the arrival of trucks with a full mast which enables loads to be placed at height. Also the introduction of the unit load carier which transports goods 'piggy back' style (Fig. 3).

Use of this equipment falls naturally into three areas: firstly, with the production environment; secondly, transportation from production to the warehouse; and thirdly, within the warehouse (Fig. 4).

The production area

Various studies have shown that the AGV can be a major tool in assisting in production, either by its use as a work platform in which it replaces the conventional assembly conveyor or for transporting 'work-in-progress' to a series of specific points in the assembly cycle. Used as a work platform, we have all seen the impressive Fiat advertisements on television a few years ago, where the AGV had a car body assembled on its back and its travel speed was synchronised to the robot arms which did the welding and assembly.

An alternative approach has also been used by car companies in the assembly of car engines. At Volkswagen, pallets containing all the spare parts required for a single engine are automatically placed on the back of the AGV. Out in the assembly hall a worker or team of workers signal their request for a Jungheinrich Teletrak to come to them. When it arrives they can initially use the pallet as a work bench to carry out any tasks which are required on the components. Then using an overhead hoist they can take the cylinder block and place it on a special jig on the front of the Teletrak. This jig will rotate the block through 360° and raise or lower it by 300mm. By this means the workmen can place the engine in the position most suitable for each operation. They, of course, have a further advantage over a conventional conveyor in that they can walk around the unit without interruption. This process brings a number of benefits regarding production quantity flexibility in that if you want to increase production, it can prove difficult to increase a conveyor system by 25%, for example.

The introduction of 25% more Teletrak units presents no problems. Job satisfaction can be improved, because not only is the work always offered at the ergonomically correct height and position, but also work can be easily varied with, for example, one man building a whole engine, or a team doing the job, or a more conventional flow-line.

As mentioned previously, the other main task in the production area is the transport of work-in-progress from workstation to workstation.

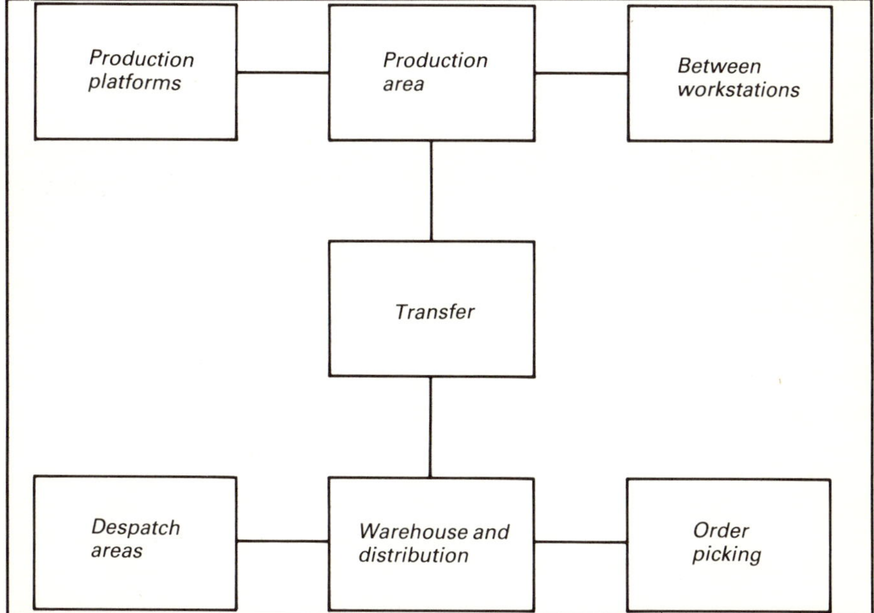

Fig. 4   The three categories of equipment use

More and more companies are now using computers to control their factory production; this brings many benefits and lends itself to the use of the AGV system. One of the major benefits of the AGV system is its controlled work pace and constant predictability which, when installed in a well organised factory, means that buffer stocks, which are so expensive, can in some cases be eliminated.

Transportation

The transportation of goods from production to the warehouse is, perhaps, the most simple and commonly used operation and the benefits are self-evident.

AGVs in the warehouse

The AGV is used as a simple pallet mover in warehouses, collecting pallets from the 'P' and 'D' stations in the racked area and distributing them to the despatch areas.

Another effective idea is on ground-level order-picking where an operator rides on an extended version of the AGV which carries a number of combi-tainers or pallets. The operator, freed of the need to guide the vehicle, can concentrate on his job of picking items.

Benefits of the system

Apart from the labour savings, there are a number of important benefits to be gained from the use of this equipment. It features smooth acceleration, it follows a predictable path, it travels by the shortest route, it does not stop for unnatural breaks (visits to the toilet, smoking, etc.) and it brakes smoothly. Its benefits, therefore, are longer component life than a conventional fork-truck which is, of course, not so thoughtfully driven.

In fact, some of the systems installed in the early sixties are still giving good service when fork-trucks installed at the same time have long since been replaced. It does not make unnecessary journeys, so battery sizes can be smaller and charging costs are lower. If it travels through an unpopulated area it does not require heating or lighting. However, perhaps the most fruitful area for cost-savings is in the area of spillage. No factory or warehouse makes accurate records of spillage costs. Remember, it is not only the cost of the damaged goods, but also the time taken to tidy up the mess and the disruption caused by unnecessary spectators. There is also the unreported spillage, where a fork-truck driver goes too quickly around a corner or brakes too sharply and his load shifts on the pallet. The driver then spends valuable time pushing the load into shape.

It is clear, therefore, that we are not looking at a straight machine for a man situation.

Environmental considerations

There are few limitations as to the conditions in which the AGV can work. If a fork-lift truck can do the job, an AGV should also be able to manage the task. Floor conditions are important, in that the equipment uses small hard wheels and for this reason the floor should be smooth. It should also preferably be concrete. Tar, for example, causes problems due to the fact that the truck travels over exactly the same path every time and will wear a groove in a relatively soft surface. Steel particles in the surface of a concrete floor will not cause problems, but steel sheets or mesh on or near the surface can cause problems because they dissipate the signal from the guidance wire.

It is possible to use AGVs outside. In fact, Jungheinrich's first installation in the UK in 1967, was almost totally out of doors.

Viability and costs

It is not possible to put a generalised cost for this type of equipment on paper. However, a simple floor installation and one AGV with automatic pick-up and put-down could be installed for approximately £35,000. Nearly all installations, however, are highly specialised to suit an individual company's criteria and so cost will rise from this point.

If we look at the costs another way, in other words on pay-back period, it is perhaps easier to generalise. In terms of pay-back it is, of course, a fact that the more shifts the company works, the shorter the pay-back period. If we consider only labour savings and the replacement value of the fork-lift truck, the average system can be paid back in 1–1½ years on a triple-shift system. On a double-shift system, the period would be 2–3 years. On a single-shift system it would need to be a fairly simple system to achieve the three year pay-back, which is the general recovery period required by industry.

The future

Growth in this sector of the material handling market has, over the years, been a slow business. However, in recent years there has been a rapid increase in interest shown by companies. There are a number of reasons for this, chief of which is of course labour costs, but also new truck designs enabling AGV manufacturers to be more versatile and also new suppliers arriving in the UK market creating in their wake fresh interest. The next few years will see an explosive growth in the field of automated guided vehicles.

Automated narrow aisle trucks

The technology which has been refined in AGVs can, of course, also be applied to more conventional fork-lift trucks. The Japanese have installations in operation using both reach trucks and counter-balance trucks. To the author's knowledge, however, no installations have been carried out in the UK for this type of equipment. The problem of finding a suitable application, probably block stacking, has so far eluded the would-be suppliers. European manufacturers, in particular Jungheinrich, have felt that there is much potential in developing the narrow aisle stackers through the AGV principle.

Fig. 5   Narrow aisle stacker

Narrow aisle trucks (Fig. 5) have been with us for some years and have, during this period, become more and more automatic up to the point where several companies now have trucks designed which are fully automatic, i.e. the driver has been removed. For those unsure of the term 'narrow aisle trucks', this means the type of machine which operates in approximately 137cm (54in.) aisles and puts its load out sideways into the racking, not having to turn through 90° to achieve this. Reach and pedestrain operated trucks are not designed for this function.

The first method of automating this type of equipment was the introduction of automatic height selection which enabled the driver to raise the forks to the correct height at the single press of a button. This was soon followed by automatic stack and de-stack, which meant, once again, at the press of a button the truck would automatically deposit or pick up the pallet from the racking with a measured, controlled movement. Initially, these trucks were controlled within their very tight aisles by guide rails, but in recent years the technology used on wire-guided AGVs was introduced on narrow aisle trucks. This, in some areas, reduced installation costs but also brought other benefits, which are not to be covered here in depth, but include improvements to hygiene, accessibility, layout, spillage and increased truck life.

A development which followed the introduction of wire guidance was a system which Jungheinrich calls 'Hot Spot'. This was a method by which the truck can be fed with a pallet location number in the form of a six or nine digit code, and the truck will then prevent the operator from entering the wrong aisle or placing the pallet in the wrong location. Also, with the introduction of data transfer by either radio or buried wire, Jungheinrich was one step from making the driver redundant. It became a relatively simple matter, therefore, to install an on-board microprocessor and mechanical actuators to pull the levers and push the buttons. In this way, the fully automated narrow aisle truck became a logical progression and not the leap in the dark which it may appear at first sight.

Automated narrow aisle trucks (ANATs) are suitable for heights of up to approximately 9 metres. It is envisaged that above that height there would be too many problems with tolerances and a stacker crane would be a more suitable alternative.

Operation

It is envisaged that the equipment will operate in a pedestrain-free area, due mainly to the fact that owing to design of this equipment, there are problems in fitting safety guards like on the AGVs. It is not hard to design the areas as being pedestrian-free which, of course, means that savings could be made on lighting and heating. Jungheinrich have also designed the equipment to operate with a 'buzz bar' supplying the electrical power. This bar runs along the racking where the stackers will draw electricity during travel down the aisles. A small battery is fitted to enable the truck to transfer quickly from aisle to aisle and this battery is then recharged while the truck operates in the aisle. By this method a multi-shift operation can be maintained without resorting to battery change or indeed needing to recharge the batteries between shifts.

Initially, ANATs were only designed for Euro pallets and used telescopic forks. Nowadays however, they are produced with the more conventional swivelling side shift and can therefore pick up perimeter base pallets comfortably. There are three feasible methods of control; that is, passing information from the central warehouse computer to the truck:

- Radio.
- Inductive loop buried in the floor at decision points.
- Inductive wire buried alongside the guidance wire.

The radio method will work in theory, but as there are too many places where it would not be suitable; this method has not been fully developed. The second method does work in a number of installations, but if the truck has problems away from the decision point it cannot call for help and has to wait until discovered. The third method is the preferred method because the inductive wire follows the path of the guidance wire and gives 'on-line' performance.

The automated narrow aisle stacker has three main advantages over the automatic stacker crane:

- The trucks are free-standing and there is no need for overhead structural work.
- Transfer from aisle to aisle is as fast as a normal truck and there are no costly, slow transfer cars.
- Installation is very simple and for this reason alteration of an existing warehouse is a viable reality.

What is the future?

In the UK, it is felt that while a few people will opt for the fully automated package, most companies will take the choice to progress by stages. They will start with a wire-guided narrow aisle stacker and perhaps add a data transfer package at a later date. Then when they are ready the final automation package can be installed.

Concluding remarks

Automated guided vehicles have been with us for many years. Their applications are many and varied, and they are being adapted to new uses all the time. They have the ability to cut costs dramatically and give us more control over our work.

ECONOMIC AND MANAGEMENT CONSIDERATIONS FOR AGV SYSTEM PURCHASE

J. Burton
Ingersoll Engineers Inc., UK

A decision to invest in an automated guided vehicle (AGV) system has to take into account many factors regarding operation, product mix, and current and future business objectives. AGV systems are, by and large, based on existing hardware and systems technology. However, they also involve a complete range of other interfaces and activities, from parts scheduling and computer interfacing to special floor specifications, all of which require time, resourcing and costing. How the end user organises and plans the project has a direct bearing upon the ease with which these systems are implemented. It is demonstrated how careful planning, clear specifications and meticulous project organisation are crucial to the successful implementation of AGV systems.

Automated guided vehicle (AGV) systems are proving capable of providing solutions to the many varied and complex materials handling problems that industries have been confronted with for many years. The suitability of AGV systems for computer integration and their ability to extend and change routings make them the ideal solution for progressive integrated manufacturing and development of JIT operation.

The competence of AGV systems to meet a variety of demands is reflected in their acceptability across a diverse range of industries, for example:

- Aerospace.
- Automotive.
- Discrete part manufacture.
- Engineering assembly.
- Food.
- Paper.
- Pharmaceuticals.

- Photographic.
- Sanitary ware.
- Textiles.

For the prospective purchaser the choice of systems currently available is large and is growing all the time. All major suppliers have successful systems in operation and the purchaser must assess which supplier's equipment is best suited to the proposed application. This can be a difficult task; like in selecting washing machines, the purchaser is confronted with a choice of systems any of which could fit the need.

However, unlike washing machines and unlike more conventional automated handling systems, the AGV system has potential beyond the role of transporting parts. It can make and act on decisions, participate in part scheduling control activities and – in short – become the most critical link in a flexible computer-integrated manufacturing system.

The AGV system is not just another method of automating material handling. Its procurement should take into consideration the fundamental role the chosen system will have within the overall integrated manufacturing and systems structure.

Objective

The objective of this paper is to offer guidance to the people responsible for planning and progressing the purchase and implementaion of AGV systems from the point when the purchaser decides an AGV system fits his needs to the installation and commissioning. It does not include advice on choice of type or make of AGV as this depends on factors outside the scope of this paper. The intention is rather to offer some guidelines on the steps to be taken prior to purchasing a system to avoid major pitfalls during the system implementation and to assist in achieving completion on time and within budget to the satisfaction of the purchaser and supplier.

The guidelines are based on Ingersoll Engineers' experiences acting with and on behalf of end user clients through various stages of contract development and implementation.

How do you know what you want?

Once the decision to invest in an AGV system has been made the next move is usually to approach suppliers and ask them for advice on their product and its suitability to the purchaser's application. In addition, seminars, books, journals and papers are all useful sources of information to provide the purchaser with an understanding of the market availability and potential. It is important and necessary for the potential purchaser to familiarise himself with all these aspects in the early stages of the project.

This, however, does not eliminate the need for the purchaser to plan for himself how he wants his system to operate and to identify its basic functions and operating parameters. It is unreasonable to expect a supplier to provide a system tailored to the purchaser's needs if the purchaser himself is unsure or confused about what those needs are.

Step one – Identify long-term needs

For any company considering the implementation of an AGV system the first step should be to plan the long-term requirements of the manufacturing facility in which the AGV is to operate. This will clearly establish the overall stages of facility development.

The AGV system's potential for flexibility and phased development for both physical and system expansion makes it shortsighted in both planning and cost terms to think only of the immediate operating requirements of the facility. Long-term plans should be laid, with phased timing corresponding to the introduction and integration of both soft and hard elements of the total facility.

We call this plan for integration the 'three-legged stool' approach because it considers the three basic elements of manufacture:

Systems.
Technology.
People.

Short, medium and long-term plans should be developed, taking into account the role of the AGV system within each phase (Fig.1).

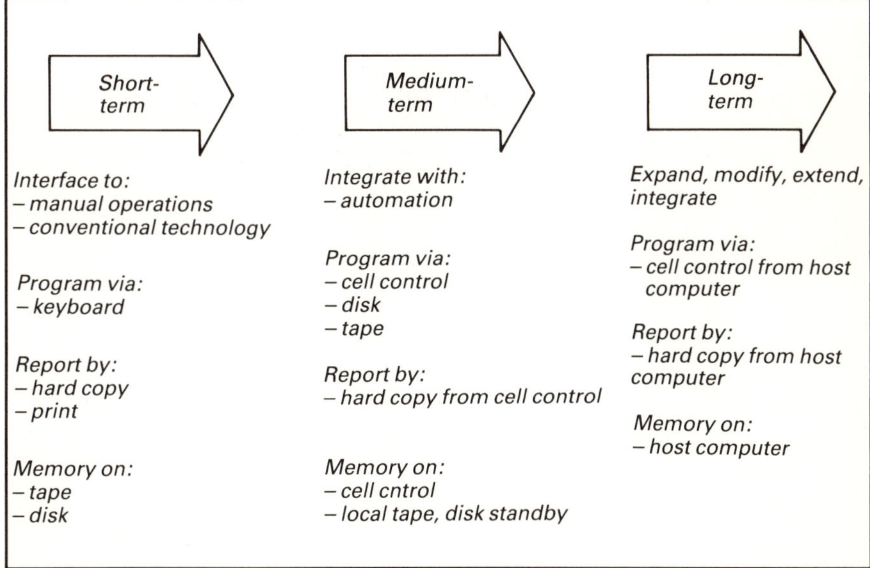

Fig. 1   Phased plans for implementing an AGV system

The level of detail should reflect the known and anticipated needs and be appropriate to the stage of development, such as:

- Anticipated changes from manned to automatic operation:
  - Machine loading.
  - Tool management.
  - Integration with automation requirements.

- Physical constraints to future expansion:
  - Space.
  - Obstacles.

- Development of CAD/CAM systems.

- Future changes in basic plant operating procedures:
  - JIT.
  - Cell technology.

- Development or changes of systems hierarchy.

- Future product models, market changes and fluctuations.

It is also important to identify the long-term needs of the people who will operate the system and make plans to meet them. People may be sceptical about the benefits of AGVs or concerned about safety aspects. Careful motivation and thorough training will help them to recognise the advantages of the new system. Results are generally better when people accept that change is necessary than when it is forced upon them.

Step two – Establish budgets and timing

The next stage of planning is to establish the budget and timing for the Phase 1 system.

The purchaser's approach to establishing cost and timing should be to define as early as possibble the scope of work and the specific responsibilities of both parties. Agreement on these points before order placement will minimise misunderstanding later. Furthermore, by defining the work content for himself the purchaser can seek to avoid costly and disruptive post-contract modifications.

Additional requirements

As we have seen with the phased planning, many elements outside the immediate provision of the AGV system supplier's scope of work will nevertheless need to be costed and will fundamentally influence the timing and effectiveness of the AGV system implementation and operation.

Underestimating the content and cost of these other elements is not uncommon. One reason may be the simplistic nature of the AGV systems in themselves. If the application is one which is suitable for a relatively standard AGV system, then the degree and implications of

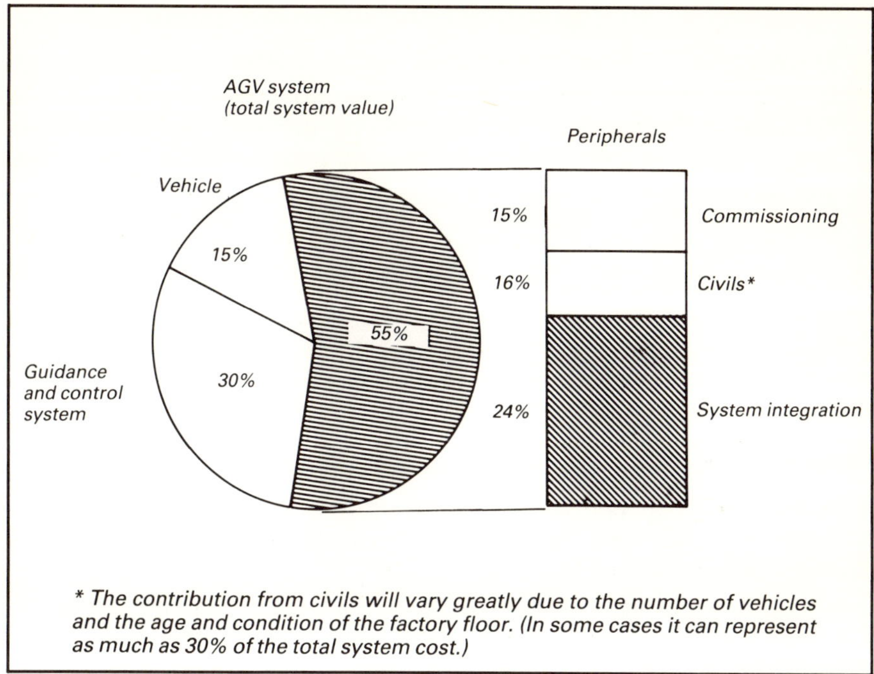

Fig. 2   Estimated percentage breakdown of typical AGV system costs

'other' elements of the cost and timing are minimised. However, with the move towards integrated manufacturing the number of applications compatible with 'standard' AGV systems is likely to decrease.

In our experience, the most commonly underestimated elements, both in time and cost, are:

- Integration – mechanical and systems.
- Commissioning and debugging.
- Special software design.
- Floor preparations.

An AGV system consists of three main elements:

- Vehicle.
- Guidance system.
- Control system.

These can roughly be proportioned in terms of cost as shown in Fig. 2, with allowance for variations according to application.

Peripheral costs

Fig. 2 shows that peripherals represent a large proportion of the AGV system's cost. A substantial proportion of this software cost will apply to special software which is often written to meet not only the special requirements of the AGV system but also to satisfy the communications and integration requirements of other systems' interfaces.

In these special systems areas the end user should consider:

- In what phase of development is the special software required?
- If the requirement comes in the next phase of development the initial system needs to be capable of updating with minimum modification, but it will not necessarily include the full operating features.
- Is the feature *really* necessary?

Simplifying the requirements down to *real* needs and phasing the development will significantly limit cost and aid smooth implementation.

Step three – Plan systems and mechanical integration

Systems and mechanical integration is a continuing activity from the concept of the facility to completion and should be monitored as such. The AGV system tends to 'sit in the middle' of the production and systems hierarchy and therefore it will integrate and interface in a number of directions.

The end user should anticipate, specify and contractually define the following:

- The scope of responsibility for providing and making the interfaces.
- The attendance of all parties on site during the installation and facility commissioning stages.

Responsibility for liaison between the parties to ensure that the integration works should rest with each supplier, but the end user will need to manage and coordinate the liaison to ensure that his overall needs are being met.

In relation to AGV system integration the end user should particularly ensure that:

- Location tolerance requirements and stopping occurrences between the AGV system and process plant are compatible and that guarding and physical interlocks are taken into account.
- The required level of autonomy is maintained between each element of the system and that 'stand-alone' operation can be resumed in the event of system or process element failure.

Step four – Assess the floor area

The floor upon which the AGVs are to travel may look flat enough, but the chances are that it is not. AGV suppliers will specify the gradients, flatness and other floor conditions necessary for ideal operation. They will also indicate the proximity tolerances of sub-floor steel and appropriate floor coating resins.

The potential work (and cost) involved in complying with these conditions should not be underestimated by the end user. Depending on the degree of work required, the cost of preparing the floor can represent up to 30% of the AGV system cost.

The end user should invite the supplier to inspect and survey the proposed area to confirm suitability or otherwise at an early stage. If rework is required, then the AGV supplier should be asked to re-inspect, survey and approve the work when it is completed.

How to prove that the system will work

There are two techniques which an end user can employ in order to prove that the system will work and that its size is correct. These are:

- Computerised model simulation.
- Functional specification.

Simulation of the proposed system via a computer model is useful to test the operation of the system against all the criteria related to such points as downtime, operation time, battery charging, effects of rework, queuing, etc. The simulation should be carried out prior to order placement and can be conducted by the potential supplier or by an independent organisation.

In the same way, a functional specification outlining the operation and requirements of each element in the system should be written. This is not a technical specification detailing types of equipment but a specification against function which outlines the system's configuration and the actions to be taken against set operating parameters.

The functional specification is the document which most closely reflects the end user's conception of and requirements for the AGV system. It will form part of the tender document against which the supplier(s) will specify the equipment provision and prepare quotations, and against which the contract will be placed.

Concluding remarks

By applying these guidelines, many of the difficulties often experienced during the introduction of high-technology AGV systems can be minimised. Each new application will have its own problems, but the ability of AGV systems to adapt to varying degrees of computerised sophistication and manufacturing flexibility gives them such potential for all levels of integration that they deserve to be considered as more than just 'another type of automated transport system'.

SOCIO-TECHNICAL CONSIDERATIONS FOR AGV SYSTEM IMPLEMENTATION

I. Persson
ACS Autocarrier System AB, Sweden

The concept of productivity can be looked upon from different points
of view. Within AB Volvo, a total approach has been made in order
to achieve an increased productivity, mainly within the final assembly
area. Due to the use of a new technique in internal handling, the
automated guided vehicle system, a new approach to the design of
assembly work has been made possible. The results after the system
had been running for some years, have been very favourable.

No single manufacturing operation which has come into effect over the
last few years, has drawn as much attention or resulted in such an
immense number of articles as the Volvo Kalmar Plant. Many people
have tried to draw conclusions from this operation in order to find
solutions as to which future steps are to be taken in other types of
industries and in other countries. The Volvo Kalmar Plant has been
considered a model for a new way of thinking, when it comes to what is
called the quality of working life.

However, Volvo recognises that its model is not the only model.
The solutions are different in different contexts, but by combining
inputs from different fields of interests, experience and knowledge, the
system continues to develop.

Over the last 10 to 20 years the trend in mass production industry
has been to develop more sophisticated engineering methods and
highly advanced hardware technology. The degree of mechanisation
has increased as a consequence, and man has become a 'cog in a big
machine'. A highly mechanised assembly line allows a minimum of
personal initiatives and thought. People are taught how to use
pre-assigned tools and given a pre-planned amount of work to perform
which leaves the degree of freedom of action utterly limited.

In a speech, given by Mr Gyllenhammar, Chief Executive of the
Volvo Group of Companies, when opening the Kalmar plant, he said:

"When dealing with a job, it is fairly easy to find that as people get
better and better educated – Sweden spends perhaps more money

per capita than any other country in the world today – their jobs get less complex. So you do simpler and simpler things with more education – and that does not make sense.

By comparison if one goes to one of our plants in the Far East, the workers there have much more complex jobs, than they have in our Swedish factories. Simply because in the Far East, they are less mechanised."

In Volvo today the great majority of jobs are mechanised. At the same time the new corporate strategy is to develop production technology and work organisation towards two ends: craftsmanship and automation. During a transition period a number of people will naturally remain in highly mechanised jobs.

So far very little of traditional final assembly work has been able to be automated.

These problems assumed such importance that Volvo engaged its resources to find out the total expenditure for these factors. The findings provided strong economic reasons for change.

Volvo then began the process of creating a new production technology. Through the adoption of new techniques and the principal change of the general organisation, personal as well as technical, an increased productivity was to be achieved. Technique, layout and people would be accepted as an entirety.

It became clear early on that solutions were not to be found only in technical modifications or changes in the administrative systems. It was necessary to look at all the problems at the same time. Volvo's major task was to find solutions so that the performance and arrangements of daily work, workplaces, etc. were designed and adapted to the demands of the workforce. This was considered the most important part of the new production philosophy – or as Volvo calls it, the socio-technical philosophy.

The first results of these intentions were presented in the final assembly area of the new plant for final assembly of cars that Volvo built in Kalmar, Sweden. By basing the production layout on AGVs, the flexibility aimed at for a changed working environment could be achieved. New methods of cooperation between assembly personnel were found through working groups with their own quality responsibility.

Expected benefits

Volvo wanted to improve productivity. This meant the need for an improved total outcome.

For economical and operational reasons the following needed to be improved:

- Flexibility of production sequencing.
- Adaptability to changes in products.

- Ability to vary production volumes.
- Labour motivation.
- Product reliability and quality.
- Response to legal requirements of noise, light, air, safety.
- Information of flow, necessary to manage and control operations.

In addition the following had to be reduced:

- The time to realise production in a new plant or in an adapted plant.
- Investments in fixed assets and design costs.
- Cash commitments to inventory.
- Vulnerability to work disruption.
- Non-productive labour and space.

Actual benefits

The Volvo Kalmar Plant

The Kalmar Plant, started up in 1974, is a final assembly plant for cars. The layout is based on the idea of teamwork; each team having its own part of the plant building. The building is designed like a star to give the teams close contact with the outside world. The material store is situated in the middle of the building to reduce the number of transport roads.

The assembly area. The assembly work is organised in two different ways (Fig. 1). In both cases, work is carried out on a stationary product, with the possibility of varying station times.

The assembly system. The assembly system is based on the assembly carriers, controlled from a central computer-based control system. The assembly carriers are not simply transport equipment, they are also carriers of information, assembly fixture and work platforms.

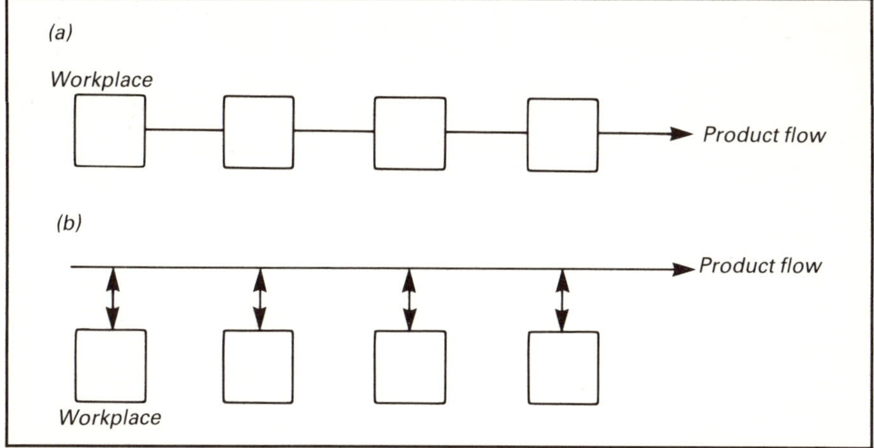

Fig. 1   *Organisation of assembly work at the Volvo Kalmar Plant: (a) straight (line) assembly, and (b) dock assembly work*

Results. Now, several years later, Volvo has formulated the results as follows:

- The total assembly time in the Kalmar Plant is less than in a conventional plant.
- Most workers are very positive towards the team organisation and have a strong feeling of belonging to the team.
- The company has delegated responsibility to the teams to design their own organisations within certain frames. In a traditional organisation all worker divisions are planned at a superior hierarchical level in the line organisation.
- At Kalmar, considerable leeway has been left to the teams to decide on the distribution of the work. What happens between the time when incoming materials are received and when complete assemblies are dispatched is largely up to the teams to decide. In order to fulfil their assignments, the team members must organise their own work. This includes, among other things, setting rules for job-switching and working ahead of breaks.
- Kalmar may be considered a qualified success since its investment costs are around 10% more than a conventional plant of equal capacity. Over half a dozen new plants, one in Holland and six in Sweden, have been built recently, based on the Kalmar experiences.

The Skövde plant

Another example where this production technology has been tried out is the Volvo Skövde-Verken Plant for final assembly of motors.

Not long after the Kalmar Plant commenced operation, this new plant was started up. The assembly system consists of a transport system based on an automated guided vehicle system and a team-orientated organisation.

In the motor assembly department, the assembly time for the local worker can be varied due to the possibilities of buffering between the worker and the work teams. After the assembly, the carrier transports the motor to the test department, where tests are carried out with the motor still on the carrier. The control for this system is entirely local, i.e. there are no integrated central control systems in the system.

Results. From this factory it is possible to make direct comparisons to see what the production and technology is worth. Before this system was installed, the same work was being done by the same workers in another room in the factory. Therefore it has been possible to compare the old system and this new system very thoroughly. Four different factors have been compared:

- Assembly time.
- The handling of interruptions.
- Flexibility in production changes.
- Social costs, advantages.

When these factors are put together and the level of advantages estimated, it can be seen that the investments are payed back in less than two years.

Another factor that is very difficult to establish, is this system's involvement with the quality concept. Most people believe that quality has improved, but why? The higher standard in quality can be attributed to such factors as:

- Working on a still-standing product.
- Better know-how from the workers.
- More observations on eath other and therefore own adjustments are made.
- Better test equipment.

Advantages from this spectra are not taken into consideration in the analysis.

Further plants based on the same concept include:

- Volvo Hällby Plant – final assembly of tractors (based on air-cushion vehicles).
- Volvo Vara Plant – final assembly of motors (based on air-cushion vehicles).
- Volvo Ghent Plant (Belgium) – final assembly of motors (based on AGVs).
- Volvo Umeå Plant – welding and assembly of truck cabins (based on AGVs).

In addition, more are in a state of design or construction.

The future

There must be a conscious effort to design organisation and assembly lines so that production is successful.

Within the automotive industry Volvo has splendid transport equipment in its conveyors. All details come in a good order, the product is controlled and the speed on the line gives the output.

Within this system Volvo has created a system of methods to measure, calculate and plan for the local workers.

The final assembly units are still labour intensive. So even if Volvo tries to automate as far as possible, it is neither economic nor possible to replace people when it comes to the final assembly of cars.

There are other aspects where Volvo's models are not effective enough.

- The custom demands on its products are different. Different product variations give different assembly times and operation will continue to vary.
- The cars will be more and more complex.
- The quality of the products must be built-in during the production process.

- The work in hand must be complete before the next consignment of work can be continued, while adjustments in a car factory are very costly.

In the forthcoming models (systems) Volvo would also like to create another form of control and guiding of the process.

Volvo would like to change its plans and control the production from the actual state in a higher degree than is possible today. To follow the product through the whole process, to allow different sequencing and to coordinate the control functions for the process, the production, the quality control and even the material handling is very important.

The automated guided vehicle systems, the people in the factory and the systems of control and supervision, have made it possible to build up a new process of production.

Concluding remarks

When analysing these new systems, it might be useful to split the most essential demands in the following way:

- To create better environment.
- To automate the transport and/or the handling of the product in work, or of the material flow to the assembly line.
- To increase the real-time information about the production process, to achieve a better control and more flexible guidance adopted to the process control.

The important thing to remember is that neither the organisation nor the equipment work by themselves. To achieve better productivity, tackle both the organisation and the equipment at the same time, so that the final solution will suit the adopted situation.

AGV SYSTEM SAFETY – NEW DEVELOPMENTS TO MEET CHANGING NEEDS

P. M. Tracey
Herga Electric Ltd, UK

Although automated guided vehicles (AGV) systems have a good safety record, the increasing use of AGVs in manned environments means that the safety aspect cannot be taken for granted. Sources of accidents and safety considerations are discussed. AGV specifications for safe operation are given, with particular attention to non-contact sensors and bumpers.

Automated guided vehicle (AGV) systems have a good safety record, probably better than most forms of automated machinery. In a survey of 37 AGV users made in 1982 in Sweden, covering more than 90% of systems existing at the time, only 10 injuries to personnel were reported. In five of these the injury was slight. In addition there were 16 cases of material damage.

However, this record is no cause for complacency. One severe accident can slow up the development of the whole industry.

Where and when do accidents occur?

Before considering how to approach the problem of safety, it is necessary to consider where and when accidents happen.

In the UK it has been found that a large proportion of accidents occur during non-standard situations. The first example is where personnel are in areas which are totally forbidden. This may be due to curiosity, because they are going to speak to a friend, or because they are taking a short cut.

Other accidents occur during breakdown situations. Typically a service engineer might be working on a breakdown. He has been told that everything must be running by the morning. However, he is still at it late at night. He is tired and his reactions are slow. He makes a wrong electrical connection and suddenly the AGV starts to move.

Impossible? Unfortunately, this is just the type of accident which has occurred on many other types of machinery.

Other accidents occur due to human fallibility. For example, if a manual operation has not been carried out correctly; perhaps a load is misplaced on a carrier or verbal instructions have not been understood.

Lastly, accidents occur due to product failure. This can be the result of the AGV not performing correctly due to a programming or component failure. It might also be due to safety devices not functioning properly.

Safety considerations for the specifier

The cost of safety must be taken fully into account at the outset. There are four basic categories to consider:

- Training.
- Procedures.
- System specification.
- AGV specification.

Training

All persons involved with AGVs should receive adequate training. These include:

- Maintenance personnel.
- Operators.
- Supervisors.
- Management.

Anybody who is entering the area where AGVs are being used should also be included, such as plant designers, contractors and visitors.

The required level of training may be small but should be properly established. This is only equivalent to issuing hard hats or protective spectacles to visitors.

Procedures

Written standard procedures regarding the safe operation of the AGV system are essential. These procedures will ensure a consistent standard and provide a check list for management. Once again, this is only parallel to standard procedures in other aspects of business management.

System specification

The type of AGV system will have a strong influence on the safety specification. Originally, AGVs were used mostly in warehouses involving a few personnel. Today, the fastest growth is in AGV systems used in assembly areas where there are many people and correspondingly higher risks.

The areas in which the AGV will operate can be divided into three alternative types as follows.

Closed zones. Some warehouse systems can be considered as closed zones with no people at all. It is then possibble to stop people entering the operating area by the use of fencing and the use of pressure-sensitive mats or light beams at entrances. Even so, it is difficult to count people entering and leaving closed zones and difficult to know if somebody has been left inside.

Mixed zones. These are areas where there may be a limited number of personnel. These may be pedestrians, people working on machines or driving manual vehicles. Restricted access should be enforced so that only trained personnel are present.

Open zones. Open zones exist wherever AGVs are working in open factory areas alongside other production processes. Historically, the open zone system is the area where the greatest number of accidents has occurred.

Well-designed systems should take into account the following:

Layout. A good layout will ensure that there are no trapping points such as with narrow aisles where pedestrians and AGVs are in the same area.

Lighting. Lighting should be to a good level with no dark patches.

Interfacing traffic. This may need similar controls to road traffic.

Warning signs. These should be clear, well lit and at the right height.

Vehicle breakdown. This must be allowed for in the design of operating systems.

Computer program integrity. This must be of a high order.

AGV specification

AGVs are now available to a very high specification. The specifier should consider the following points with regard to safety.

Warning lights. These are most desirable on moving vehicles and should be visible all-round.

Audible signals. These are useful and important at blind corners.

Brakes. These must be of high integrity.

The load. The load must be secure so that it will not move under emergency braking or cornering.

The computer program. This must be of high integrity. The hardware must not be affected by voltage variations or electrical interference.

Colour. A bright colour is easier to see than a dark one.

Stop buttons. The emergency stop buttons must be accessible from all sides of the vehicle.

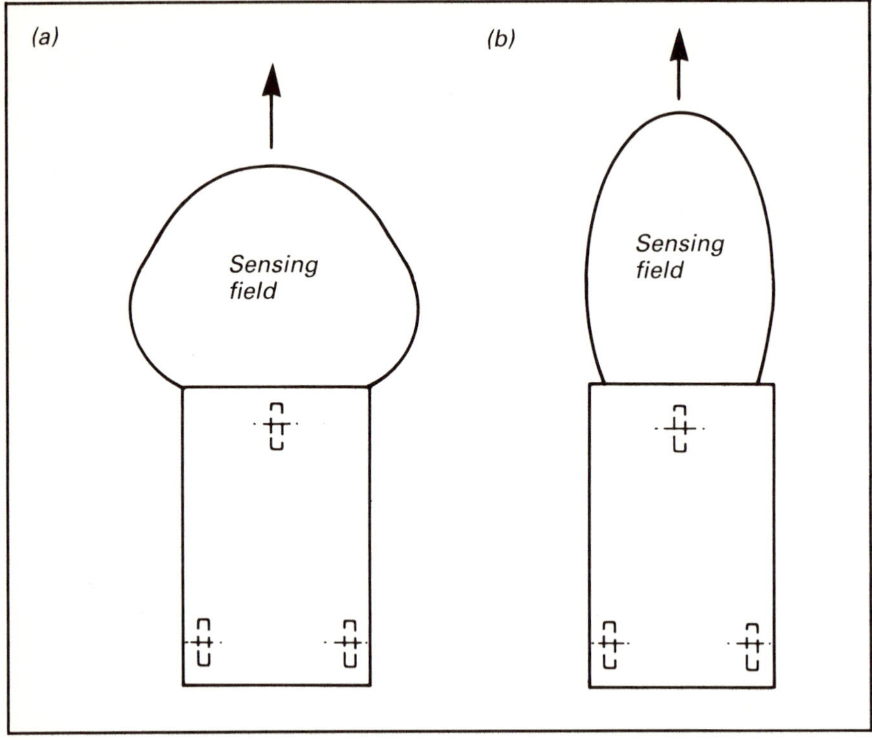

Fig. 1   AGV with (a) wide-angle and (b) narrow-angle non-contact sensing

Non-contact sensing devices. Non-contact sensors such as infrared or ultrasonic detectors have the advantage of being able to sense a considerable distance in front of the path of the vehicle. They are normally used as a first line of defence to slow the machine down. However, they have certain disadvantages:

* Some systems are affected by environmental noise such as from pneumatics and magnetic or radio interference.
* The sensing field can be over a wide or narrow angle in front of the vehicle, as shown in Fig. 1. If it is very wide it will give some protection against cornering and crabwise movement of the AGV. However, it may also be affected by other AGVs passing in the opposite direction and by objects on the side of the path of the vehicle. A narrow-angled beam will avoid these problems but can still create difficulty when an AGV approaches a wall before cornering and when it is arriving at a loading point. It will also not give protection when the vehicle is cornering, as shown in Fig. 2.

Contact sensors. Contact sensors such as pressure-sensitive bumpers do not have the disadvantages of non-contact sensors but are limited in their size and as a result in their sensing field.

After a contact sensor has operated, the vehicle must stop before it can cause an accident. In the worst condition, a loaded AGV going at a certain speed will have a maximum associated stopping distance. This is related to the point in time when the contact sensor operates, through the time taken by the control system and the braking system to the point when the vehicle comes to a halt. Other factors to take into account are floor condition and sloping floors.

The contact sensor should have built-in overtravel to allow for this stopping distance. In other words, the distance between operation of the sensor and it finally being squashed flat must be greater than the maximum stopping distance of the AGV. It is also desirable that the sensor should not have sharp edges which will cause cuts or bruising if pressed in this way through human contact.

In many systems the non-contact sensor acts as an early warning to reduce speed. Contact sensors are then used to stop the AGV. In this situation, it is important to ascertain that the non-contact sensors are truly fail-safe.

The height of the sensor will depend upon the design of the AGV. However, it appears that a maximum band of between 50mm (2in.) and 305mm (12in.) above the ground is acceptable.

It is particularly important to consider sensors along the sides of the AGV. The British Health and Safety Executive is particularly concerned about the possibility of accidents if an AGV should unexpectedly turn or move crabwise. For example, a person may be standing out of the proper path of the vehicle or what he recognizes to be the proper path and it may suddenly come sideways at him, perhaps pinning him against a wall. The steering mechanism of an AGV gives far greater crabwise motion than, for example, a motor-car. This is shown in Fig. 2. Consequently it is important that the sensors on the sides of a vehicle should also be designed to give the required

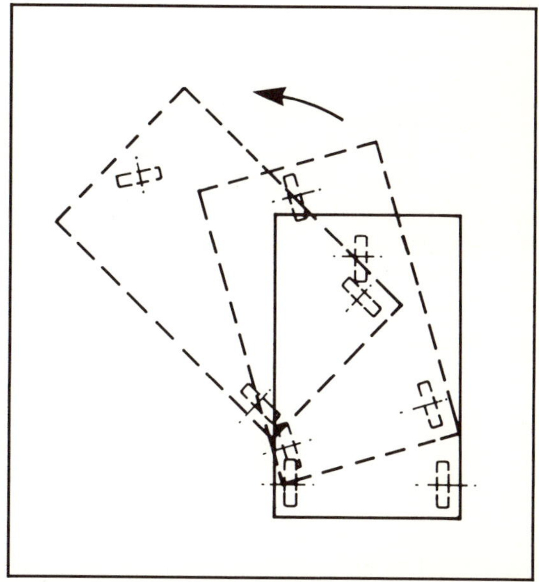

Fig. 2 Crabwise movement when cornering

overtravel. This overtravel may be limited by the aisle width or the design of the docking facility.

Many AGVs in use today use a suspended skirt as a bumper. There are several disadvantages to this design:

- It is normally held up by wires which are connected to interlock switches inside the AGV. This type of protection is not favoured by many authorities because it is very easy for an operator to change the adjustment or to fix a clip on the wire where it enters the AGV, thus preventing it from operating the switch when the skirt is pressed.
- An alternative design uses a photo-electric cell on the AGV reflecting off a disc on the inside of the skirt. Reliable operation depends on the alignment of the reflecting disc staying accurate. It may not operate if the bumper is pressed head on so that the reflective disc stays at the same angle to the beam. It is also necessary to ensure that the photo-electric control is truly fail-safe.
- Some skirts are affected by vibration, such as on rough floors.
- Most skirts can give only a limited degree of protection along the sides.
- On most designs it is possible to stand inside the skirt. This is undesirable to say the least but may be necessary for personnel who are servicing the AGV. More importantly, it is very undesirable on AGV assembly carriers where operators will be standing very close to the vehicle.
- The skirt is very susceptible to damage, particularly by sharp objects such as the forks of other fork-lift trucks.

On the sides of AGVs, electric contact strips have been tried. These have three disadvantages:

- They normally have open contacts which close when actuated. Although it is possible to monitor the control circuit associated with contact strips, it is not possible to monitor the contact strip itself and this is the item which is most susceptible to damage. Thus it is not possible for the safety circuit to detect if the strip is in good condition or if it has become contaminated.
- The contact strip has limited overtravel.
- Contact strips are easy to damage.

New pressure-sensitive bumpers and sensing edges are now available which overcome these criticisms. The design uses a new intrinsic fibre optic sensor, which works on the following principle: By bending glass fibre at regular predetermined intervals, a phenomenon known as micro-bending loss is achieved. In fact, a displacement of only 0.03mm will switch off 95 per cent of the light which is passing through. The sensing fibre can be up to 50m long and requires very little pressure to operate. The glass fibre is used in a loop with a light emitter at one end

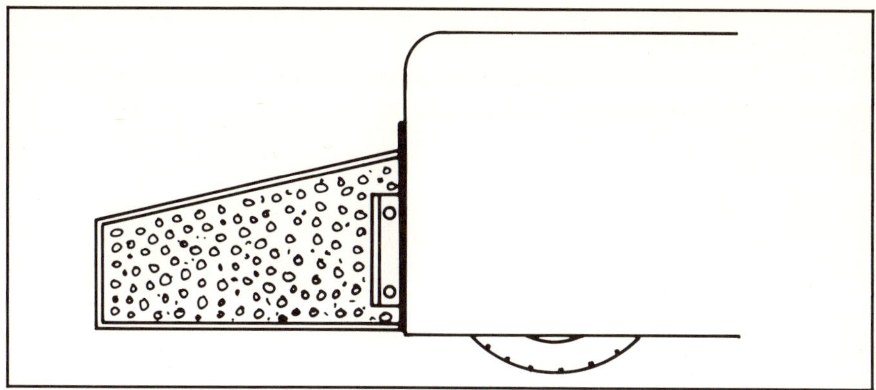

Fig. 3   Cross-section of 'Herga' bumper

and a photo-detector at the other. The fibre optic system can then be compared with a photo-electric control on a guarding system such as a press brake.

The control circuit is pulsed code monitored to check the complete system including the fibre and all the components in the control every 5ms. Any failure in any part of the system will either cause the system to shut down safely immediately or allow it to continue to operate safely and then prevent the system restarting.

When applied to AGVs, the sensor is built into the bumper as shown in Fig. 3.

The sensor is fixed to a mounting plate built onto the front of the AGV and is itself covered by an actuating plate. The foam core can be made to almost any shape and is coated with a neoprene or polyurethane finish. The bumper can be designed to allow overtravel

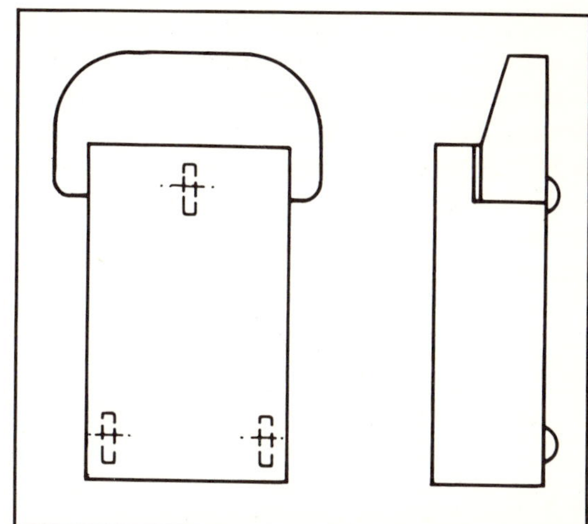

Fig. 4   Wraparound 'Herga' bumper

of up to 460mm (18in.) after the sensor has actuated. A further advantage is that the soft foam will not bruise or cut someone who is squashed by it. The bumper can wrap around the front of the AGV as shown in Fig. 4. This gives additional protection when cornering.

The bumpers can also be produced to replace electric contact strips along the sides of the AGV, and again they can be made to any shape with the desired amount of overtravel.

Alternative controls are available to meet differing safety standards in different countries and applications. For the manufacturer, this means that one design of bumper can be used for all markets. The only thing which needs changing is the control.

The new type of bumper is particularly suitable for AGV assembly carriers which are operating in close proximity to people. This is because it is not possible to stand inside it. It will operate when someone stands on it.

The new bumper can be adapted to meet most special requirements: for example welding is taking place and there is a danger of weld spat, protective skirts can be produced which fit over the bumper in the same way as protective clothing.

Appearance. For products to sell well, appearance is extremely important. A streamlined AGV is going to attract a buyer's attention much more than an 'old lady holding up her skirt'. Bumpers can be fashioned to a variety of shapes to assist the industrial designer. It is important to consult with the manufacturer first to ensure that a shape is feasible from the bumper manufacturer's point of view.

Standards. In spite of the desirability of common world standards, it appears that the USA, Japan, Germany and the UK to name a few are all working on their own standards.

Most published standards at present are extremely general and do not take into account new uses for AGVs where many people are involved.

The new ANSI standard in the USA has now been published for general comment. It provides very good guidance but avoids detailed recommendation on many aspects such as the integrity of control circuits.

Many organisations, particularly in the automobile industry, have written their own standards. This has generally been due to the lack of up-to-date national regulations. The danger is that a number of different standards will result in extra cost for the industry as a whole.

2

Design

The major aspects of AGV design are examined in the opening paper. Then follows a comparative evaluation of guidance techniques giving advantages and disadvantages of each. A similar approach is used to discuss the merits and demerits of centralised and decentralised control systems. The final paper considers the most important parameters in the overall design of an AGV system, including the guidepath layout and analysis of load movements.

AGV DESIGNS

P. McEllin
SKF Engineering Products Ltd, UK

The main short-term trends in automated guided vehicle (AGV) development are discussed. The key areas considered are control (on-board and off-board), mechanical design, maintenance and diagnostics, power supply, downgraded operation, vehicle safety and load transfer.

Automation technology has been around for many years, but it is only now becoming popular amongst industry as a practical alternative to conventional systems. The rate of increase in its use is in line with a greater proportion of standard and proven software reaching an advanced stage of development, and therefore with a decreasing risk factor. Industry in general is becoming more systems orientated; understanding of computer-based systems has increased immeasurably over the past ten years. Also, the demands upon industry are changing, and modern manufacturing systems are being asked to respond quickly to changing market conditions, to cope with ever-increasing product varieties and shorter product life cycles. Intense competition from a generally contracting European manufacturing capacity demands the efficiency from low to medium batch production that is normally only available from mass production.

The response from industry to these pressures has been linked to the economic rate of increase in availability of computer-based systems. Without these systems the general trends towards non-dedicated investment and manufacturing flexibility would not be economically viable.

The high growth rate currently being experienced in the AGV market results from the prevailing market conditions, demand for technology, availability of products to satisfy this demand and the economics of introduction. The demand created naturally results in a competitive market, and the numerous suppliers now active in this field compete and develop their products in order to gain greater market acceptance.

Acceptability has fuelled the advances in AGV technology, which now provides a user-friendly and reliable transport system. The advantages offered by AGVs are many:

- Capability of full computer control and also the ability to integrate the system into an overall production unit host management computer. This provides the possibilities of:
  – Optimising the efficiency of the handling system.
  – Reducing or eliminating labour.
  – Operating unmanned shifts.
  – Ensuring consistent product quality by the removal of reliance on skilled labour.
  – Removing product transit damage due to reliance on human factors.
- Ability to cope with complex or random workpiece or product routing, including varying production rates.
- Viability for the automatic movement of low to medium volume unit loads, including transport over reasonably large distances, without high investment dedicated conveyor systems.
- The flexibility to adapt to changes in product design and factory machine layouts without further major capital expenditure.
- Ability to extend the machine layout, number of operating stations, etc., at low cost.
- Ability to introduce further capacity into the existing system.
- Ability to have a variety of vehicles performing different functions while operating on the same circuit.
- Freedom to plan factory layouts, without the restrictions normally imposed by dedicated conveying systems.
- Product transfer between parallel and series arranged machine groups.
- Balancing of production flow during varying operation cycle times by increasing supplementary workstations for the longest operations.
- Individual breakdown or interruptions in either the process or handling system do not disturb the whole system.
- Power source is independent, giving the ability to operate the transport system independently.
- The AGV control system will allow fully automated, semi-automatic or direct manual control. This is particularly important in the event of failure of some element of the production system, in order to ensure the continued availability of the handling system.
- The AGV system has the capability of generating a great deal of up-to-date information, the integration of which into the management system makes a significant contribution to the total flow of data required for production and inventory control, administration, purchasing, etc.

- Unobstructed floor, which facilitates cleaning and maintenance, etc., and contributes to a clean and pleasant working environment.
- Environmentally unpleasant jobs can be done in special enclosures, while retaining production flow and releasing operators for other tasks.
- Low noise level.
- Non-human intervention ensures consistency in the operating safety of the system.

What then are the main short-term trends in AGV development (as opposed to the long-term ideas of free-ranging systems, etc.)? These trends are considered below, with reference to a particular AGV – the Babcock FATA Microtruk.

Control

Control philosophies have to date progressed from those which have centralised intelligence to those with decentralised intelligence. To draw an analogy, if you ask a driver to deliver goods to another factory do you deliver a fresh instruction to the driver at each decision point along the route in order to guide his 'patch', or do you give the driver a map and let him decide his own route, taking into account local conditions such as traffic congestion? The greater the intelligence at a lower level, the less is required centrally and the greater the possibility of simplifying system requirements. This means that the vehicles may be applied in a wider number of applications without, in many cases, centralised control being necessary at all.

Typically the control system of these vehicles basically consists of a microprocessor. Programs may be stored up to a predetermined number either in RAM or EPROM memory. In RAM, the programs can be written (via either a portable keyboard or the off-board control and communication system) and carried out immediately or cancelled or modified. In EPROM, programs are written permanently using equipment external to the AGV.

The two alternatives are not exclusive; that is, it is possible to have some programs in RAM and some in EPROM.

EPROM on-board control

In practical terms this means that should the circuit requirements be simple, it may be possible to store all the various permutations of routes within the vehicle's permanent memory. These programs may then be actioned in one of the following ways:

- From a key pad on board the vehicle.
- From a key pad off board the vehicle via the controller and communication line.
- From a host system via the controller and communication line.
- By a bank of photocells.

The programs are loaded only in the initialisation phase. Once present in the truck's internal memory, they are started by a simple instruction such as a program number.

The software objectives are achieved by using an easy-to-use programming language that considers the truck functions and installation variables. This user-friendly method of programming enables operators to learn to write and load programs quickly.

The language is composed of a sequence of ACTIONS (functions to carry out) and EVENTS (awaiting external signals dependent upon the installation). Fig. 1 shows a programming example.

Off-board control

Off-board control is necessary when:

- The AGV requires interfacing up to a higher level.
- More than one vehicle is operating on the circuit (i.e. traffic control is necessary).
- The possible number of route permutations is beyond the memory capacity of the AGV micro.

In the majority of cases off-board control will consist of a PLC, although for systems that require a high level of data handling microcomputers may be used either in conjunction with or in place of the PLC.

The use of a PLC for this task offers certain advantages due to the simplicity and cost of the software involved and the familiarity of many organisations with this level of programming. This is particularly useful for system modifications, for which the user is not then dependent upon the supplier.

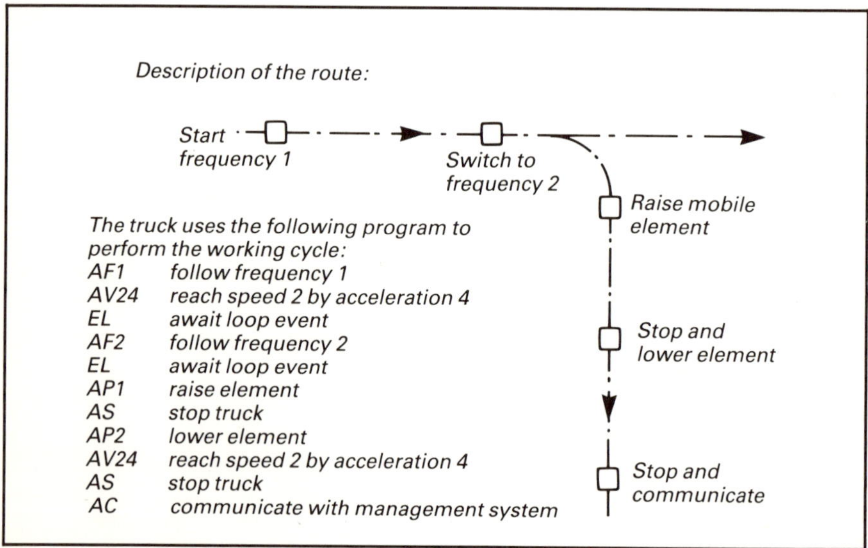

Fig. 1 Programming example

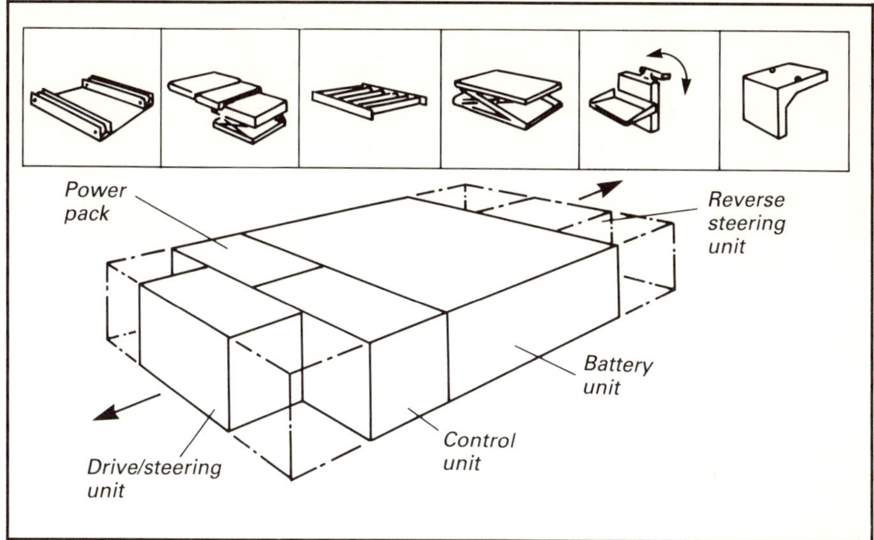

Fig. 2   Vehicle concept: the Babcock FATA Microtruk

The trend towards higher on-board intelligence is thus likely to continue, making user software simpler to produce.

Mechanical design

Due to an expanding variety of applications for this technology it is becoming increasingly difficult to offer products to suit wide differences in mechanical interfaces within a standard range of equipment. Vehicle design is thus becoming modularised, which will allow the rapid reorganisation of vehicle layout using standard modules within a customised chassis design. This may be uprated along with the drive units according to payload requirements.

The basic vehicle modules are the drive, microprocessor, power control and battery (Fig. 2). Additional attachments, which may be of standard or special nature, are used for sequencing from the on-board processor.

Maintenance

Accessibility and ease of maintenance are of key importance in automated systems within which downtime must be minimised. The term user-friendly then extends from the operational/control area to the maintenance and fault-finding aspects.

The standard range of drive units is mounted on independent subframes which are easily removable for maintenance purposes. All major electrical connections are via multi-pin connections, again allowing for the easy removal of subassemblies.

The microprocessor and power control electronics are mounted in sliding drawer units, which ensure easy access and, if required, are quickly detachable for maintenance. These units are shock-mounted and sealed against the ingress of dust and moisture. For high temperature working environments, air conditioning may be supplied to regulate the operating temperature. Additionally, each of the major electrical components is of a plugable nature. All wiring is contained within the RHS chassis design and terminates in bulk head connectors and multi-pin units. This means that for construction purposes standard looms may be used, and once installed these are completely protected from damage.

Diagnostics

The diagnosing of faults on complex high-technology equipment can be time consuming without the aid of a diagnostic system.

The first line in diagnosing a Microtruk fault is a study of the 12 LEDs on the operator's control panel. These LEDs have two modes of operation. First, during normal operation they give indications that confirm normal operation of particular functions. Monitoring of these LEDs as the truck progresses through its program gives confirmation that the particular operation is being performed.

In the event of a fault operation being detected, the amber ERROR light is lit, the truck stops, and one or more LEDs are lit. In this situation the LEDs have different meanings, given by the outer legends on the panel.

If during the program execution the pocket VDU is left plugged into the truck, then when an error situation occurs, the cause of the error will also be printed out as a plain English message on the terminal. However, this only occurs once, at the time of the fault.

In the event of a fault occurring that does not give an LED indication or that does not enable the truck even to start operating, a fault-finding diagram is used, with sequential tests at each stage.

Batteries

Two basic types of batteries are commonly used for AGVs: traction and automotive. The most suitable type is selected depending upon each specific application.

Traction batteries

These units, traditionally used in fork-lifts, have the highest energy capacities combined with very robust construction. Because they are designed for a finite number of charge/discharge cycles (approximately 1500), each cycle should utilise as much of the available charge as possible (usually 80% of the battery capacity). On completion of this discharge, they should be taken out of service and recharged in a controlled manner to 100%. After recharging they are again put into service.

Therefore to obtain continuous (i.e. 24 hours a day) service from an AGV with traction batteries it is necessary to have two batteries, one charging and one on the truck. The batteries are exchanged at the end of their charge capacity.

Automotive batteries

Such batteries based on commercial vehicle units have low to medium energy capacities and are more suited to light-duty AGVs. Because of their construction, they are capable of accepting a charge at any time and, when fitted to AGVs, allow recharging while the vehicle is stationary.

The ideal application for these batteries is with AGVs used for assembly line work, where the vehicles are stationary during fitting times. During these times, the AGV couples automatically to a power supply and charges the battery. It is quite possible to maintain an equalisation charge, say weekly.

New alternatives

Recently battery technology has advanced to provide further alternatives. Deep-cycle, sealed, maintenance-free power cells provide attractive advantages over standard traction batteries. These cells may be stored in any position and do not give off explosive gases or corrosive electolyte like conventional lead-acid power systems.

Dry cell traction units, although comparatively expensive, now enable traction batteries to be opportunity charged. This enables systems to operate on a 24-hour basis without the requirement for any manual intervention or maintenance and removes the need for additional batteries.

Downgraded operation

Provision is made for the operation of the vehicle in the event of system failure at various levels.

At the lowest level the vehicle may be operated by a pendant control which allows it to be manoeuvred off its guide wire in any direction. A facility is also provided to operate any attachments which are on board (e.g. pallet transfer mechanisms). This pendant control bypasses the control electronics on board the vehicle so that in the event of vehicle failure it is still manoeuvrable to maintenance areas, etc.

Above this level the same pendant control may be used in conjunction with the steering antenna, which allows the vehicle to follow the required floor frequency as selected.

In systems that require off-board control for linking into a higher level of management computer, the failure of this host facility is of particular concern to system operators who still require to maintain limited output. The decisions on AGV journeys then become manual functions, input from an off-board key pad situated in the central control facility. The system will operate in the same manner as if the instructions had been downloaded directly from the host.

A strong trend in today's FMS is to keep intelligence at the lowest possible levels and as independent as possible of host systems. The AGV control system may then be responsible using simplified algorithms determining pallet positions within storage buffers, and mapping the status of transactions throughout the system. The system map may contain detailed information regarding the status of the components within buffer storage. Therefore in the event of host failure the operators have the requirements upon themselves simplified (i.e. they are required only to instruct the AGV to buffer). The PLC decides where and updates the system map. Once the host system is back on stream the PLC updates the map and full level operation is quickly restored.

PLC failure, although unlikely, is potentially a problem when system output is mandatory. Excluding the luxuries of hot or cold stand-by systems, the AGV system is still operable in a lower form.

Vehicle safety systems

With loaded driverless vehicles weighing up to several tonnes, operating in normal industrial shopfloor environments at speeds of up to one metre per second, there have to be stringent safety precautions to prevent collisions. It is obviously a requirement of both supplier and user to make AGV systems as safe as possible.

Babcock FATA fits a two-stage safety device on the Microtruk which covers the path in front of the vehicle (also behind if the AGV is reversible). The first stage of the device comprises longer range (1-2m) ultrasonic sensors that scan just beyond the width of the truck path. If an obstacle is encountered, the AGV starts a program slow down and can also emit an audible warning. If the obstacle remains in position, by the time the AGV reaches it the speed will have already dropped to a crawl. A spring safety bumper on the vehicle cuts all drive power and applies the emergency brake when the obstacle is finally touched.

If an obstacle appears in front of the truck and strikes the bumper, the emergency brake is immediately engaged, bringing the AGV to a halt within the travel distance of the bumper. In this way there is no possibility of the obstacle being pressured by the body of the truck.

Transfer attachments: Stillage or tray storage

Roller beds are the most effective and economical method of transferring trays. However, storage can be expensive if it is necessary to power off-board units individually and combine these with PEC (photoelectric cell) interlocks. A cost-effective method uses an actuator-operated friction-driven roller on board the AGV to power up the unpowered off-board rollers. This may be bidirectional and multi-level if required.

Larger trays or stillages may be handled using side-loading AGVs in preference to end-loading units, which use considerable floor space. Two-way extending forks provide the most flexible solution, combined with a mast or lift table for multi-level storage.

The vehicle payload is transferred via polyurethane castor wheels with drive traction transmitted via spring-loaded drive modules, adjustable for pressure according to payload. This provides a constant traction force despite undulations in the floor surface and combines a constant ride height with stability.

The AGV layout for a bidirectional vehicle contains two drive/steering units, each of which operates completely independently of the other. The positioning of these central to the AGV chassis ensures precise and repeatable tracking throughout the vehicle path, which is particularly important for machine shop environments where layouts are usually confined. Dual drive systems also offer significant advantages in such environments for traction purposes during oil spillage, etc.

Pallet-machine transfer interface

This involves the transportation of machine tool pallets complete with fixturing and components between machining centres, storage and fixturing stations and often wash and inspection stations, etc. Common requirements are also for the AGV to provide the transportation for tool management systems and swarf removal, etc., this often being a dual function.

The Babcock FATA approach to AGV design is to use whenever possible the standard Microtruk chassis design and modify the application module for each project's particular needs. These take a variety of forms, i.e. single/twin saddle, and with or without the responsibility for pallet transfer. There are also many dimensional and payload differences between each application, but similar design criteria apply:

- There is a requirement for precision docking.
- There is a requirement for interlock security.
- There are practical considerations for swarf and coolant spillage.

Precision docking

The positional tolerance of AGV docking without using control system integrity is within ± 5mm in both planes. It is therefore necessary to provide a mechanical means of achieving the finite limit for machine-pallet transfer. Two basic systems are offered by Babcock FATA, both of which rely upon cones for accuracy. These cones may be different combinations of cones and flat height tolerance plates and be positioned either in the floor or above the floor on the machine tools or pallet stands. The philosophy remains exactly the same: the AGV is provided with a lift table around which is suspended a framework supporting the transfer mechanism or pallet guide rails. When the lift is in the raised position this framework becomes locked but upon lowering is allowed to float for settlement onto the cones.

A variety of mechanisms are then used in order to deliver or retrieve the pallet at the station.

Interlock security

A major concern in AGV systems is that a pallet not be wrongly delivered to a machine tool, which could result in considerable damage. A number of security methods are available to alleviate this problem:

- When the AGV control system is issued with a transportation order it may also be issued with a pallet identification code. When the AGV reaches its destination, this code may be read from a reader on board the AGV and transmitted back through a local communication point specific to the destination. The AGV control may then verify the journey and transmit a commence transfer instruction. This method has the advantage of limiting the number of readers required, in contrast to off-board code reading systems, which although able to handle larger quantities of information are required at each machine.

- When the AGV docks at a machine tool, limit switches in the base of the female cones on board the AGV ensure that the transfer device is properly seated. Failure in this respect results in a possible three docking attempts controlled by the on-board software, after which a request for operator intervention is sent.

  With the transfer mechanism correctly seated a photocell handshake interlock operates to request either deposit or retrieval. The machine checks its own station and only responds if the expected criteria are satisfied. The transfer may then commence.

- In the pallet transfer to stations such as buffers or fixture stations, the same level of integrity need not be applied; indeed it may be argued that it is not important that pallets may be wrongly positioned other than at machines because any faulty transactions would eventually be highlighted at the machine stations. However, the normal method of ensuring that a pallet is delivered to a station of the correct status and that the AGV is positioned to carry out this transaction is a photocell on-board the AGV at the station.

Swarf and coolant considerations

Practical considerations must be allowed for the possible spillage of coolant and swarf. This is particularly a problem with some large hollow castings, which despite wash and dry facilities on machines may still contain pockets of coolant. Drip trays are necessary with, in certain cases, extending trays which do not allow drippage onto the floor areas during transfer. These should be centrally drained to a reservoir on board the AGV for manual or automatic periodic draining.

A COMPARATIVE EVALUATION OF AGV NAVIGATION TECHNIQUES

P. Boegli
Schindler-Digitron AG, Switzerland

The vast majority of automated guided vehicles in operation today follow an inductive guidewire buried in the floor. Other guidance techniques, using optical/chemical/magnetic guidepaths, dead reckoning, position referencing, inertial navigation, environmental imaging, etc., have been developed, laboratory tested and, to varying degrees, implemented in industrial installations. After a brief outline of the guidance techniques available, the relative merits and demerits of each technique, according to various technical and economic criteria, are detailed. Having shown that no single technique is in all respects ideal, it is concluded that the best currently practical solution is based on a judicious combination of several techniques.

As its name implies, an automated guided vehicle (AGV) system encompasses a navigation system of which one part defines a network of guidepaths that the automated guided vehicles (AGVs) have to follow. Another part residing within each vehicle enables it to determine its position relative to that guidepath so that it can follow it.

Navigation systems can be divided into two groups. One group, using a *physical* guidepath, contains the well-known inductive guide-wire technique which is used in the majority of AGV systems today. Other physical guidepaths are rails or passive lines painted or stuck onto the floor. The other group uses *virtual guidepaths* – that is, guidepaths that must first be derived from indirect indicators by the AGVs. Examples are inertial navigation systems, systems using beacons for establishing a position reference frame, dead reckoning vehicles or computer vision systems that allow AGVs to navigate in much the same way as humans do.

Another possible distinction is between static and dynamic guide-path definition. In the static case there is exactly one guidepath between the end-points of an AGV travel. If there are several possible routes, one of which is selected at travelling time, it is a case of dynamic guidepath definition. In the extreme, guidepaths can be rerouted by infinitesimally small increments.

The following discussion will deal with static guidepath systems in an industrial environment. Due attention will be paid to the necessity to rearrange guidepaths as a result of shop-floor reorganisations.

Design objectives of a navigation system

A navigation system in the context discussed here should meet the following objectives:

- Guide the vehicle reliably from A to B within the working surroundings (e.g. a factory shop-floor) and with a precision sufficient to avoid the known obstacles (walls/pillars/machines/ reserved areas for temporary materials storage, for other traffic/ ramps/etc.). In practice many of these obstacles remain fairly static during the life-cycle of an AGV system. There should be no intrinsic limit to the distance that can be covered by the navigation system.
- Provide for precise manoeuvring of the AGV in the vicinity of load pick-up and deposit stations. The exact location and execution of these stations may vary as a consequence of shop-floor reorganisations.
- Provide a base for an efficient traffic control possibly involving a large number of vehicles.
- Fulfil safety requirements for protection of people, production infrastructure (including the AGV itself) and other material.

It goes without saying that these main objectives must be met at minimum system cost for the owner-user of an AGV system.

It is in the light of the above objectives that the following seven principal guidance techniques will be compared and evaluated:

- *Mechanical guidepath* using some form of rails, directly or indirectly (i.e. through a steering mechanism) to guide the vehicle. (This technique, being of minor importance in the modern AGV systems scene is considered mainly for reference purposes.)

- *Inductive guidepath* as the 'classical' AGV guidance system using a floor-embedded wire carrying alternating current.

- *Optical/chemical/magnetic guidepath* using a passive stripe on the floor that is detectable by proximity sensors (magnetic) or optically. Of the latter a principle is known where a 'chemical' guidepath is excited with ultraviolet waves and re-emits in the visible range, thereby vastly improving the contrast ratio[1].

- *Dead reckoning* using tables of distance and heading information stored in the on-board microcomputer enabling the AGV to extrapolate its movement from a known reference point and without any physical guidepath.

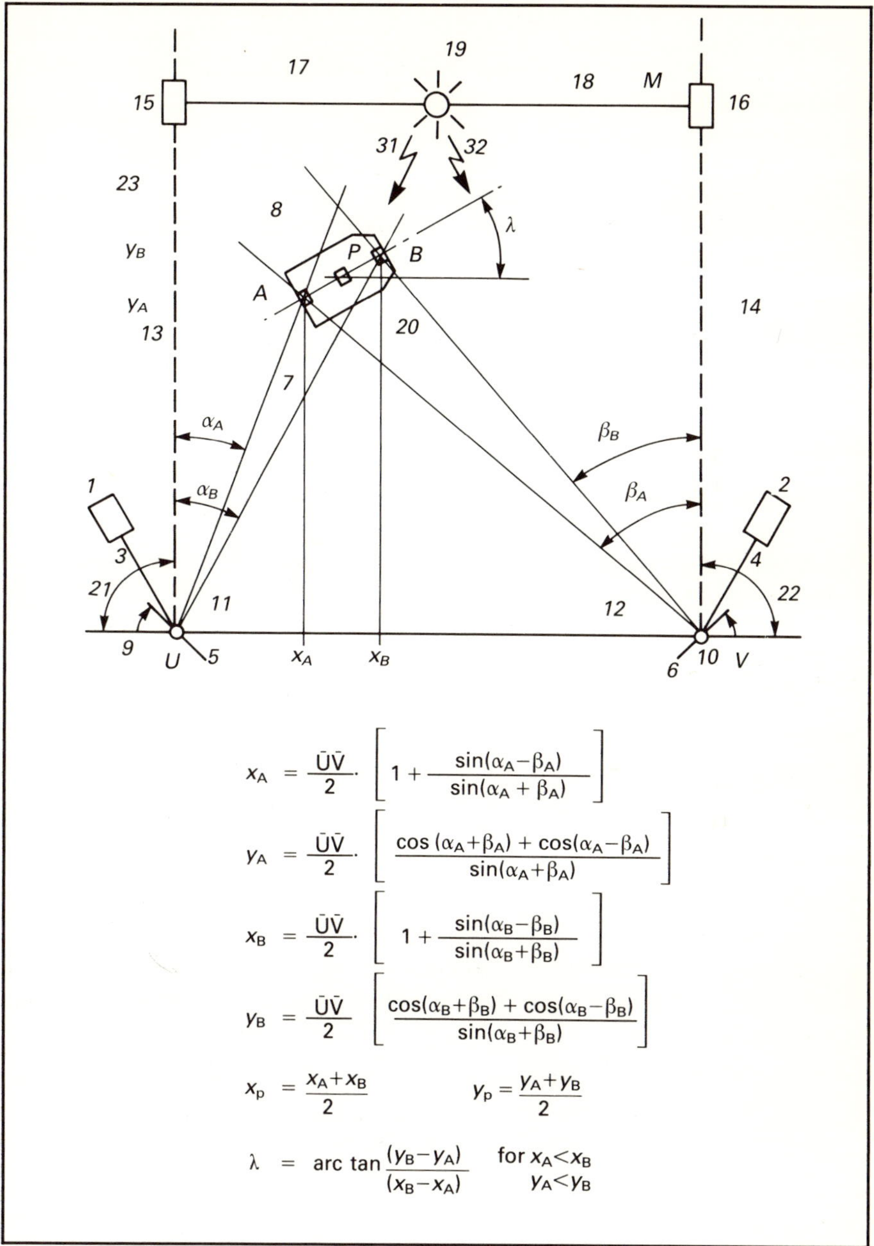

$$x_A = \frac{\overline{UV}}{2} \cdot \left[1 + \frac{\sin(\alpha_A - \beta_A)}{\sin(\alpha_A + \beta_A)} \right]$$

$$y_A = \frac{\overline{UV}}{2} \cdot \left[\frac{\cos(\alpha_A + \beta_A) + \cos(\alpha_A - \beta_A)}{\sin(\alpha_A + \beta_A)} \right]$$

$$x_B = \frac{\overline{UV}}{2} \cdot \left[1 + \frac{\sin(\alpha_B - \beta_B)}{\sin(\alpha_B + \beta_B)} \right]$$

$$y_B = \frac{\overline{UV}}{2} \left[\frac{\cos(\alpha_B + \beta_B) + \cos(\alpha_B - \beta_B)}{\sin(\alpha_B + \beta_B)} \right]$$

$$x_p = \frac{x_A + x_B}{2} \qquad y_p = \frac{y_A + y_B}{2}$$

$$\lambda = \arctan \frac{(y_B - y_A)}{(x_B - x_A)} \qquad \begin{array}{l} \text{for } x_A < x_B \\ y_A < y_B \end{array}$$

Fig. 1 Navigation system using two stationary mounted rotating beacons, together with formulae to derive position and heading

- *Beacon system* as outlined in Fig. 1 for the purpose of performance assessment. Other systems (e.g. with beacons mounted on the AGVs and stationary beam receivers) are possible[2] but are not considered here as it is felt that the precision mechanics of the beacons do not stand up well to the vibrating environment of an AGV.

The system presented here is depicted in Fig. 1. It consists of two precisely focused (e.g. laser) beacons (1, 2) scanning over the area to be controlled (14) in a time-linear sweep by means of rotating mirrors (5, 6). An omnidirectional transmitter (19) emits a synchronising signal when a beam of a beacon passes over one of two photo-receivers (15, 16). An AGV can compute the angles α_A, α_B, β_A and β_B by evaluation of the time relationship between the synchronising signal received via (P) and the passing of the two beacon signals received by its photo-detectors (A, B). Its position and heading can then be derived by evaluating the formulae written under Fig. 1. A more detailed description is to be found elsewhere[3].

- *Inertial navigation systems* measuring translational and rotational acceleration of the AGV and deducing its position and heading relative to a reference point exclusively by double integration of the respective accelerations. Mixed systems as described elsewhere[4] are not considered here because it is felt that the flexibility an inertial system offers is most valuable near the docking stations. But this is where the mentioned system needs recalibrations by absolute means.

- *Direct imaging systems* where optical or ultrasound images of the surrounding working area are used on-board the AGV to deduce its position within that working area.

Evaluation criteria

As a consequence of the design objectives for a navigation system a number of performance criteria can be derived allowing a relative performance evaluation of the various guidance technologies. These are in an arbitrary order:

- *Range* – How far can an AGV be guided by the technology considered? For most navigation systems there is a stepwise degradation in cost efficiency and/or performance at distinct lengths of guidepath. The first such step is assessed.

- *Accuracy* – By how much will the track of the AGV deviate from the position of an ideal guidepath? Tolerances in laying down the guidepath as well as system-inherent tracking errors are taken into account.

- *Flexibility* – How much effort (and cost) is necessary to move a guidepath in order to adapt to changed operational requirements? The assumption is made that the topology of the guidepath network remains unaltered for the reasons outlined below.

In any AGV system, irrespective of its navigation system the topology of the guidepath network must be known in order to direct the vehicle(s) from starting location to destination, to avoid collisions between vehicles, to monitor and control traffic flow, to optimise throughput, etc. With the exception of the most primitive rail systems this topology is stored in one or several computers. Whenever the topology of the guidepath network is changed the necessity arises to redefine (i.e. reprogram) the *logics* of the AGV system to a greater or lesser extent. This effort far outweighs in practice the effort necessary just to move the guidepath proper by some tens of centimetres.

- *Reliability* – What is the probability for a malfunction of the navigation system considered in industrial use? What are the mechanisms degrading or even inhibiting the proper function of the system?

- *Controllability* – What are the means of traffic control (i.e. start/stop initiation, control of guidepath switches) intrinsic to the navigation system? Is there any means for data communication between the automated guided vehicles and a supervisory central control system inherent in the navigation system under consideration? If yes, what is the performance level to be expected?

- *Vehicle equipment cost* – What is the cost of the navigation equipment needed aboard the AGV based on an estimation of its complexity and maintenance requirements? As absolute figures would depend heavily on an actual implementation of the navigation system under consideration, only an estimate of its relative position can be given.

- *Stationary equipment cost* – What is the cost of the navigation equipment infrastructure needed (i.e. floor installation, control cabinets, electronics and/or optronics for guidepath definition, etc.) based on an estimation of its complexity and maintenance requirements? As absolute figures would again depend heavily on an actual implementation of the navigation system under consideration, only an estimate of its relative position can be given.

Performance assessment

For all of the seven navigation systems mentioned above a relative 'merit profile' is shown in Fig. 2. The reasoning behind each is outlined in the following.

Mechanical guidepath

Range – Excellent. The cost of a rail system is strictly proportional to the length of the installed rail and performance is not influenced by the dimension of the guidepath network.

(a)

Eval. Criterion / Rating	Poor	Acceptable	Medium	Good	Excellent
Range					△
Accuracy					△
Flexibility	△				
Reliability				△	
Controllability		△			
Vehicle equipment cost				△	
Station equipment cost		△			

(b)

Eval. Criterion / Rating	Poor	Acceptable	Medium	Good	Excellent
Range				△	
Accuracy				△	
Flexibility			△		
Reliability					△
Controllability					△
Vehicle equipment cost				△	
Station equipment cost			△		

(c)

Eval. Criterion / Rating	Poor	Acceptable	Medium	Good	Excellent
Range					△
Accuracy					△
Flexibility				△	
Reliability			△		
Controllability	△				
Vehicle equipment cost			△		
Station equipment cost					△

(d)

Eval. Criterion / Rating	Poor	Acceptable	Medium	Good	Excellent
Range	△				
Accuracy		△			
Flexibility					△
Reliability				△	
Controllability	△				
Vehicle equipment cost				△	
Station equipment cost					△

(e)

Eval. Criterion / Rating	Poor	Acceptable	Medium	Good	Excellent
Range			△		
Accuracy			△		
Flexibility					△
Reliability		△			
Controllability			△		
Vehicle equipment cost			△		
Station equipment cost		△			

(f)

Eval. Criterion / Rating	Poor	Acceptable	Medium	Good	Excellent
Range			△		
Accuracy		△			
Flexibility					△
Reliability			△		
Controllability	△				
Vehicle equipment cost	△				
Station equipment cost					△

(g)

Eval. Criterion / Rating	Poor	Acceptable	Medium	Good	Excellent
Range			△		
Accuracy					△
Flexibility					△
Reliability			△		
Controllability		△			
Vehicle equipment cost	△				
Station equipment cost				△	

Fig. 2 Merit profiles for: (a) mechanical guidepath, (b) inductive guidepath, (c) optical/chemical/magnetic guidepath, (d) dead-reckoning, (e) beacon system, (f) inertial navigation system, (g) direct imaging system

Accuracy – Excellent. Guidance accuracies in the sub-millimetre range can be achieved if necessary, although in practice the effort needed is only rewarded in special applications as the cost rises sharply with more stringent accuracy requirements.

Flexibility – Poor. Rails have to support the weight of the vehicle and/or guidance forces and therefore have to be mounted rigidly. Furthermore many actual constructions are sunk into the floor to make the travelling area of the AGVs usable by other traffic, which means that a considerable effort is necessary if it is required to rearrange the guidepath.

Reliability – Good. With rail in the classical railway sense an excellent figure can be achieved easily. If there is only a mechanical guidance slot sunk in the floor, then in working with a finger-guided vehicle the problem of clogging of the slot has to be considered.

Controllability – Acceptable. The technique of building remotely controlled switches is well known for all kinds of rail structures. The problem of jamming needs attention. Start/stop control is possible (e.g. by mechanical catches) but usually has rough operation. Smooth velocity control requires considerable expense.

Vehicle equipment cost – Excellent. The functions of load carrying, movement and directional control can be combined within the same constructional elements. Mechanical guidance systems are the only ones needing no electronic servo control for steering.

Stationary equipment cost – Acceptable for modest accuracy requirements, but rising for high-accuracy systems. Switches and other mechanical motion control equipment tend to be expensive.

Inductive guidepath

Range – Good. The length of an inductive guidewire loop is limited only by the drive capability of the alternating current amplifier used. With appropriate equipment several hundred metres can easily be spanned by a single loop.

Accuracy – Good. Accuracy is limited by distortion of the electro-magnetic field caused by metallic objects in the vicinity of the guidewire as well as by tolerances in cutting the wire-slot into the floor.

Flexibility – Medium. To move an existing guidepath a new slot has to be cut into the floor and the guidewire loop reconnected. Provided that the special floor-cutting equipment required is at hand the job can be done in a short time and with modest effort.

Reliability – Excellent. As the guidepath is embedded in the floor it is almost ideally protected from environmental influences. The electronics to supply the alternating current and to measure the resulting electromagnetic field can easily be made up to the most exacting reliability requirements.

Controllability – Excellent. Switches are implemented most easily by using several distinct frequencies or by switching on or off the appropriate guidance frequencies. The same method can be used for start/stop control. More intricate data communication needs can be satisfied by modulating control frequencies in the wire loops.

Vehicle equipment cost – Good. The technology for detection of an electromagnetic field of a guidewire and the processing of the resulting signals is well known today and the necessary components are readily available.

Stationary equipment cost – Medium. Although a well-known, stable and therefore cost-efficient technique, the expenses for floor-cutting and sealing and for the control cabinets add up to medium stationary equipment cost.

Optical/chemical/magnetic guidepath

Range – Excellent. The cost of an optical, chemical or magnetic guidepath is proportional to its installed length. Performance is not affected thereby.

Accuracy – Very much dependent on the working principle applied. Simple magnetic or optical sytems suffer substantial errors due to poor selectivity of the sensors used, especially when their ground clearances cannot be kept constant. In principle, optical or chemical navigation systems can be made very accurate thanks to the precise focusing possible with optical wevelengths. The construction of high-precision sensors does have its price but costs remain reasonable.

Flexibility – Good. The guidepaths considered here are mostly attached to the surface of the floor by painting or adhesive bonding. They can therefore be removed relatively easily and a new stretch of guidepath applied to a new location should the necessity arise.

Reliability – Medium. As the guidepath is mostly attached to the floor surface it is vulnerable to environmental influences (e.g. wear, mechanical damage, dirt). Careful fixation by adhesives of high shear resisitance, attention to minimal thickness of the guidepath, use of protective covers and, last but not least application of intelligent sensors all help to get the reliability problem under control.

Controllability – Poor. As optical, chemical and magnetic guidepaths are all passive in their nature there is no inherent means of switch control or start/stop control. All types of dynamic data exchange between AGVs and the stationary control system must therefore be performed via dedicated data communication equipment (e.g. a radio link).

Vehicle equipment cost – Good to medium. Magnetic guidepath sensors are available at low cost. Although the complexity of an intelligent optical or chemical sensor exhibiting a high resistance against dirt on the guidepath is higher the costs remain within reasonable limits.

Stationary equipment cost – Excellent. The considered passive guidepath materials boast the lowest cost per unit length of all the physical guidepath systems.

Dead reckoning

Range – Poor. Dead reckoning is essentially an open loop system (i.e. there is neither a physical nor a logical guidepath defined in absolute coordinates). The AGV has its own position at the start of the dead reckoning travel as the only reference. Errors of the odometric system are introduced through wheel slippage and floor unevenness inevitable in industrial environments. After a certain distance guidance accuracy falls to unacceptable levels, limiting the useful range of dead reckoning to some ten metres.

Accuracy – Acceptable. As described above, accuracy is strongly interdependent with range. Used within reasonable range limits the accuracy of dead reckoning is sufficient even for docking purposes.

Flexibility – Excellent as there is no physical guidepath to handle whenever a piece of track has to be rearranged. Only the software tables in the vehicle microcomputer have to be altered.

Reliability – Good. There are no dedicated guidance components detracting from the intrinsic vehicle reliability figure but floor quality plays an important part in reliability assessment.

Controllability – Poor. As there is no stationary installation at all no communication is possible via the navigation system. Start/stop and switch control have to be done before initiating a dead reckoning travel.

Vehicle equipment cost – Good. Dead reckoning requires a microcomputer, a high-quality servo control system and an odometric sensor to be installed in the vehicle. As these elements are incorporated in most modern AGVs anyway there is little additional cost necessary.

Stationary equipment cost – Excellent (Zero cost!)

Beacon systems

Range – Medium. With beacons accuracy is directly related to range. With medium requirements for guidance accuracy and using high-precision optical sensors and time-bases, rotating beacon systems can be used over more than a hundred metres in line of sight. Fixed stationary beacons allow ranges in excess of that figure, albeit only for straight-ahead travel.

With all beacon systems an unobstructed line of sight between the beacon and its respective receiver(s) is mandatory. In many industrial environments this requirement may well be the ultimate limit of range or preclude the use of a beacon system at all.

Accuracy – Medium. With rotating beacon systems as considered the exacting specifications required for docking purposes can only be fulfilled within some tens of metres from the beacons.

Flexibility – Generally excellent, but can be severely retricted by the operating environment, such as when obstacles restrict sectors of sight.

Reliability – Acceptable. The precision needed in the optical equipment to achieve a sufficient guidance accuracy is considerable. Drift effects can therefore heavily affect long-term reliability. External light sources such as safety light curtains, optical limit switches, welding arcs, etc., can be a source of spurious malfunctions.

Controllability – Medium in the direction from beacon to beam receiver as a light beam can be modulated with a very high data rate. The data packets can only be very short, though, as there is an immediate trade-off between accuracy of position determination and beam divergence necessary for longer communication time slots. In the inverse direction there is usually no communication possibility within the beacon system.

Vehicle equipment cost – Medium, assuming that the (rotating) laser beacons are mounted stationarily and that the AGV contains two omnidirectional beam receivers to allow computation of heading.

Stationary equipment cost – Acceptable. The use of laser beacons with very small beam divergence (max. 20 arc seconds) and highly constant angular velocity is mandatory. Their number must be large enough to reach the complete operating area of the AGV system.

Inertial navigation systems

Range – Medium. As a pure inertial navigation system derives its knowledge about its position solely from gyros and accelerometers it is prone to drift, which in turn necessitates the periodic recalibration by another, absolute navigation system.

Accuracy – Acceptable. The vibration spectrum encountered on AGVs makes the job of the inertial sensors very difficult. This leads to navigation errors precluding direct docking operations without special measures after average travel distances and heading changes.

Flexibility – Excellent. As there is no stationary installation for describing the guidepaths only AGV internal software tables have to be adapted whenever the operational environment changes.

Reliability – Medium. No exact figures are available to the author but it is assumed that the maintenance effort required to keep precision gyros working under vibration cannot be neglected. With commercial availability of laser gyros this situation is bound to change.

Controllability – Poor. There is no means inherent in inertial navigation systems for switch control, for start/stop control or general data communication between AGVs and any supervisory control system.

Vehicle equipment cost – Poor. With today's technology the cost of a two-dimensional inertial platform allowing stable and precise guidance of an AGV over several hundred metres is considerable.

Stationary equipment cost – Excellent. No stationary equipment required.

Direct imaging systems

Range – Good. The range is limited only by the on-board memory required to store the necessary details for AGV position determination.

Accuracy – Excellent. Guidance errors are roughly proportional to the distance to the nearest obstacle (i.e. they get smaller approaching a docking station).

Flexibility – Excellent. Guidepath description resides only in the AGV memory. It can be put there easily by 'teach-in' under manual control. With that approach, however, it will be unavoidable that each vehicle will follow a slightly different path. Special precautions have to be taken that the AGV does not take the human trainer as part of its usual image of surroundings. In addition, a means for deleting parts of the teach-in process spoilt by operator errors must be provided.

Reliability – Medium. The process of computing the actual position of the AGV must be tolerant to minor deviations of the actual image of the surroundings from the stored image. Otherwise the AGV might be misled by workers or fork-lifts moving around or pallets with work in progress standing at various places. The images must thus not only be 'seen' but also to some exent be 'understood'. This level of pattern recognisation intelligence is still a matter of research and not of commercial availability. It is felt that it might be a weak point in direct imaging sytems for a long time to come.

Controllability – Acceptable. Stationary optical signalling devices are imaginable to transmit variable data from a supervisory control system to the vehicles, thereby using the navigation system (e.g. LED panels at specific locations displaying a dot pattern code). For data communication in the inverse direction additional equipment is needed.

Vehicle equipment cost – Poor. Today's imaging sensors are unreliable in industrial environments, not sufficiently precise or too expensive.

Cheaper and better CCD-cameras will change this situation in the foreseeable future, though. Another factor, is the computational power needed aboard the vehicle for image processing. The hardware aspect will diminish with every generation of (micro) computers but high software cost will be with us for some time to come.

Stationary equipment cost – Good. Some provisions will be necessary to ensure a stable appearance of the working environment of the AGV that is 'comprehensible' to the image processing software.

Concluding remarks

It has been shown that no single navigation system scores top ratings in all respects considered. Therefore the question arises whether a combination of two or more systems possesses all desired qualities. To this end Fig. 3 has been created in an attempt to give a performance vs system cost relation between the various guidance systems. As can be seen, the 'modern' technologies (i.e. beacon, inertial navigation and direct imaging) are all near the top end of the cost scale. A combination of any of these systems would raise the system cost to prohibitive levels and must be ruled out on this basis.

As the purely mechanical guidepath does not need the electronics required by all other navigation systems and because the mechanical construction of the vehicle must be optimised for the rail, it does not usually lend itself easily to combinations with other guidance principles. What remains is a combination between the well-proven

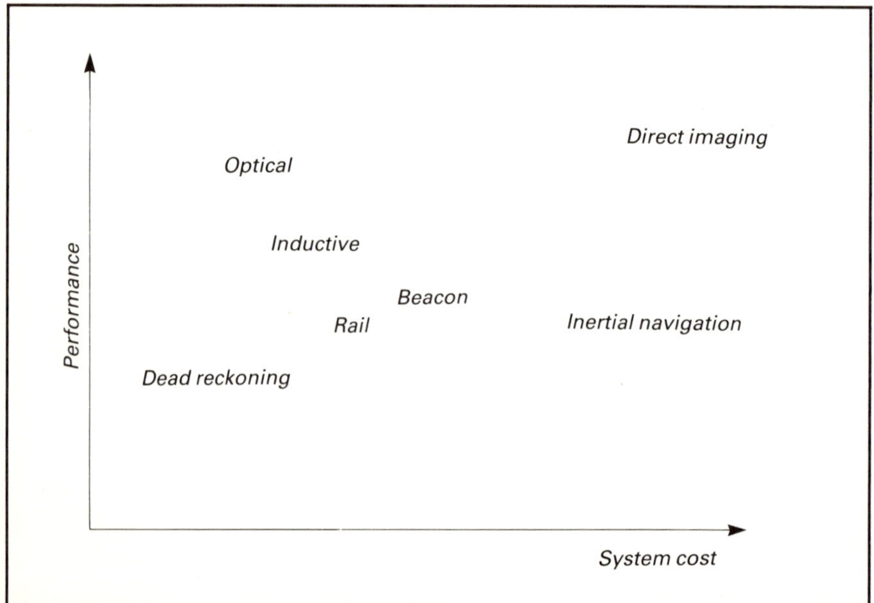

Fig. 3   System cost-effectiveness comparison

Fig. 4 *Merit profile combined inductive guidepath (△), optical guidepath (□) and dead reckoning (○) system*

Rating / Eval. Criterion	Poor	Accept-able	Medium	Good	Excel-lent
Range					□
Accuracy					□
Flexibility					○
Reliability					△
Controllability					△
Vehicle equip-ment cost			□		
Station equip-ment cost			△		

inductive guidepath with its reliability and controllability advantages and the dead reckoning technique for its flexibility within confined zones. Where improved range and accuracy together with useful flexibility is of importance an optical or chemical guidepath can be applied in place of or in combination with the aforementioned systems. The resulting merit profile in Fig. 4 clearly shows the superiority in terms of performance *and* cost obtainable with the suggested system combination and with today's industrially available technologies.

References

[1] DeBruine, C. J. 1979. Line follower vehicle with sensing head. European Patent No. EP 0 012 554, December 1979.

[2] Premi, S. K. and Besant, C. B. 1983. A review of various vehicle guidance techniques that can be used by mobile robots or AGVs. In, *Proc. 2nd Int. Conf. on AGVS*, 7–9 June, 1983, pp. 195–210.

[3] Heiz, U. 1983. Einrichtung fuer orsbewegliche, ebenegebundene Objekte zur Selbstbestimmung ihrer Lagekoordinaten und Richtungswinkel. Swiss Patent No. EP 84 110 861.6, October 1983.

[4] Katoo, H., Fujiwara, K., Yoshikazu, A. and Hyogo, J. P. 1982. Roboterfahrzeug. German Offenlegungsschrift No. DE 9 3435117 A1, April 1982.

CENTRALISED AND DECENTRALISED CONTROL OF AGVs

H. Lindgren
SattControl AB, Sweden

Decentralised and centralised system solutions to AGV control are described. The advantages and disadvantages of the different solutions are discussed.

Today most automated guided vehicle (AGV) systems are designed with wire-guided AGVs, which follow a wire buried in the floor. The wire carries a current which generates an electromagnetic field that the AGV can detect and follow. Such wires are supplied with different frequencies, enabling junctions and crossings in the AGV route layout. To drive from one point to another the AGV follows a given frequency or set of frequencies.

To control the materials flow there must be some form of central computer which gives assignments to the AGVs in an optimal way. The system must include functions which avoid AGV collisions, set priorities at crossings, and interface to external units (e.g. pick-up and deposit stations, doors, fire alarms, etc.). Two different system solutions are described in this article.

Decentralised control

The decentralised control system is built in three levels: central control, distributed traffic control and AGV control. The block diagram of Fig. 1 shows a decentralised AGV control system.

Central control

The central control is performed by a mini- or microcomputer. At this level is decided which available AGV should take which assignment (i.e. pick-up) and where idle AGVs should wait until assignments are available. The central computer communicates with an administrative computer, controls automatic charging, compiles statistics, controls system restart, etc. These decisions are made using information from the administrative computer, from terminals connected directly to the

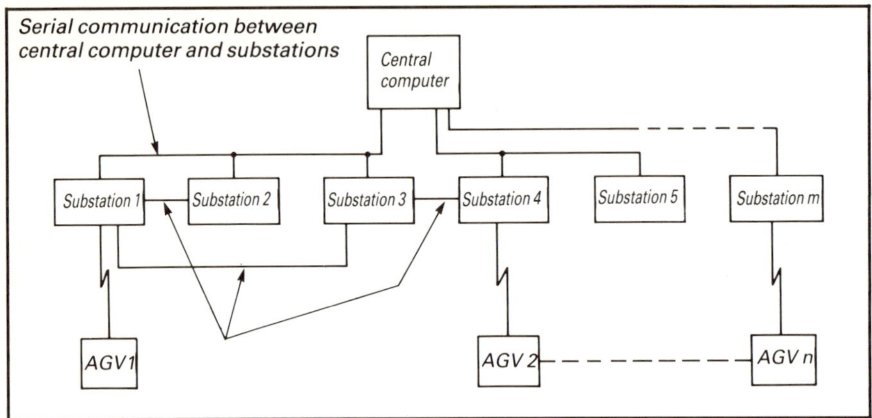

Fig. 1   Decentralised system

central computer, and from currrent system information. The latter is received via substations. When a decision is made to use a particular AGV for an assignment the information is sent to the substation which controls the section of the route layout where the AGV is.

Distributed traffic control

The distributed control is performed in microcomputer-based substations mounted around the AGV route layout. Each substation controls a section of the layout. The substations contain frequency generators, communication loop interfaces, digital input and output interfaces, and communication interfaces to the central computer and other substations.

The substation directs the AGV along the route to its destination. This is done at communication loops on the AGV route layout. Each substation has a map of its section of the layout and uses this when giving route directions to the AGV.

Traffic control is performed by the substation. It will not allow an AGV to continue unless the route ahead is clear. The substation also checks whether the AGV is loaded or not.

Via the communication loops the substation checks the AGV status and destination, and based upon this information it issues route directions. The information received from the AGV is sent from the substation to the central computer and forms the basis of statistics, error messages, etc.

When the AGVs drive from one substation area to another the different substations communicate with each other. Signals such as 'can AGV continue?' are sent from one substation and the receiving substation must reply with 'clear to enter' for the AGV to continue its route.

The substation also interfaces between the central computer and peripherals such as conveyors, transfer units, etc. Digital output signals can be set either by the central computer or directly by the substation (e.g. when coordinating loading and unloading of the AGV). Digital input signals are transmitted to the central computer. Input signals can be used for assignment creation in the central computer, and for the coordination of loading and unloading at substation level.

AGV control

The onboard AGV microcomputer transmits the AGV status to the substation via the communication loops. The AGV receives a new route direction from the substation in the same way.

The route direction that the AGV receives is a set of instructions describing how the AGV drives from the present communication point to the next. The description may contain a series of steps with different functions (e.g. change steering frequency, change AGV speed, change direction, etc.). The following is a typical set of instructions:

Step 1: Follow frequency one at full speed.
Step 2: Follow frequency two at half speed.
Step 3: Position at conveyor.
Step 4: Unload pallet.
Step 5: Report route direction completed to substation.

Centralised control

The centralised control system is also built in three levels: central control, assignment control and AGV control. The block diagram of Fig. 2 shows a centralised AGV control system.

Central control

As for a decentralised control system, the central control is performed by a mini- or microcomputer. At this level the AGV assignments are created and passed to the assignment control computer. Other functions at this level are communication with other computers, compilation of statistics, system restart, etc.

These decisions are based upon information from administrative computers, terminals and current system information. Current system information is received from the assignment control computer.

Assignment control

The assignment control is performed by a mini- or microcomputer. This can either be a separate computer or the same one as used for central control. The assignment controller also includes equipment necessary to communicate with the AGVs. This may be done in many ways, such as radio link, infrared light, ultrasonic sound, via the steering wire, etc. For simplicity, a radio link is used as an example in the following.

At this level decisions are made regarding choice of AGV for the

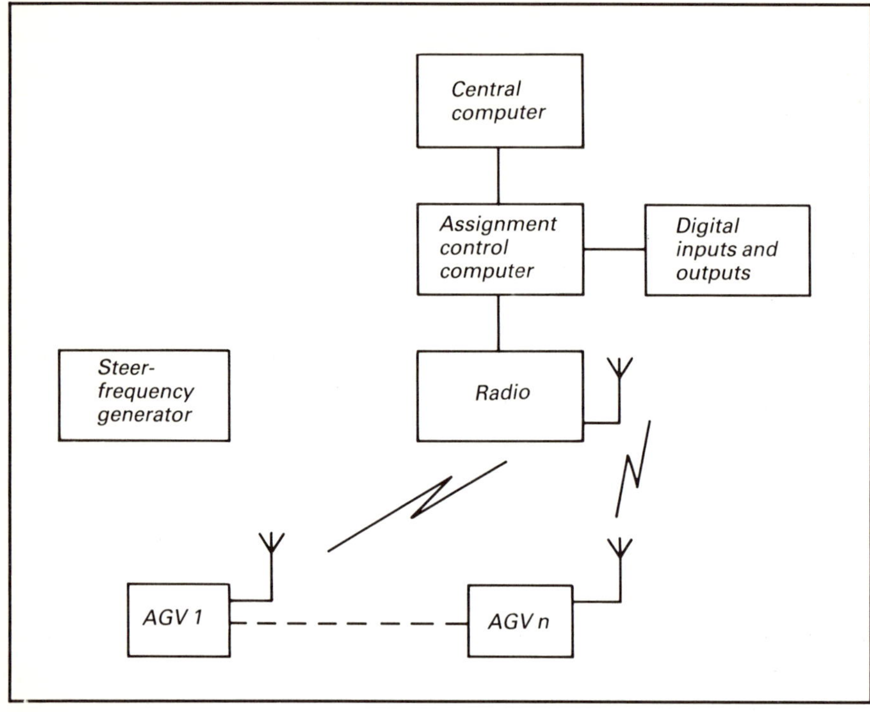

Fig. 2   *Centralised system*

next assignment and choice of wait position for idle AGVs. Functions for communication with the control computer, automatic charging and handling digital signals are also included at this level. In addition, there are the functions for the AGV communication via radio equipment.

To be able to give route directions to the AGVs the assignment computer has a description of the whole AGV route layout (map). The computer also keeps a record of each AGV's current position. There are two ways in which an AGV can record its position:

- Suitable positions in the AGV route layout are coded with an absolute code. The AGV reads these codes and can thereby report its current position.
- Suitable positions in the AGV route layout are marked. The AGV knows, from what it has previously done, which marked position it arrives at.

In between the coded or marked positions the AGV can measure distances by itself, and report these intermediate positions to the computer.

When a drive instruction is given to an AGV, the AGV knows how it should drive (which frequency, speed, etc.) to the next sub-destination. At the next sub-destination the AGV gets a new drive instruction, and so on. If the AGV does not get a drive instruction it will wait. This may occur when the route ahead is blocked.

The assignment computer acts as the contact between units such as conveyors, sensors, etc., supplying information such as where loads are waiting to be picked up. The computer uses this information for checks before an AGV is sent out.

AGV control

The microcomputer onboard the AGV has, via its radio equipment, continuous contact with the assignment computer. The AGV receives information the whole time from the computer concerning whether or not it should continue. This procedure eliminates the risk of AGVs colliding with each other at crossings, etc.

Advantages and disadvantages

Table 1 shows a comparison between the two systems described. Both the centralised and the decentralised control system require intelligent AGVs (i.e. onboard microcomputers).

Advantages of the decentralised solution

In the decentralised system, information regarding the locations of AGVs is more secure as each AGV reports at each communication point. This is because each communication point in the installation is unique and therefore the AGV itself does not need to measure distances or detect anything to establish its current position. This means that it is very simple to reinstate an AGV in the system after,

Table 1 Comparison between decentralised and centralised AGV control system

Function	Decentralised system	Centralised system
Assignment creation	Central control	Central control
Choice of AGV for an assignment	Central computer	Assignment control computer
Traffic control	Substations	Assignment control computer
Digital input and output signals	Substations	Assignment control computer or via distributed interface units
Interfaces between AGVs and conveyors	Substations	Assignment control computer or local signals between AGV and conveyor
Communication with AGVs	Inductive loops in floor which are connected to the substations	Continuous via radio link between assignment control computer and AGVs
AGV position detection	Absolute via the communication loops	Via some absolute positions and distance measuring
Layout description (map)	Substations	Assignment control computer

say, servicing or repair. The AGV can be placed anywhere in the installation, because the central computer finds out directly where the AGV is located, via the substation. The central computer can then either assign the AGV the previous assignment (if it was unfinished) or a new assignment.

Another substantial advantage is that each substation covers a section of the AGV layout, which makes it simple to describe a route map in each substation. Furthermore, the amount of data handled by the substation programs is manageable. The map contains, for each communication point controlled by the particular substation, the drive instruction for the AGV to reach the next communication point on its way to the final destination. The map also includes a description of which points, crossings, etc., have to be free for the AGV to be able to continue.

Each substation controls a limited area, which means that it is possible to test each section of the AGV layout locally. This can be done before the total layout has been installed and before linking the substations to the central computer. This means, also, that the installation can be set up and tested in stages while production continues.

In the decentralised solution it is easy to deal with the local interlockings for pick-up positions, deposit positions, doors, etc. Duplication of signals is not necessary as the signals required at, say, a deposit position by a roller conveyor would be required anyway for the control of that conveyor.

Fault finding in an operational installation is also easier in the decentralised solution because of the local test facility. In the local substation, status information can be read from the LED displays on the panel, or a terminal can be connected to the substation to obtain even more information. This status can be obtained locally while an engineer is observing the actual movements of the AGVs.

The decentralised solution does not impose any limit on the number of substations and AGVs (i.e. the size of the AGV system). A further advantage is that the cable route between fixed sensors and the substations is relatively short.

As traffic control is completely handled at substation level (the central computer only has to decide the final destination), a smaller central computer is required than in the centralised solution.

Disadvantages of the decentralised solution

A decentralised solution mostly means communication with the AGVs via inductive loops laid in the floor. In large or complex installations the number of inductive loops can be large. This means that a lot of track cutting in the floor has to be carried out to connect the inductive loops to the substations. Each substation controls a number of inductive loops, so in a large installation many substations may be required.

Each substation has to be linked to the central computer and, where no natural cable routes exist, this could present a problem. In the same way, each substation needs to be linked to substations controlling adjacent sections of the AGV route layout.

Advantages of the centralised solution

An important advantage of the centralised solution is that it requires less floor cutting (only tracks for the steering cables are needed). However, calibration points need to be fixed, in some way, around the AGV layout. Substations and the associated cabling are also not required.

When expanding the system, the only changes required are the extension of the steering cables, installation of new calibration points, and reprogramming of the route map held in the central computer.

Via the continuous communication between the central computer and the AGV, the central computer can monitor the AGV status at all times and provide detailed error messages and statistics.

Disadvantages of the centralised solution

With the use of a radio channel a limited number of AGVs can use each frequency. For example, if each AGV needs to communicate with the central computer once per second, using 2400 bauds on the data transfer rate, then a maximum of ten AGVs may use each frequency.

Permission from the relevant authorities is required for radio communication and approval for each frequency used has to be obtained. This could be a problem in some areas where many frequencies are already in use, and may take a long time. In particular, it may be difficult to get more than one frequency, which is necessary for a system with a large number of AGVs.

The basic cost of radio equipment may mean that a simple AGV system with a small number of AGVs may not be cost-effective. The equipment needed in each AGV for radio communication increases the cost of each AGV by approximately 10%. Furthermore, depending upon the environment (e.g. cold store, etc.), the antenna equipment required may be very complicated. Also, interference with the radio equipment may present a problem in certain types of industry (e.g. with welding machines).

In a centralised solution the central computer needs to be quite powerful for a larger AGV system in order to handle the large volumes of data. The map needs to cover the whole route layout, which makes it very hard to handle.

Digital input and output signals in centralised solutions need to be either connected directly to the central computer or distributed via local interface units. In the first instance it means a lot of cable installation; in the second instance, decentralised interface units with serial communication (comparable to substations). A method of

Table 2   Summary of the most important advantages and disadvantages

Function	Decentralised system	Centralised system
AGV position detection	+	−
Layout description (map)	+	−
Installation during production	+	−
Interfaces for pick-up and deposit positions	+	−
Fault-finding	+	−
Many AGVs in the system	+	−
Track cutting	−	+
Wiring for communication	−	+
Communication to AGVs	−	+
Small expansions of the system	−	+

avoiding too much signal cabling is to use local signal exchanges directly between the AGV and the fixed unit. This could be done, for instance, by a directional lamp on the AGV and a photocell on the fixed unit. The problem with this is that every AGV has to be adjusted to every fixed unit (e.g. in an installation with 10 AGVs and 20 fixed units this means that 200 adjustments have to be carried out).

In a centralised solution there are usually only a few fixed positions where an AGV can be put back into the system after, say, servicing. At each of these positions the operator needs to specify the location of the AGV via an onboard terminal.

Table 2 gives a summary of the most important advantages and disadvantages of the two systems.

Concluding remarks

As we have seen above, both decentralised and centralised AGV control systems have their advantages and disadvantages. When selecting the type of AGV control system, factors such as materials flow, available floor space and transport distances involved have to be taken into consideration.

A decentralised system will normally prove to be the best solution in an installation where at least one of the following points is significant:

- High materials flow.
- Many AGVs.
- Complex AGV layout.

In the same way, a centralised control system will normally be the best solution when there is a large but simple AGV layout with low materials flow and few AGVs.

DESIGN PARAMETERS AND SPECIFICATION OF AN AGV SYSTEM

U. Jansson
Schindler-Digitron, UK

This paper outlines the various aspects to be considered by the prospective buyer when considering the purchase of an AGV system. Its purpose is to highlight and discuss those areas vital to the success of such a project including the transport matrix, simulation and battery charging concepts. It assumes that the justification of the system has already been dealt with.

To design and specify an AGV system in detail involves a long list of activities and parameters to be clarified. The importance of each parameter varies depending on the type of application.

In a standard transport application (e.g. high bay warehouse) with a high capacity demand, parameters such as layout design, control system and computer simulation are of utmost importance.

Whilst in a flexible manufacturing system (FMS) application where a 4000kg machine pallet with a diameter of 1650mm is to be transferred with an accuracy of ±0.4mm, the physical and logical interface between the machines and the AGV come to the fore.

The purpose of this presentation is to give a general description of the most important parameters and in which steps an AGV system is designed. To limit the scope of this paper it is assumed that the logistic function of the transport is well analysed and that the potential to improve the efficiency with an AGV system has been defined. Fig. 1 outlines the basic design procedure of an AGV system.

Load analysis

The first parameters to analyse are, of course, the load itself and the pallet system used. All variants need to be defined as regards:

- Load dimensions.
- Maximum load weight.
- Centre of gravity.

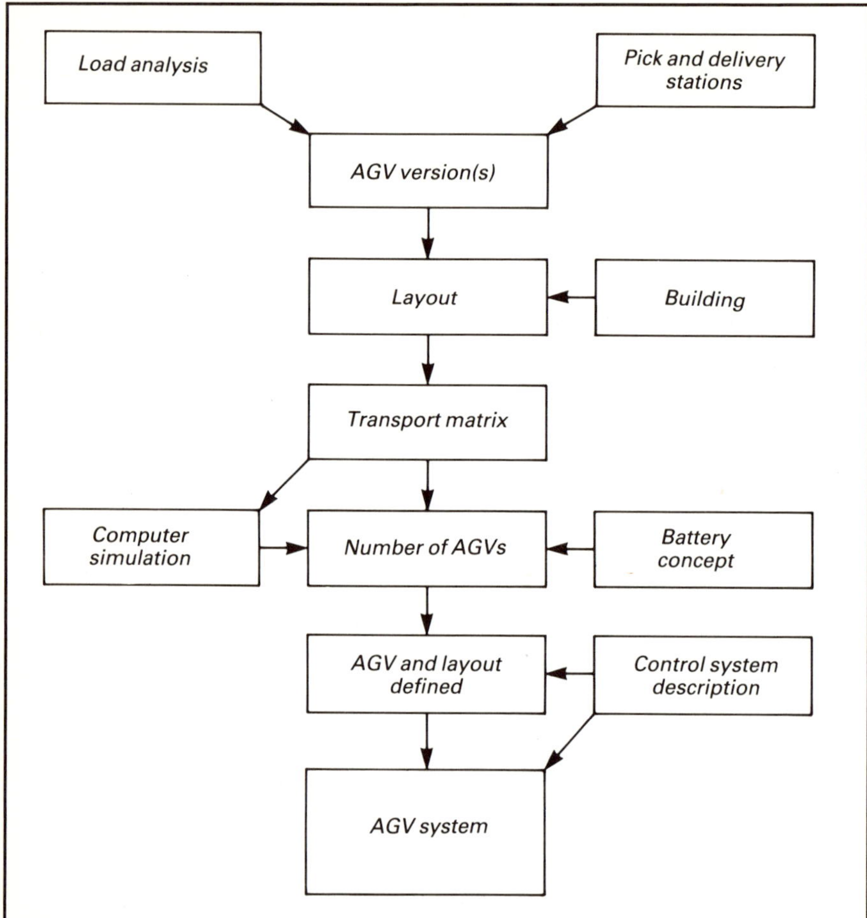

Fig. 1 Design procedure of an AGV system

- Pallet design and dimensions.
- How is the load secured (pallet frames, plastic stretch-wrapping, etc.)

To avoid special vehicles, which are expensive and have longer delivery times, it is recommended that the next step is to investigate the possibility of reducing the number of variants and to use a standard pallet system. To justify financially a mixture of special vehicles is often very difficult, especially if the pallets are also supposed to be stored in an efficient way.

Pick and delivery (P/D) stations

The next activity is to define the various pick and delivery stations required. Even here there is a need for standardisation. The pick and delivery stations (P/D stations) can be divided into two groups: passive and active.

Examples of passive stations are fixed racks (one or more levels) and floor storage; active stations: conveyors, lifts and machine pallet changers.

The active stations require a more complex exchange of logical signals between the AGV control system and, for example, the conveyor control system; whilst a passive station at most requires a presence signal.

It is here important to define the height of the P/D station and what positioning accuracy is required. A proper definition of both the logical and physical interface is vital for the detail design of the control system and the AGV itself.

AGV type

Based on the definition of the loads and the P/D stations, the most suitable AGV can be picked. However, a large number of different AGVs are available; the main differences are in:

- Maximum load.
- Maximum load dimensions.
- Moving pattern (one-way, symmetric with rotation, parallel movement, etc.).
- Maximum speed.
- How the load is carried.
- Level of intelligence on board.
- Control system concept.

Often a number of good alternatives exist and the building, the layout, the transport flow and the control system have to be considered before a final recommendation can be made.

Building and layout

If it is an existing building, the following have to be clarified:

- Floor condition (inspection of wear and other irregularities).
- Maximum floor load allowed.
- Depth of reinforcing bars within the floor.
- Level in differences, ramps, etc.
- Location of door openings (height and width) and other layout limitations (e.g. cable installations, ducting, etc.).

For a new building, the AGV supplier's recommendations for floor design should be discussed with the building contractor as early as possible. With the basic building conditions clarified the layout work can begin. Factors to be considered are:

- Width of the existing aisles.

- Area requirements for loading/unloading.
- Number of parallel P/D stations and/or conveyor buffer positions.
- One- or two-way traffic.
- Number of communication points.
- Single spur or double spur.
- Alternative routes to prevent bottlenecks.
- Location of battery chargers.
- Location of service area and test layout.
- Location of control cabinets.

Transport matrix

Based on a flow scheme a transport matrix can be produced. The matrix is essential for the calculation of the number of vehicles required.

The matrix should not only reflect the transport frequency between the various layout destinations, but should also define the variations (max./min.) during the time of operation, the distances, the relevant AGV specifications (speed, load type and load transfer time), empty transports and average speed. Fig. 2 shows a simplified example of a transport matrix.

The transport matrix can be used to provide a theoretical calculation of the required number of vehicles, and the layout can be analysed and

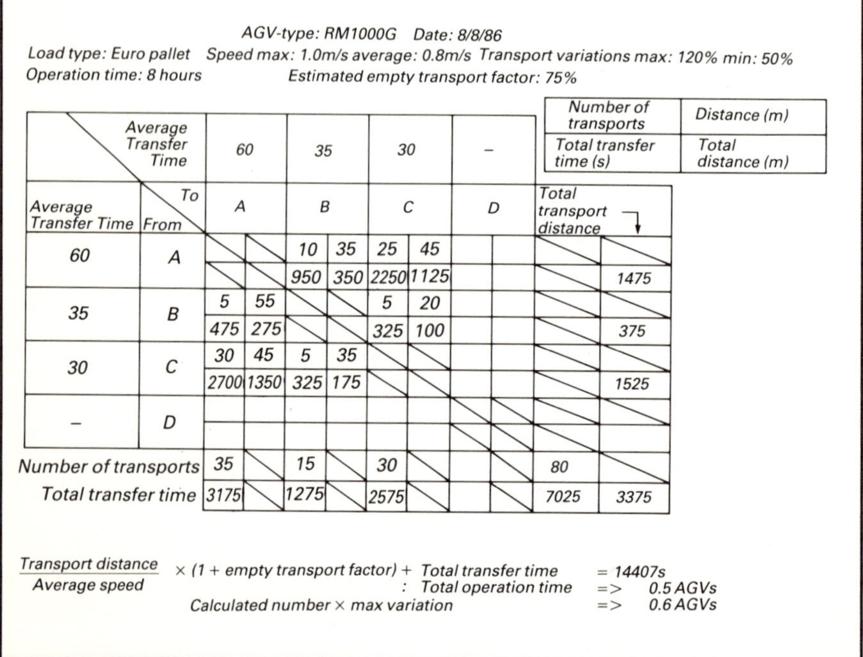

Fig. 2   *Example of transport matrix calculation*

adjusted as required. This type of calculation is often accurate enough for the offer phase; however, for the detailed design a computer simulation is strongly recommended.

Computer simulation

With a manual calculation the interplay between the various pieces of equipment is not reflected and it is difficult to get a realistic view of a complex AGV system. Other papers in this book (Section Three) explain the intrinsic details of computer simulation; however, some of the reasons for carrying out a computer simulation from an AGV supplier's point of view are given below:

- To find and adjust bottlenecks in the system during the planning phase.
- To guarantee the capacity and the number of vehicles required.
- To create a system in balance, which can cope with future expansion.
- To simplify the programming of the control system by defining the correct control strategy.
- To create a useful document for the final capacity test of the completed installation.

Nowadays there are many simulation packages available on the market. There are also quite large differences in the control systems of the AGV suppliers. It is therefore important to remember that only a simulation package, which in the true sense reflects the control strategy of the chosen AGV supplier, can be relied upon as a basis for guaranteed performance.

Battery concept

Design of an AGV system also includes decisions about the most suitable battery concept. The following factors are important:

- Total time in operation per day and week.
- Breaks available for battery charging.
- The variations in transport demands per AGV during a day or shift.
- Automated charging or battery exchange.

For one-shift applications the standard lead-acid battery is most common. For more than 10 hours' operation there is a point where either the AGV system has to have an over-capacity for charging during normal operation or the battery sets have to be charged during the operation. To reduce the manual operations to a minimum the fully automated systems of today normally have automated docking to the battery charger.

The alternative to lead-acid is nickel-cadmium (NiCd) batteries. Their advantage is that they can be charged in a short time. The disadvantage is the low maximum quantity of energy an NiCd battery carries. It has, therefore, to be recharged very frequently. However, in applications with multi-shift operations and a relatively low percentage of transport time, i.e. FMS applications and flexible assembly systems, NiCd batteries have proven to be successful.

Control system

So far only the physical design of an AGV system has been described. However, parallel to this the 'soul' of the AGV system, the control system has to be formed. The design and the structural basis of the control system are to a great extent the decisive factors in whether or not an AGV system will be a success.

It is important that the software is developed for a real-time environment, i.e. with short response times, data security, data processing and quick re-start of the whole equipment level after downtime. The control system does not have to be complicated. For many applications with point-to-point transport and a manual disposition, a single local control cabinet is sufficient. For applications with a complex layout and a high throughput a more comprehensive structure is required. Each local control cabinet controls a part of the layout and is connected to a central master controller for overall coordination. With comprehensive traffic rules and a disposition which is automatic and dynamic, optimal utilisations of the vehicles can be attained.

However, today more than ever, with the introduction of concepts like CIM (computer integrated manufacturing) and production automation, the designer of the AGV control structure cannot only look upon the needs of the transport functions; other conditions and desires have to be considered:

- Other equipment and functions.
- Connection to the existing systems.
- The company's strategy for computers and communication.
- Clear interfaces between suppliers.
- Extendability/introduction in stages.
- Internal or external software development.

Theoretically different base solutions can be discerned. Practically, the result often is a hierarchical logical structure. The ideal solution is a network in which each subsystem unit can communicate with each other independent of computer make.

Concluding remarks

To design an AGV system is very much a question of physical and logical system integration. At the same time the trade of automated material handling develops very quickly and the initial design phase becomes more and more important.

In this context there are two pieces of advice for future AGV users:

- Budget for a proper paid study. In an early stage form a strong partnership with one of the suppliers, where the state-of-the-art knowledge is.
- Do not forget the users. A perfect theoretical plan has little value in the commissioning phase if the final users have not had their say, been educated, or been motivated.

3
Operation

Effective operation is clearly dependent on good design – the two cannot be easily separated. Thus the first three papers deal with operational characteristics such as the number of vehicles, capacity and availability of the system, flexibility and control issues, such as dispatching rules to give satisfactory vehicle utilisation as well as good handling service. But the second paper in particular points back to the dependency of operational factors on the design of the basic layout. The last two papers both illustrate how computer simulation techniques are used to evaluate AGV networks.

CAPACITY, AVAILABILITY AND FLEXIBILITY OF AGV SYSTEMS

S. Karlsson
AB Bygg-och Transportekonomi (BT), Sweden

The rapid development of AGV systems means that technical innovations are often incorporated before their effect on system capabilities is fully appreciated. Relations between technical parameters and important system parameters – transport capacity, availability and flexibility – are presented. These relations offer a basis for evaluation of new technical features.

Product and system development for automated guided vehicles (AGVs) is very rapid and innovative. However, technical features are sometimes implemented without due consideration of their influence on system capability. It is obvious that development is useful only if it means increased system capability. Thus to be able to evaluate the importance of design ideas and basic technical parameters it is essential to analyse their effect on important system parameters like transport capacity, availability and flexibility. This paper presents some general relations between basic technical parameters and system capabilities for typical AGV applications in warehouses and production.

Capacity

Computer simulation is very often used when analysing complex systems, and stochastic simulation with a computer is, of course, a useful tool when working out capacity calculations for AGV systems. However, in order to get a deeper understanding of the properties of the system, mathematical and analytic methods are often superior to simulation. The calculation method described here, has proved to be very useful when evaluating layouts and technical variables with respect to capacity. The characteristic properties are shown in the 'capacity diagram' (Fig. 1).

Along the X-axis is transport flow (number of pallet movements per hour); the Y-axis is used for average cycle time (i.e. the average time an AGV is allocated for each transport operation, including time for

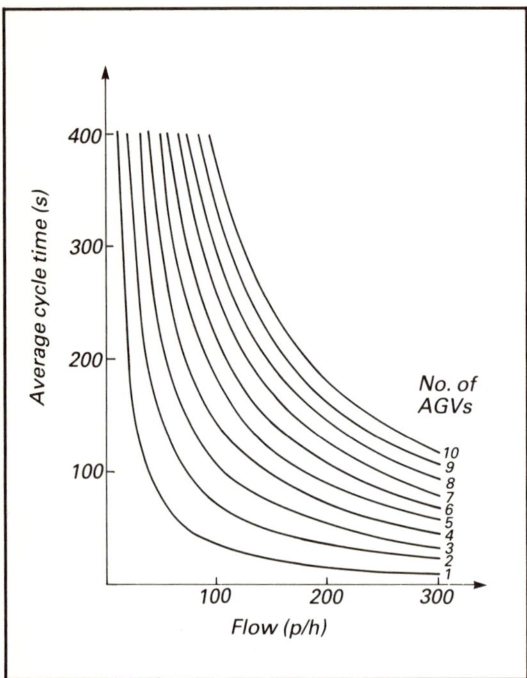

Fig. 1   Capacity diagram: average cycle time vs pallet flow

waiting, driving empty, etc). The diagram shows curves, which gives the ratio between flow and average cycle time when using 1, 2, 3, 4, 5, 6, 7, 8, 9 and 10 vehicles respectively. So far the diagram is common for all kinds of cyclic transport systems.

When analysing a specific AGV system it can be a good idea to first search for the part of the layout that has the lowest capacity (the 'bottleneck'), which can be illustrated by a vertical line in the diagram. If, for example, a pick-up-station is to be used for 43% of the transports and for 30 seconds at each pick-up operation, the pick-up station will have a maximum capacity of 120 pallets/hour (p/h) and will set a limit for the entire system at 280p/h (Fig. 2a).

If waiting times due to queues are not considered in the analysis, the cycle time for a specific system can be represented by a horizontal line in the diagram. However, in practice waiting times will appear at P/D-stations (pick-up and delivery) and in crossings. These waiting times increase as flow increases. They can be calculated by methods using queue theory as discussed below.

Fig. 2a includes a 'system characteristic curve' (curve a) for a typical automatic warehouse application with relatively short transport distances. The capacity requirement for this installation is 150p/h. The diagram shows that six AGVs are necessary for the required flow. It is also clear how the utilisation of each further AGV decreases due to queues. The first AGV will manage 35p/h, while the sixth only adds 26p/h to the system capacity.

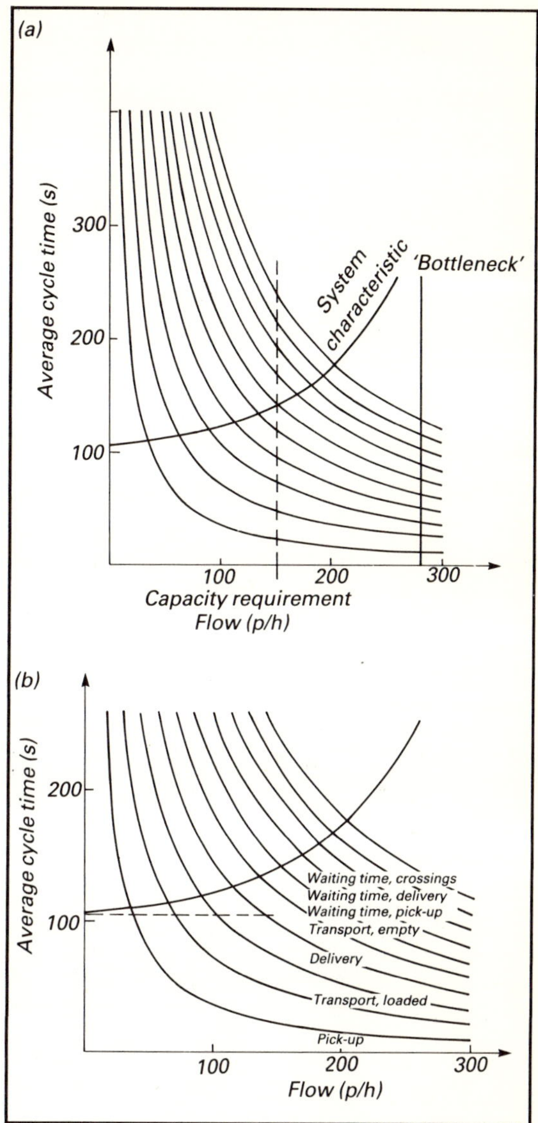

Fig. 2 Capacity diagram for example system

Fig. 2b shows in greater detail how the cycle time is divided at 150p/h:

Pick-up time	20s
Transport time, loaded	41s
Delivery time	20s
Transport time, empty	23s
Waiting time, pick-up	11s
Waiting time, delivery	10s
Waiting time, crossings	16s
Total cycle time	141s

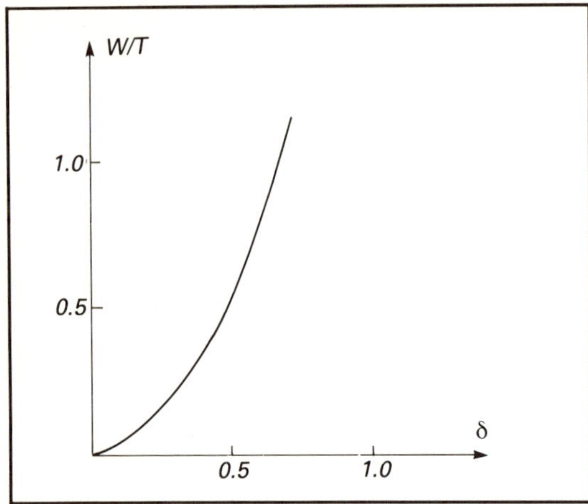

Fig. 3   *Station waiting time index vs utilisation factor*

Calculation of average transport times for loaded AGVs is easy to carry out if the layout and distribution of assignments between different P/D-stations is known. To calculate transport times for empty AGVs it is also necessary to know the strategies for assigning idle AGVs. The waiting times are calculated by summarising what each P/D-station and each crossing add to the total cycle time. The method is as follows. First calculate the time (T) the P-station is occupied for each pallet to be picked up (time to drive from waiting position to pick-up position plus time to load plus time to get destination from warehouse control system). The waiting time (W) as a function of T is given by

$$\frac{W}{T} = \frac{\delta}{2(1-\delta)}$$

where δ is the utilisation factor ($\delta \rightarrow 1$ when the P-station is 100% occupied). This is illustrated in Fig. 3.

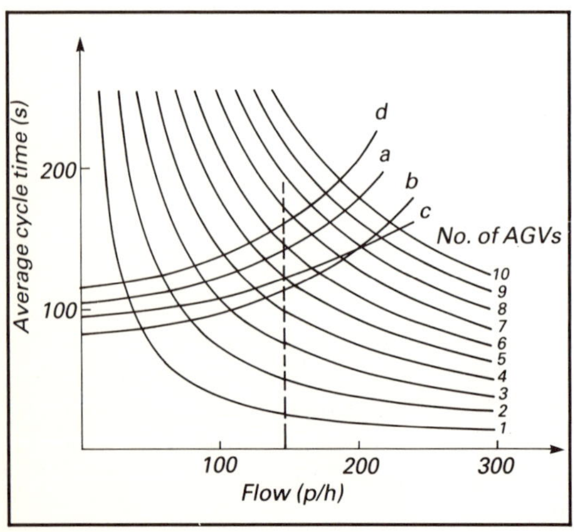

Fig. 4   *Effect on system capacity of various technical parameters*

The method described above is used to determine how basic technical parameters influence the system capability. The following are examples of application of the method.

Increase AGV speed by 50% (see Fig. 4, curve b). Increased speed will of course have an influence on transport times, but also to a certain extent on waiting times, as the utilisation factor for P/D and crossings will decrease. The diagram shows that by increasing speed by 50% one can reduce the number of AGVs from six to five and still manage the required flow. The capacity with five AGVs will be 155p/h (i.e. transportation work per AGV will increase by 24%). Alternatively, six AGVs will manage a flow of 174p/h (i.e. the total transportation work will increase by 16%).

It is obvious that speed is more important for the system capability in installations with longer distances. The majority of all AGV applications in warehouses and production will have a 15–30% increase in capacity when AGV speed is increased by 50%.

Reduce P/D-time by 25% (see Fig. 4, curve c). This will also influence waiting times at P/D-stations. From the diagram it is seen that five AGVs would be enough to manage the required flow. System capacity with six AGVs is 170p/h (i.e. 13% increase in transport work).

In addition, the diagram shows that if a future expansion to eight AGVs or more is considered the reduction in P/D-time is more important than the increased speed. Generally in applications with short distances (as is normally the case in warehouses) it is very important in layout design work to make sure that utilisation factors for P/D-stations are kept low. A good rule of thumb is never to utilise a P/D-station more than 50%.

Increase time driving empty by 40% (see Fig. 4, curve d). With insufficient strategies for assigning AGVs it is easy to increase the time for driving empty. In this example, 9 seconds on average per transport would mean a 40% increrease. Then six AGVs would manage only 139p/h (i.e. 93% of required capacity). Adding an extra AGV would give a capacity of 155p/h, and the transport work per AGV would be only 89% of what was first calculated (curve a).

Battery capacity and charging technique

The above-described method is very useful when analysing capacity for the most busy part of the day, for which most systems are designed. To have all AGVs available during the busy hours, the batteries should be charged at other times. Charging during production hours is often done when the daily production time is 9–18 hours (if shorter, charging can be done at night; if longer, battery replacement is normally required).

To be able to optimise charging during production hours it is necessary to know the required capacity in the near future. If this knowledge is not available there is a big risk that AGVs immediately after being sent to charging will be required because a new batch of pallets has been received. Thus the most important basis for controlled battery charging is a production plan. With such a plan there are opportunities to achieve further improvements in the system by using an advanced charging technique based on accurate measurements of charging content.

Availability

When analysing the availability of an AGV system it is necessary to look at the system components and also at the other systems cooperating with the AGV system. Take the example of an automatic warehouse (see Fig. 5). The system consists of components in series and in parallel. Parallel components can normally replace each other when one is out of order.

The availability for the entire system is:

$$\eta = \eta_{crane} \times \eta_{control} \times \eta_{AGV} \times \eta_{conveyor}$$

Note that the AGV system is not able to function if the warehouse control system is out of order (the complete system is not available). Thus it is important to analyse the complete system when calculating the availability of automatic systems. If this is done properly early in the project it is possible to design the system for high availability by suitable layout design, by avoiding overly centralised control systems, by doubling certain components, etc. On the other hand, if the availability analysis is not carried out properly, even a system constructed with very reliable components can have a low availability. In fact, in most cases good maintenance and a suitable spare part stock are even more important for system availability than reliable components.

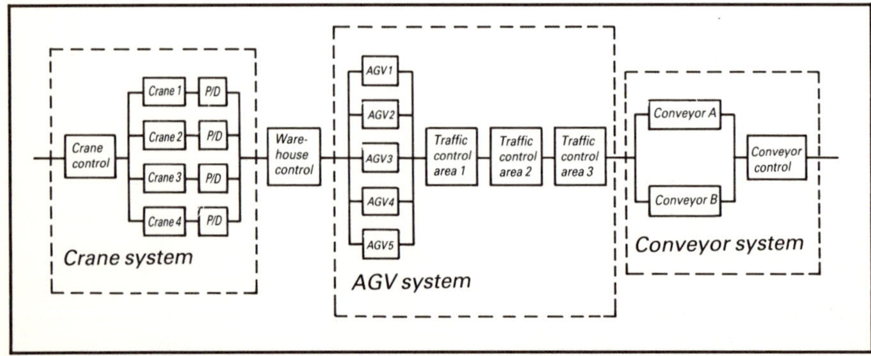

Fig. 5 Availability diagram of an automated warehouse

Now back to the example. The most important components of the AGV system are traffic control and the AGV guidewire network. Traffic contol is in most cases built up using very reliable electronic equipment. The most important parameter for this is the time to correct errors. Traffic control can have a negative influence on system availability if a long time is needed to localise and correct errors that occur very rarely. Equipment for traffic control therefore has to be designed with good troubleshooting aids and built up in such a way that defect units can be replaced very quickly.

Guidewire networks are extremely reliable. After studying different principles for automatic truck systems, this author is convinced that wire guidance will be the leading system for at least the next five years because of its high availability.

The AGVs do not have as much influence as the other components on system availability. If one AGV is out of order, normally the system is still available. To fulfil capacity requirements, however, in most cases it is important to have all AGVs available during busy hours. Thus when designing AGVs, as opposed to manual trucks, more effort should be directed toward reliability, troubleshooting aids and easy replacement of parts. For a good service an AGV must have an availability of 97–98%.

In summary, the two most important factors to achieve a high system availability are:

- System design – this should be carried out with due care and attention.
- Service – get the best!

Flexibility

Generally, automatic systems are less flexible than manual systems and AGV systems are more flexible than conveyor systems. Thus it is not surprising that one's opinion about the flexibility of AGV systems depends very much on the terms of reference. Here, 'flexibility' is defined in accordance with the following considerations:

- The general system concept should be easy to adapt to different applications.
- The system should be designed for 'flexible' utilisation of its components.
- The installed system should be easy to change.

Although most AGV systems are relatively flexible in all three respects, layout changes are very rare in most systems after production start-up. The reason could be that flexibility is not needed or that the systems are too difficult to change. This author believes that the second reason comes closer to the truth. Even though it is relatively easy to make changes in and extensions of a system, the user often encounters difficulties. His knowledge of how the system is designed is limited and he will often have problems in specifying software changes both in traffic control and in warehouse control (e.g. strategies for assigning AGVs).

Furthermore, it is often said that automatic systems without wires are more flexible than systems with wire guidance. This author does not agree. With present technology for computer programming it is much more complicated (and expensive) to reprogram control systems than to move wires in the floor. It will accordingly be more difficult to change a wireless system if a programming technique has not first been developed.

Therefore, to increase flexibility in an AGV system, the following is recommended: Develop programming languages for AGVs and AGV control that can be understood by the system user. It should be possible for the customer, or the service engineer responsible for the system, to do reprogramming and start-up after a change in the guidewire layout. Examples of languages which meet this requirement are CARMOD (Carrago modular layout definition system) and CAL (Carriage assembly language). With these it is also possible to define the entire system with terminal dialogues.

Concluding remarks

Here, the importance of capacity, availability and flexibility have been discussed separately. In reality they are of course not indepedent of each other. Activities intended to improve one parameter can have a negative influence on the others. Thus, finally, it is appropriate to give some idea of how to optimise AGV system capability with respect to all three parameters:

- Develop well standardised products and software packages that can be configured in different ways to cover a wide range of applications. For example, load handling on the AGVs should easily be adapted to different loads without the need to redesign other AGV functions.
- The system design work should be carried out very carefully. Mistakes in the project planning phase are very difficult to correct afterwards.
- Make sure that every system after installation is backed up by a professional service organisation supported by good troubleshooting aids.

OPERATIONAL CHARACTERISTICS OF AGV SYSTEMS

W. L. Maxwell and J. A. Muckstadt
Cornell University, USA

Designing an automatic guided vehicle system is a complex task. Besides hardware considerations, the design engineer should assess the impacts on facilities' layout, material procurement policy, and production policy. Some analysis techniques and a methodological framework for specifying the operational characteristics of an automatic guided vehicle system are presented.

When planning a layout for a manufacturing facility, a large number of factors must be considered. This paper is concerned with how manufacturing systems' layouts are affected when movement of material – components as well as finished products – is accomplished using an automatic guided vehicle (AGV) system. Normally the engineer concentrates first on determining where individual departments, loading and unloading docks, storage facilities, and aisles will be located within a facility. The layout of each department determines, among many other things, where components will enter the department (drop-off points for a material movement system) and leave the department (pick-up points for the system).

If driver-operated lift-trucks are used to perform the pick-up and drop-off activities, then considerable attention is placed on locating departments and the pick-up and drop-off points within departments so that the total travel distances, taking into account flow quantities, are kept as low as possible. Planning aids, such as CRAFT[1,2], are often used to assist the engineer in developing a layout; others can be found elsewhere[3].

When an AGV system is used to move material, the layout problem becomes considerably more complicated. For the purposes of this paper, an AGV system is one in which identical vehicles move loads without an accompanying operator. Movement is accomplished using some type of guide-path in or on the floor, on-board computers or other control devices, local traffic control devices, and possibly network control computers. For these types of systems, in addition to the facility and departmental layout issues mentioned, the engineer must also consider track layout alternatives. There are, of course, mechanical and electrical problems associated with the type of equipment selected. The equipment may be a self-propelled tugger that pulls trailers, a self-propelled platform, or a towline system. The system may interface with an automatic storage and retrieval system. Examples of these types of hardware can be seen elsewhere[4].

However, no matter which type of system is selected and no matter what type of control system is used, the engineer must address certain questions concerning track layout. For example, the designer must ask: How will the aisles be used? What will the direction of movement be in each aisle? Should movement be permitted in both directions? Should there be bypasses at pick-up and drop-off points? If so, how long should they be? Should there be spurs? If so, what should their lengths be?

In addition, control issues for the system must be studied carefully, so that the system will operate effectively. Loading and unloading functions must be accomplished with minimal delay; otherwise, congestion will occur. This in turn causes more vehicles to be needed and the utilisation of each vehicle to decrease. Thus the system must be designed so that the loading and unloading functions occur in a timely manner rather than at someone's leisure.

A dispatching system must be designed to ensure proper vehicle utilisation. Components and raw materials must be moved from receiving docks to storage facilities to production lines according to needs as they arise. Finished products of subassemblies must be moved from each department to storage areas or shipping docks. The system may also be used to move waste materials from the manufacturing area. These functions cannot be carried out effectively unless considerable thought and design effort has gone into planning the vehicle dispatching and control system.

The extent of the design effort for the AGV system depends on the flow quantities, distances travelled, and the timing of the requirements for the vehicles. However, unless there is a reasonable amount of activity per unit of time, there is probably little need to have an AGV system at all, unless there are some overriding environmental or safety concerns.

One purpose of this paper is to show how the design of an AGV system can determine the minimum number of required vehicles. Determination of the optimal number of vehicles is extremely difficult when considering detailed time-phased pick-up and drop-off requirements, pick-up and drop-off area floor space capacity, and track congestion (one vehicle blocking the path of another). A large-scale integer program can be formulated including all these factors (and solved for certain small problems). Rather than pursuing this approach, which the present authors feel is intractable at this time, this paper describes a simple time-dependent model to find the minimum number of vehicles. The authors' experience in designing real systems has proven the worth of this approach.

The paper's second goal is to present other analysis tools that can be used to evaluate the time-dependent behaviour of an AGV system. In particular, a procedure is described for dispatching vehicles and it is shown how to measure the blocking time caused by congestion and the size of storage areas.

The design aids presented can be used, with very slight modification, for planning a layout when lift-trucks are used for transporting material.

The problem

It is assumed that the designer has a particular layout in mind for a system and wants to determine how many vehicles are needed to support that design. An example of a proposed track layout is shown in Fig. 1. The track layout can be thought of as a network. The vehicles move through the network between *nodes* (labelled A, B, C,

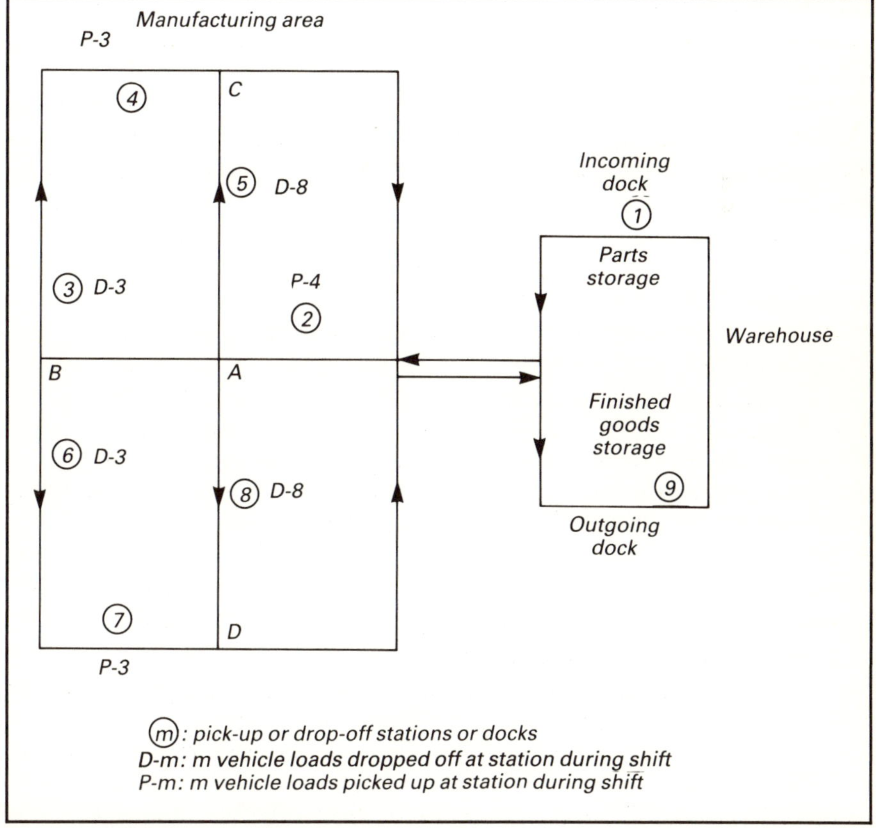

Fig. 1   Layout for example problem

D, E in Fig. 1) on directed segments, which correspond to a guide-path connecting one node with another. The nodes correspond to the intersection points of the various segments of the guide-path. In Fig. 1 the segments, denoted by ordered pairs of nodes, are (A,B), (A,C), A,D), (B,C), (B,D), (C,E), (D,E), and (E,A). The segment (E,A) involves travelling from the manufacturing area, through the warehouse, to diversion node A. Here, a vehicle entering node E from segments (C,E) or (D,E) is not allowed to move directly to node A; one might wish to consider allowing this type of movement as an alternative, however.

Each segment may have some member of pick-up or drop-off stations located on it, as illustrated in Fig. 1. For control purposes, segments are often subdivided into a sequence of non-overlapping and contiguous zones. The zones are usually designed so that no more than one vehicle is permitted within a zone at a time. Computer-based control systems often have some type of zone control to prohibit collisions. Zone control is also used to control access to intersections such as nodes C, D, and E in Fig. 1.

The system being analysed is assumed to have the following characteristics:

- Vehicles move in only one direction on any segment. (Some systems which have elaborate and more costly control systems allow the direction of flow to alternate. For simplicity, it is assumed that flows always go in one direction.)
- Vehicles do not pass each other on a segment. There are no spurs or bypasses in the system. This assumption makes the final analysis harder since congestion is reduced when spurs or bypasses exist. (This assumption can be dropped without affecting the primary method of analysis.)
- A zone control system is used to prevent collisions, particularly at intersections.

Also, the following data are assumed to be available:

- Load and unload times are known for each location.
- Travel speeds between stations for loaded and unloaded vehicles are known so that travel times are easily computed.
- The amount required to be moved from one station to another is known. It is assumed that there are integer vehicle loads and that vehicles are always dispatched to pick up or drop off complete loads. In other words, if one vehicle load must be moved from station i to station j, then one vehicle is used to accomplish this task. Splitting of loads between vehicles is not allowed. This assumption is crucial since splitting loads or carrying part of one load combined with part of another load substantially increases the combinatorial aspects of the problem. Let v_{ij} be the number of vehicle loads that must be moved from station i to staion j during a shift. If delivery

requirements vary from shift to shift during a day due to different production line manning schedules or different raw material dock arrival schedules, then the following analysis must be done for each shift.

Given the set of assumptions and available data, a mathematical model can be formulated to find the minimum number of vehicles needed to move the required vehicle loads. The model presented is time independent; that is, there is no attempt to account for the times at which loads become available for shipment or are needed at the appropriate stations. Since time is essentially ignored, the model is also unable to account for blocking or congestion in the system. These aspects of the problem are investigated in the penultimate section.

The model

First, observe that at least a certain amount of time is required to carry out the movement of material. If $\bar{t}_{ij}$ measures the time to load a vehicle at i, move it from i to j, and unload it at j, then $\sum_{ij} v_{ij} \cdot \bar{t}_{ij}$ units of vehicle time must be available. Let t_i be the load time at station i, u_j the unload time at station j, and t'_{ij} the time for a loaded vehicle to travel from i to j. Then $\bar{t}_{ij} = t_i + u_j + t'_{ij}$. Also, observe that if movement is to take place between two stations it will be best accomplished using the shortest time path because blocking is not considered in the model described in this section. Observe, too, that $\sum_j v_{ij}$ vehicles are needed at station i to move the material, and that $\sum_i v_{ij}$ vehicles arrive at station j during the shift. For stations which are not storage points for vehicles, the total vehicle flow into the station within the shift must equal the total vehicle flow out. If a station is a place at which vehicles can accumulate, then there is a possibility that more may come in than go out. It is also assumed that some number of vehicle trips are available for dispatch at the beginning and end of the shift at certain stations.

The model's objective function will measure the total travel time for empty vehicles moving between stations. By adding this amount to $\sum_{ij} v_{ij} \cdot \bar{t}_{ij}$ the total vehicle time required during the shift can be computed. Thus if this total is H hours, and h hours are available on the shift per vehicle, then $[H/h]$ vehicles are required to carry out the required material movement plan, where $[x]$ is the smallest integer greater than or equal to x.

To formulate the problem the net vehicle flow into each station is determined. The net flow for station i is $\sum_j v_{ji} - \sum_j v_{ij}$. Since there may be a requirement to have a certain number of vehicles at a station at the end of a shift or to have some vehicles available at the beginning of a shift, the net flow for station i must be adjusted to reflect these possibilities. For example, if f_i vehicles are available at the beginning of the shift and g_i vehicles are required at the end of the shift, the net flow for station i is:

$$\text{NF}(i) = \sum_j v_{ji} - \sum_j v_{ij} + f_i - g_i$$

For the problem to be well defined, $\sum_i NF(i) = 0$. Thus the problem is to determine how to allocate the vehicles that are available ($\text{NF}(i) > 0$) at station i to satisfy the deficits at other stations j ($\text{NF}(j) < 0$) so that the total travel time for moving empty vehicles is as small as possible.

Let

$$a_i = \begin{cases} \text{NF}(i) & , \text{if NF}(i) \geqslant 0 \\ 0 & , \text{otherwise} \end{cases}$$

$$b_i = \begin{cases} \text{NF}(i) & , \text{if NF}(i) < 0 \\ 0 & , \text{otherwise} \end{cases}$$

and let t_{ij} be the shortest travel time from station i to station j when a vehicle is unloaded, and x_{ij} the number of empty vehicle trips that should be made from station i to station j during the shift.

The problem is to find the values for the variables x_{ij} that minimise

$$\sum_i \sum_j t_{ij} x_{ij}$$

subject to

$$\sum_j x_{ij} = a_i, \text{ if the net flow for station } i \text{ is non-negative}$$

$$-\sum_k x_{ki} = b_i, \text{ if the net flow into station } i \text{ is negative} \quad \text{(SP)}$$

$$\text{and } x_{ij} \geqslant 0$$

It is easy to see that the above problem is a simple transportation problem of linear programming[5,6]. The solution indicates how many vehicle trips should be made with empty vehicles between stations i and j. Since this problem is a transportation problem, all variables will have integer values.

Consider an example that illustrates these ideas. It is assumed that a layout has been specified (the track layout and the positioning of the pick-up and drop-off stations are known). Only the issue of vehicle requirements must be considered. Specifically, it is assumed that the aisle configuration, the proposed positions for the pick-up and drop-off stations, and the proposed track layout – including the direction in which vehicles can travel – are as shown in Fig. 1. The aisle between the warehouse and manufacturing area has two guide-paths, one for each direction.

The basic travel time data required are the times to move an empty vehicle from a node to the nearest station on each segment from that node (or to the node ending the segment if there are no stations in a segment), the travel time for an empty vehicle from a station on a

Table 1   Travel time data for layout of Fig. 1

From	A	A	A	B	B	5	8	3	6	C	D	4	7	E	9	1	2
To	B	5	8	3	6	C	D	4	7	E	E	C	D	9	1	2	A
Time(s)	52.0	80.0	56.0	52.0	52.0	48.0	48.0	144.0	132.0	210.0	210.0	48.0	36.0	126.0	166.0	180.0	40.0

Table 2   Travel times between stations (min)

FROM	TO Station								
Station	1	2	3	4	5	6	7	8	9
1	–	3.0	5.4	7.8	5.0	5.4	7.6	4.6	6.0
2	10.6	–	2.4	4.8	2.0	2.4	4.6	1.6	6.0
3	11.4	14.4	–	2.4	16.4	16.8	19.0	16.0	8.8
4	9.0	12.0	14.4	–	14.0	14.4	16.6	13.6	6.4
5	9.0	12.0	14.4	16.8	–	14.4	16.6	13.6	6.4
6	11.0	14.0	16.4	18.8	16.0	–	2.2	15.6	8.4
7	8.8	11.8	14.2	16.6	13.8	14.2	–	13.4	6.2
8	9.0	12.0	14.4	16.8	14.0	14.4	16.6	–	6.4
9	2.6	5.6	8.0	10.4	7.6	8.0	10.2	7.2	–

Note: The intersection at node E in Fig. 1 exists only for merging of the input segments to the warehouse. After entering the manufacturing area a vehicle must return to the warehouse before again entering the manufacturing area

segment to the next, and the travel time for an empty vehicle from the last station on a segment to the end node of the segment. For the layout of Fig. 1 these times are given in Table 1.

From these basic travel times, the minimum travel times between stations are calculated (see Table 2). (As stated, the shortest travel time route is used since blocking is not considered in this stage of the analysis.) For this problem these times can be determined easily. For more complicated problems these times will be found using a shortest-route algorithm[5]. For simplicity it is assumed that the minimum loaded travel time, t_{ij}, is the same as the minimum unloaded travel time, t_{ij} for track layouts involving ramps or elevators, two separate travel time tables are necessary for the analysis.

The number of vehicle loads moving from a 'FROM' station to a 'TO' station is given in Table 3. It is seen that a total of 22 vehicles loads must move from station 1 to other stations and that a total of 10 loads must be sent from other stations to station 9. The net flows for each station are given in Table 4; these are used to construct the right-hand-style values in the transportation model. The transportation network corresponding to this problem is shown in Fig. 2. An optimal solution to this problem is given in Table 5. Note that there is more than one optimal solution to this problem.

Table 3   FROM–TO Vehicles chart (loads per shift)

FROM Station	TO Station								
	1	2	3	4	5	6	7	8	9
1			3		8	3		8	
2									4
3									
4									3
5									
6									
7									3
8									
9									

Table 4   Movement of unloaded vehicles

	Station number									
	1	2	3	4	5	6	7	8	9	Totals
Total TO	0	0	3	0	8	3	0	8	10	32
Total FROM	22	4	0	3	0	0	3	0	0	32
Net flow	−22	−4	3	−3	8	3	−3	8	10	0

Table 5   An optical solution for the example problem

FROM Station		3	5	9	9	6	8
TO Station		4	1	2	1	7	1
Number of empty vehicles moved		8	4	6	3	8

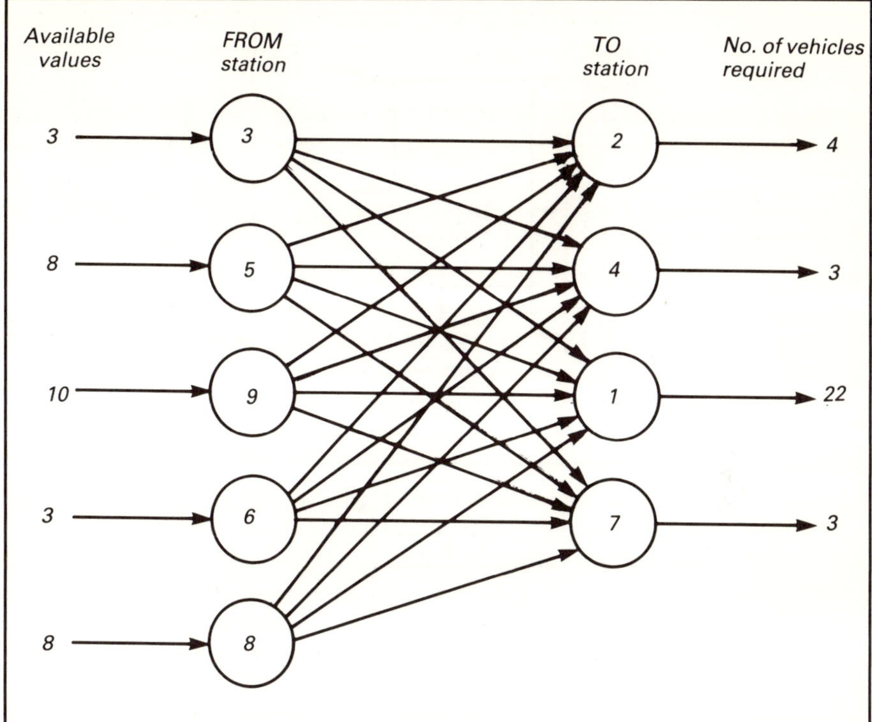

Fig. 2   Transportation network for the example problem

To find the minimum number of vehicles, the total loaded travel, unload, and load times are first computed. Assuming that each pick-up at station 1 and 2 and drop-off at station 9 takes 5 minutes and all other pick-ups and drop-offs each require 20 minutes, this is

$$\sum_{ij} v_{ij}\, \bar{t}_{ij} = 919\text{min}$$

From the solution of prroblem SP, the minimum total unloaded vehicle travel time is

$$\sum_{ij} v_{ij}\, x_{ij} = 195.8\text{min}$$

Assuming 430 minutes per shift per vehicle, the minimum number of vehicles is

$$N = \left\lceil \frac{114.8}{430} \right\rceil = 3$$

Thus at least three vehicles are required. To this point, the dispatching rule needed to implement the solution has not been considered. Consequently it has not been possible to determine the extent of blocking that might occur. However, with the design of the track system as shown in Fig. 1, little congestion can occur when a reasonable dispatching rule is followed. Three vehicles are all that are needed for this problem. The next section presents a dispatching rule and a method for calculating the blocking associated with it and the layout.

As mentioned, the layout in the example has certain desirable characteristics that should help minimise vehicle blocking. First, note that to minimise the unloaded vehicle travel distance the drop-off stations should be located with respect to the pick-up stations such that for each station located on the segment the total drop-off activity for that and preceding stations on the segment is at least as great as the pick-up activity. For example, for segment (B,C) in Fig. 1 there is drop-off activity at station 3 and pick-up at station 4. The pick-up at station 4 can be accomplished using the three vehicles used to drop off material at station 3. If the drop-off/pick-up activity were reversed between these stations, empty vehicle travel time would increase. If the stations are not arranged properly, the minimum number of required vehicles can easily increase.

The concept of station orientation can be formalised as follows: Let $S(a,b)$ be the set of stations on segment (a,b). For the example problem these sets are given in Table 6.

Let $W(a,b)$ be the number of vehicles entering segment (a,b) during the shift that stop on that segment for some pick-up or drop-off activity plus the number of empty vehicles available on this segment at the beginning of the shift. Thus

$$W(a,b) = \sum_{j \in S(a,b)} \left(f_j + \sum_{i \notin (a,b)} (v_{ij} + x_{ij}) \right)$$

Observe that $W(a,b)$ also measures the number of vehicles exiting segment (a,b) plus the number of vehicles required on this segment at the shift's end. Consequently

$$W(a,b) = \sum_{i \in S(a,b)} \left(g_i + \sum_{j \notin (a,b)} (v_{ij} + x_{ij}) \right)$$

Next, let $M(a,b)$ represent the minimum number of vehicles required on this segment (a,b) during the shift, assuming vehicles could reverse direction on the segment. Then

$$M(a,b) = \max\left(\sum_{j \in S(a,b)} \left(f_j + \sum_{i \notin (a,b)} v_{ij} \right), \sum_{i \in S(a,b)} \left(g_i + \sum_{j \notin (a,b)} v_{ij} \right) \right)$$

Table 6   Station assignment to segments

Segment, (a,b)	A,B	A,C	A,D	B,C	B,D	C,E	D,E	E,A
Stations in $S(a,b)$	–	5	8	3,4	6,7	–	–	9,1,2

Table 7   Tabulation of W(a,b) and M(a,b) for the example problem

Segment, (a,b)	A,B	A,C	A,D	B,C	B,D	C,E	D,E	E,A
W(a,b)	0	8	8	3	3	0	0	22
M(a,b)	0	8	8	3	3	0	0	22

A segment (a,b) is said to be balanced if $M(a,b) = W(a,b)$. An equivalent definition is that the number of empty vehicles that enter and stop on a segment,

$$\sum_{i \notin S(a,b)} \sum_{j \varepsilon S(a,b)} x_{ij},$$

and the number of empty vehicles exiting a segment whose trips begin on the segment,

$$\sum_{i \varepsilon S(a,b)} \sum_{j \notin S(a,b)} x_{ij},$$

are not both positive. If the vehicle pick-up and drop-off stations are properly oriented, then the segment is balanced and empty vehicle travel time is minimised.

For the layout of Fig. 1, the data for delivery quantities given in Table 3, and the optimal solution given in Table 5, the values for $W(a,b)$ and $M(a,b)$ are displayed in Table 7. These data indicate that all of the segments are balanced for the layout given in Fig. 1.

A second observation is that some aisles should basically be transit aisles while others should be work aisles. For example, in Fig. 1 the centre aisle leading from the warehouse across the building should be primarily a transit aisle. Every vehicle leaving the warehouse must travel on this path. Thus if there is considerable drop-off or pick-up activity on the centre aisle there is a possibility for considerable blocking to occur. This, in turn, can increase the vehicle requirements. Station 2 on the centre aisle in the layout of Fig. 1 is, based upon this second observation, poorly located. It should be moved to segment (A,C) or (A,D), which, in turn, might indicate that the relative locations of stations 2 and 5, or 2 and 8, should be changed.

A third observation is that the layout should be designed so that the workload should be balanced among the work aisles as much as possible. The work should be balanced in two ways so that blocking can be minimised. First, the total work time (pick-up and drop-off activity time) on each segment should be balanced. Second, the number of vehicles that load or unload on each work aisle should be balanced. Of course, as much work should be accomplished as close to the warehouse, in this example, as possible. But, as a general guideline, the workload should be assigned to stations to achieve a balanced load so that blocking time can be minimised. The trade-off between blocking time and travel time can be estimated for different layouts by using methods developed in the following section.

Dispatching rules and time-dependent analysis

To this point it has been shown how to calculate the minimum number of vehicles needed to satisfy the system's material movement requirements. The number of vehicles actually needed may exceed this minimum, because time-dependent effects, such as blocking and battery changing, have not been considered. Also, it is impossible to use the static model developed in the last section to evaluate space requirements for each station. To estimate the effect of blocking and the need for space a dynamic model is needed. The heart of such a dynamic model is a set of rules used for dispatching the vehicles; that is, the rules for determining the sequence in which various routes will be visited.

The dispatching rule used depends on the environment being controlled. Different situations can require the use of different rules. The rule presented here has been designed for a manufacturing setting with the following characteristics.

First, the manufacturing area is used primarily to assemble finished products or major subassemblies. These units are produced on a paced line so that the output occurs at approximately a constant rate. Therefore the number of vehicle loads entering a pick-up station at the end of an assembly line during each hour of a shift is almost constant. The finished product is moved over the AGV system from these pick-up stations to either a central storage facility or an outgoing truck dock.

Second, the components used in the assembly operations are delivered to the different assembly lines using the AGV system. A load of components is taken from storage and placed on a vehicle which is then dispatched to the proper drop-off station on the floor. The need to deliver loads from the warehouse to the shop-floor is related to the rate of production. Since this rate is roughly constant, the consumption rate of components also remains approximately constant. Here, this is assumed to be the case. Ideally, deliveries of components to the beginning of each line should be spaced to correspond exactly with the consumption rate. Similarly, pick-up of finished product should be spread evenly over the shift. As will be seen, this is not always possible to effect, so that storage space requirements normally exceed the volume corresponding to a single vehicle load.

Obviously many dispatching rules can be developed for the situation described. The one proposed here attempts to assign vehicle trips to routes so that the times between visits (drop-offs or pick-ups) at each station are spread uniformly across the period of time vehicles operate during the shift.

The dispatching heuristic proposed depends on the concepts of a route, a trip, and a trip number. A route is a path beginning at a station and ending at the same station, with a designation of the stations visited for pick-up or drop-off activity. A trip is a particular instance of a vehicle moving over the path of a route. The solution of problem SP combined with input data v_{ij} can be used to determine R, the number of different routes, and for each route r, $m(r)$, the number of trips using route r. If

$$M = \sum_{r=1}^{R} m(r)$$

a total of M trips must be made by the N vehicles during the shift. Each of the M trips will be assigned one of M trip numbers.

The heuristic dispatching method is now considered and its use illustrated in the example problem. It is assumed that the routes are sorted and numbered so that $m(r) > m(r+1), r = 1, \ldots, R - 1$.

Dispatching procedure

Step 1: For each route, calculate a relative vehicle trip number for each trip of the route, assuming that the first trip over the route has a relative trip number of 1. Let $n_r(k)$ be the relative trip number of trip k on route r; then

$$n_r(k) = <\frac{M}{n(r)}(k-1) + 1>, k = 1,2,\ldots,m(r)$$

where $<x>$ is x rounded to the nearest integer.

Step 2: Assign trips to trip numbers. Begin by assuming that no trips have been assigned a trip number. For routes $r = 1, \ldots, R$, in that order, do the following:
    (a) Find the first unassigned trip number. Denote this by j(r).
    (b) For $k = 1, \ldots, m(r)$ assign visit k of route r to trip number $1 + (n_r(k) + j(r) - 1) \bmod M$. In this step, a trip of route r may be assigned the same trip number as a trip of a previously considered route.

Step 3: Resolve trip to trip number assignment conflicts. For trip number $j = 1, \ldots, M$ do the following, beginning with trip number 1. If more than one trip is assigned to trip number j, then a conflict exists. When a conflict exists, find the largest route number having a trip assigned to trip number j. Reassign that trip to the unassigned trip number that is nearest to trip number j.

Step 1 spreads the trips of a route as evenly in time as possible. It is motivated by the assumption of a constant production and component supply rate. Step 2 results in a trip to trip number assignment that is at least feasible for the first trip of a route.

Step 3 is motivated by two observations. The first is that Step 2 can result in an infeasible trip to trip number assignment. The second is that conflict resolution has minimal effects on storage space requirements since the routes are considered in the reverse order of their frequency of use.

To illustrate this heuristic dispatching procedure, as well as to demonstrate how dynamic analysis of blocking and space requirements can be done, the example of Fig. 1 is used. Based upon the loaded vehicle requirements stated in Table 3 and the solution to problem SP given in Table 5, there are five routes ($R=5$). Table 8 lists these routes, ranked in decreasing order of $m(r)$, the number of trips on route r. Note that 175.2 minutes of slack time exists with three vehicles. If there is less than 175.2 minutes of blocking time, then three vehicles can be used to perform the pick-up and drop-off requirements.

The first step of the algorithm indicates the relative trip number assigned to each trip of each route, assuming the first trip on each route occurs on the first trip number. Using the formula in Step 1, the results displayed in Table 9 are obtained. Clearly this solution cannot be implemented since, for example, there are five routes assigned to the first trip number.

The second step of the algorithm attempts to preserve the spacing between trips assigned to each route on Step 1. The starting trip number for each route is assigned so that it does not conflict with any previously assigned trip number. The spacing between trips remains the same as it was in Step 1. The results of Step 2 are given in Table 10. It is seen that some trips (trips 15, 16, 23 and 24) have two routes assigned to them while others (trips 13, 19, 20 and 26) have none. If conflicts occur, then trip assignments are altered in the third step so

Table 8   Routes and number of trips per route for the example problem

Route, r	Number of trips, m(r)	Stations visited on route r
1	8	1–5–1
2	8	1–8–1
3	4	2–9–2
4	3	1–3–4–9–1
5	3	1–6–7–9–1
	$M = 26$	

Table 9   Relative trip number assignments to routes

Route	1	2	3	4	5	6	7	8	9	10	11	12	13	14	15	16	17	18	19	20	21	22	23	24	25	26
1	x			x				x			x			x			x				x			x		
2	x			x				x			x			x			x				x			x		
3	x							x						x							x					
4	x									x								x								
5	x									x								x								

x indicates that the route is assigned to that trip number

Table 10   Trip number assignments to routes after step 2

Trip number

Route	1	2	3	4	5	6	7	8	9	10	11	12	13	14	15	16	17	18	19	20	21	22	23	24	25	26
1	x			x				x			x			x			x				x			x		
2		x			x				x			x			x			x				x			x	
3			x							x						x							x			
4						x							x						x							
5							x													x						x

x indicates the trip number has been tentatively assigned to the route

Table 11   Final trip assignment for the example problem

Trip number

Route	1	2	3	4	5	6	7	8	9	10	11	12	13	14	15	16	17	18	19	20	21	22	23	24	25	26
1	x			x				x			x			x			x				x			x		
2		x			x				x			x			x			x				x			x	
3			x							x						x							x			
4						x							x							x						
5							x												x							x

that a feasible schedule results. The results of Step 3 are given in Table 11. The changes in the trip assignments in Step 3 are made so that increases in inventory storage area requirements are estimated to be as small as possible. This is accomplished by recognising that a change of one trip number for a route visited m_1 times should ideally increase the inventory less than for a one trip number change for a route visited m_2 times, when $m_1 < m_2$. The term 'ideally' is used because the exact timing of the vehicle departures and arrivals to and from each station are not known. Thus, although under ideal circumstances – an exactly equal spacing of arrival/departure times – this minimises the increase in inventory storage area requirements, a change may be made using this proposed procedure that actually increases space requirements.

To determine the blocking effects corresponding to this dispatching sequence, its implementation is simulated. This is a deterministic simulation which tracks the movement of the three vehicles throughout the system. The results of the simulation are given in Table 12. The arrival and departure times for each station are given for each trip. It is assumed that all three vehicles are available for loading at station 1 at the beginning of the shift. The schedule reflects the assumption that vehicles are dispatched from the warehouse as soon as possible.

The results indicate that it is possible to complete the required number of trips during the shift using three vehicles. This can be seen by noting that the departure time from station 9 allows all vehicles to arrive at station 1 prior to time 430. In fact, there are 113.2 minutes of excess time available. Also, 62 minutes of blocking occur. Thus over one hour of vehicle time is lost due to waiting. Since excess time exists for the three vehicles, blocking does not cause an increase in the number of vehicles needed to meet the requirements. However, if only 400 minutes had been available per vehicle during the shift, then the

Table 12 Station arrival and departure times for each trip

Trip no.	Veh. no.	1		2		3		4		5		6		7		8		9		Block-ing
		A	D	A	D	A	D	A	D	A	D	A	D	A	D	A	D	A	D	
1	1	0	5	8	8					10	30							36.4	36.4	0
2	2	0	10	13	13											14.6	34.6	41.0	41.0	5
3	3	0	10	13	18											19.6	34.6	41.0	46.0	25
4	1	39	44	47	47					49	69							75.4	75.4	0
5	2	43.6	49	52	52											53.6	73.6	80	80	0.4
6	3	48.6	54	57	57	53.4	79.4	81.8	101.8									198.2	113.2	0.4
7	1	78	83	86	86							88.4	108.4	110.6	130.6			136.8	141.8	0
8	2	82.6	88	91	91					93	113							119.4	119.4	0.4
9	3	115.8	120.8	123.8	123.8											125.4	145.4	151.8	151.8	0
10	2	122	122	125	130											131.6	145.4	151.8	156.8	13.8
11	1	144.4	149.4	152.4	152.4					154.4	174.4							180.8	180.8	0
12	3	154.4	159.4	162.4	162.4											164	184	190.4	190.4	0
13	2	159.4	164.4	167.4	167.4	169.8	189.8	192.2	212.2									218.6	223.6	0
14	1	183.4	188.4	191.4	191.4					193.4	213.4							219.8	223.6	3.8
15	3	193	198	201	201											202.6	222.6	229	229	0
16	2	226.2	226.2	229.2	234.2											235.8	235.8	242.2	247.2	0
17	1	226.2	231.2	234.2	234.2					236.2	256.2							262.6	262.6	0
18	3	231.6	236.6	239.6	239.6											241.2	261.2	267.6	267.6	0
19	2	249.8	254.8	257.8	257.8							260.2	280.2	282.4	292.4			308.6	313.6	0
20	1	265.2	270.2	273.2	273.2	275.6	295.6	298	318									324.4	329.4	0
21	3	270.2	275.2	278.2	278.2					280.2	300.2							306.6	306.6	0
22	3	309.2	314.2	317.2	317.2											313.8	338.8	345.2	345.2	0
23	2	316.2	316.2	319.2	324.2											325.8	339	345.4	350.4	13.2
24	1	332	337	340	340					342	362							368.4	358.4	0
25	3	347.8	352.8	355.8	355.8											352.4	377.4	383.8	383.8	0
26	2	353	358	361	361							363.4	383.4	385.6	406.6			411.8	416.8	0

results of the simulation would indicate that an additional vehicle may be needed. To determine if another is needed in this case, the numbers in Table 12 are examined more closely to see if some of the blocking time could be avoided.

When making the calculations, it was assumed that vehicles travel to stations using a shortest travel time route. This may lead to unnecessary blocking. For example, when vehicle 2 makes trip 10 it is delayed 13.8 minutes. By making this trip over a route going past station 5 rather than station 8, vehicle 2 increases its travel time by 0.4 minutes. However, its progress is not blocked before its arrival at station 9 at time 138.4. At that time vehicle 1 is being unloaded at station 9 and consequently would block the departure of vehicle 2 until time 141.8. Thus, if this change is made in the routing of this trip, vehicle 2 becomes available for dispatch 10 minutes sooner than indicated in the schedule shown in Table 12. Similar calculations can be made for trips 3 and 23. Of course, an evaluation of the effect of making routing changes requires completing a new simulation. After these calculations have been made the feasibility of using three vehicles to complete the required trips within the available time can be checked.

The above analysis shows that, even if it is not possible to prove that a feasible sequence exists using the first simulation results, it may be possible to find a feasible solution by choosing longer paths for certain trips. The evaluation of the alternatives is carried out using the simulation. In addition to routing changes, it is also possible to consider alterations to the dispatching sequence. Blocking may be reduced by interchanging the assignments of some vehicle trips. For example, blocking can be reduced by interchanging the assignments of vehicle trips 9 and 10.

To this point the discussion has concentrated on evaluating the dispatching sequence's effect on vehicle blocking. There is another important effect that can be measured; namely, the storage capacity requirements at each pick-up or drop-off station. Since it has been assumed that production rates are constant, the rate at which inventory accumulates at drop-off stations is constant. Fig. 3 indicates how inventory levels fluctuate over an eight-hour shift for stations 3 and 5 when the dispatching sequence in Table 11 is followed. It is seen that the amount of inventory in these stations changes considerably during the shift. The maximum inventory level considerably exceeds one vehicle load. For example, the maximum level at station 3 is about three times the amount delivered on a vehicle trip. The dispatching schedule may be adjusted to reduce peak storage requirements. This may be accomplished by inserting idleness into the schedule at certain points. Departures from the warehouse can be delayed so that time between pick-ups and drop-offs can be made to be more uniform. The example used here also illustrates the effect of battery (or mainte-

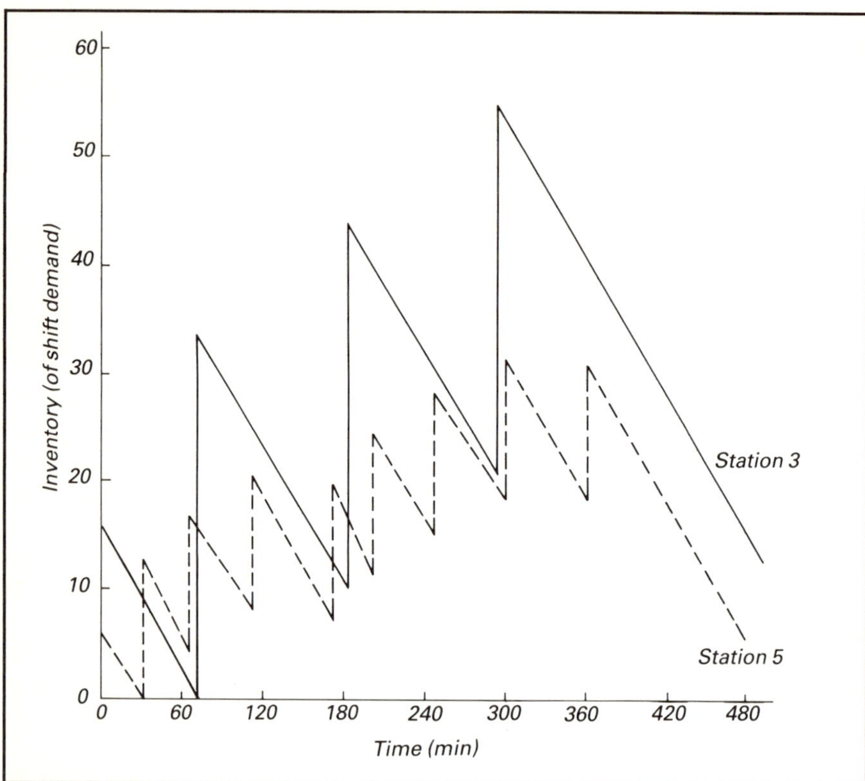

Fig. 3   Fluctuations at the inventory level at stations 3 and 5

nance) changes on all vehicles at the same time during a shift. By spacing these throughout the shift, it is possible to reduce both the shift's initial inventory level at some stations and blocking. Different rules for inserted idleness and alternative battery change or maintenance strategies can be evaluated using a deterministic simulation.

Concluding remarks

First, a process for designing system layouts and the supporting AGV system is considered. This design process, which is illustrated in Fig. 4, is iterative. It is repeated as additional understanding of the fundamental issues is obtained for a particular application; by no means is it a sequential, one-pass process.

The first stage in each iteration is the gathering of the relevant data on vehicle speeds and required inter-departmental flows. The flow values should be scrutinised carefully for consistency; one may also need to deal with multiple shift operations and/or different future scenarios for the flow values.

The second stage involves the design of the department layout, the track layout, and the location of the pick-up and drop-off stations. In the section above dealing with the model some guidelines were suggested that are useful when establishing this physical layout. The second stage also involves designing the station-vehicle interface as well as the control mechanisms for pick-up and drop-off. This interface could be automated or manual. In either case it requires estimating the times for the physical transfer to/from the vehicle and estimating the waiting times to effect such transfers. These time estimates are found using IE work measurement techniques and queuing analyses respectively.

The third stage involves conducting an analysis using techniques such as those presented in this paper. In this stage the minimum number of required vehicles is determined as well as the vehicle routes and the number of trips over each route. A dynamic evaluation of vehicle requirements is then conducted to determine whether or not the suggested number of vehicles can carry out the movement requirements. Once this has been established, additional analysis is conducted to determine a dispatching schedule that evaluates both blocking and space requirements. The result may be to increase the number of vehicles so that space requirements can be reduced. These types of trade-offs can be measured using the simulation method suggested in this paper.

With the focus on productivity, there is increasing interest in the use of AGV systems to transport material between work-stations. Equipment manufacturers have concentrated their development efforts on hardware issues and computer monitoring of movement and location; by and large they do not provide services to address the issues raised here. For example, see reference [7].

The authors hope that they have brought to the attention of prospective users of an AGV system those operational (as opposed to hardware) design issues that must be considered. This paper has

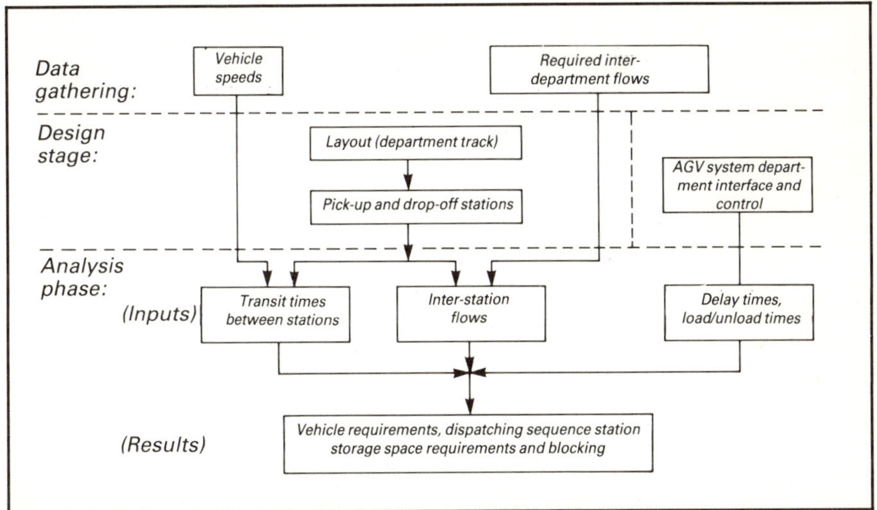

Fig. 4   Design process for system layout and AGV system

presented one approach that can be used to address some of these issues. Many issues remain, however, as a challenge to the research community.

Acknowledgements

The authors greatly appreciate the many helpful comments of Dr Richard Francis and the encouragement of the Departmental Editor, Dr Thom Hodgson. This paper extends some of the work of Dr William L. Maxwell presented elsewhere[8].

References

[1] Armour, C. and Buffa, E. S. 1963. A heuristic algorithm and simulation approach to relative location of facilities, *Management Science*, 9(2): 294–309.

[2] Francis, R. L. and White, J. A. 1974. *Facility Layout and Location: An Analytical Approach*, Prentice-Hall, Englewood Cliffs, NJ, USA.

[3] Tompkins, J. A. and Moore, J. M. *Computer Aided Layout: A User's Guide*, AIIE Monograph 119, AIIE, Norcross, GA, USA.

[4] Qunilan, J. C. 1980. The great AGVS race, *Material Handling Engineering*, 35(6): 56–64.

[5] Hillier, F. S. and Lieberman, G. J. 1980. *Introduction to Operations Research*, Holden-Day, San Francisco, CA, USA.

[6] Hitchcock, F. L. 1941. The distribution of a product from several sources to numerous locations, *Journal of Mathematical Physics*, 20: 224–230.

[7] Material Handling Institute 1980. *Considerations for Planning and Installing Automatic Guided Vehicle Systems*, AGVS Document No. 101. Material Handling Institute, Pittsburgh, PA, USA.

[8] Maxwell, W. L. 1981. Solving material handling design problems with OR, *Industrial Engineering,* 13(4): 58–69.

CHARACTERISATION OF AGV DISPATCHING RULES

P. J. Egbelu
Syracuse University, USA
and
J. M. A. Tanchoco
Purdue University, USA

Hardware failures notwithstanding, the ability of an automated system operating according to promised potential is dependent upon the operational control measures in force. Some heuristic rules for dispatching automated guided vehicles (AGVs) in a job shop environment are presented. The rules are useful for assigning priorities to workstations requesting the services of a vehicle for material pick-up. The likely effects of these rules on the performance of a job shop are postulated. Simulation results to demonstrate the effects of these rules are also presented.

Automated material handling systems, though more flexible and capable than their counterpart (non-computer-controlled systems), do pose more serious and challenging operational control problems. These control problems increase with the level of system automation. The manner by which these problems are resolved determines the operating effectiveness of the total system, as typified by an automated guided vehicle (AGV) system.

In a manufacturing environment consisting of several machine centres performing different machining functions, a typical part or unit load visits several centres before its machining requirements are satisfied. A unit load continues to circulate in the shop between work centres until it receives its last service. It is this transition of unit loads or parts that generate the vehicle dispatching (task assignment) problem in an AGV system. On the completion of a delivery task, a vehicle is reassigned to another mission immediately if there is any unattended handling task in the facility. Otherwise, the vehicle is set idle and continues to remain idle until a handling task appears, at which time it may be reassigned. If at the time a vehicle is released from a task (i.e. when it completes a task), there is only one outstanding (unattended) mission in the facility, then the vehicle task assignment problem may be trivial. On the other hand, if multiple loads are awaiting pick-up simultaneously at different locations in the facility, then the vehicle task assignment presents a serious operational control problem that involves matching the vehicle to the task. This problem also appears in another form. When multiple vehicles are idle (unassigned) simultaneously and a pick-up task arises, appropriate criteria must be established for selecting an idle vehicle to perform the task (i.e. matching task to vehicle). In a job shop environment where there is no recognised flow pattern of unit loads, vehicles in an AGV system can be dispersed throughout the network or concentrated in a region at any particular time. Vehicle distribution in the network is an operational control problem. The selected control measures, by which vehicles are assigned tasks, can affect material flow, buffer storage requirements in the departments, machine utilisation, and vehicle effectiveness.

The vehicle dispatching decisions fall into two categories. The first category is a decision involving the selection of a vehicle from a set of idle vehicles to assign it to a unit load pick-up task generated at some part of the facility. This class of decisions involves a single work centre and one or more vehicles, and is generally the result of a request from a work centre for vehicle service. The decision is to determine the appropriate vehicle to assign to the pick-up task. The secondary category of decisions involves the selection of a work centre from a set of work centres simultaneously requesting the service of any vehicle, a decision which usually involves a single vehicle and multiple work centres. The vehicle involved has just completed a delivery task and requires reassignment to another pick-up task. At the time the vehicle became available, several pick-up tasks exist in the shop at various departments. The decision is to prioritise the departments and to dispatch the vehicle to the department with the highest priority.

In this paper, the first class of problems is referred to as 'Work-centre-initiated task assignment (dispatching) problems', while the resulting rules to resolve them are referred to as 'Work-centre-initiated task assignment (dispatching) rules'. The second class of problems is identified as 'Vehicle-initiated task assignment (dispatching) problems'. The corresponding methods for problem resolution will be referred to as 'Vehicle-initiated task assignment (dispatching) rules'. Details of each rule subclass will be presented shortly.

Background work

Most available research reports on AGV systems are hardware orientated rather than operational[1,2]. Technical publications on hardware developments are protected by corporate confidentiality and proprietorship. Material available on AGV system operations are mostly success stories relating to specific applications[3-6]. Fortunately,

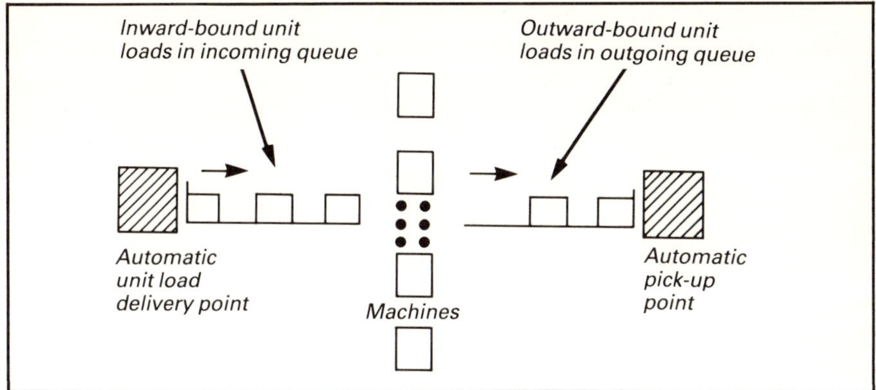

Fig. 1   Schematic representation of a work centre

tremendous insight can be gained on AGV dispatching through work done in physical distribution and transportation. The problem of controlling in-plant industrial vehicles is merely a microscopic representation of the problems present in physical distribution. The vehicle dispatching problem was first written on by Dantzig and Ramser[7]. Since then, various forms of the problem have been identified in public transportation[8,9], airlines [10,11], mining[12], and manufacturing[2,13], to cite only a few.

There is some recent work on vehicle dispatching directed specifically to the control of automated guided vehicles (AGVs). Maxwell[14] and Maxwell and Muckstadt[2] address the problem of determining an optimum schedule for dispatching vehicles that result in minimum vehicle fleet size. Integer programming was employed for seeking the optimum solution, and the problem environment was a paced assembly operation in manufacturing setting. The application environment of the work in Egbelu[1] and Egbelu and Tanchoco[15] was a job shop. The shop was assumed to be dynamic in the sense that jobs are continuously admitted into the shop. The study objective was to characterise vehicle dispatching rules in a job shop environment with a constrained layout, and simulation was the analysis tool. Other work on vehicle dispatching in a manufacturing setting includes that by Blair[16] and Russell and Tanchoco[13].

A work-centre-initiated task assignment problem

Fig. 1 is a representation of a typical machining centre in an AGV system. It consists of one or more machines, an incoming unit load queue, and an outgoing unit load queue. For simplicity, consider the machines to be identical and the queues to be full. Load stands in front of the queues facilitate the automatic transfer of unit loads to or away from the load deck of the transporter. Depending on the part processing rate of a department, unit loads are drawn from the incoming queue, processed, and released into the output queue at some rate. The deposition of a unit load into the output queue also initiates a request by the department for an unassigned vehicle for the immediate removal of the load just deposited. Several heuristic rules can be employed for assigning priorities to vehicles for dispatching. These include the following.

Random vehicle (RV) rule

Under this rule, the pick-up task is randomly assigned to any available vehicle in the shop. This is done without regard to the relative location of the vehicles to the point where the vehicle is needed.

Nearest vehicle (NV) rule

Under this rule, the distance of the vehicle demand point from every available vehicle is computed. Using the travelling speed of the vehicles, the vehicle whose travel distance is computed to be the shortest is dispatched to the vehicle demand point. This is equivalent to dispatching idle vehicle j such that

$$d_j = \min \{d_i\} \text{ for all idle vehicles } i$$

where d_i is the distance of the vehicle i from the vehicle demand point.
   The computation of d_i is according to the following expression:

$$d_i = d'_i + \sum_{k=1}^{J-1} d(n_k, n_{k+1})$$

where J is the number of nodes along the shortest path travelling away from vehicle i to the point where the vehicle is demanded. (See the Appendix for the representation of an AGV system guide path as a network.) $d'_i = 0$ if the vehicle i is currently at a node point. Otherwise, it is the distance of vehicle i to the next node on its path (i.e. if it is between two nodes). $d(n_k, n_{k+1})$ is the distance between the adjacent nodes n_k and n_{k+1} such that n_k precedes n_{k+1}. Dsitance computation between nodes is as shown in the Appendix.
   Rather than dispatching vehicles based on the shortest distance rule, the dispatching decision could be made based on the shortest travel time. The decision is to dispatch vehicle j such that

$$d_j/s_j = \min\{d_i/s_i\} \text{ for all idle vehicles } i$$

where

$$s_i \text{ is the travelling speed of vehicle } i$$

For a congested traffic network, the shortest travel distance path is not necessarily the shortest travel time rule.

Farthest vehicle (FV) rule

This is an antithetical rule to the NV rule. Using the definition in the last section, the vehicle j satisfying the condition

$$d_j/s_j = \max_{\forall i} \{d_i/s_i\}$$

is dispatched to the vehicle demand point. This rule does not provide any direct usefulness as a viable dispatching rule, but it does show the system designer what effect unnecessary vehicle travelling distance could have on system handling effectiveness.

Longest idle vehicle (LIV) rule

This rule assigns the highest dispatching priority to the vehicle that has remained idle the longest among all the idle vehicles. This is equivalent to dispatching vehicle j such that

$$t_j = \max_{V_i} \{t_i\}$$

where

$$t_i = T_c - T_i$$

and where T_c is the current time or the time the vehicle dispatching decision is to be made and T_i is the time vehicle i was last set idle, $T_c \geq T_i$ for all i.

The advantages of this rule are its workload balancing effect on all participating vehicles.

Least utilised vehicle (LUV) rule

This rule takes advantage of the fact that, in most computer-controlled AGV systems, time-persistent statistics on the utilisation of each vehicle are maintained. For idle vehicle i, if U_i denotes its mean utilisation up to the time the vehicle dispatching decision is to be made, the decision rule is to dispatch vehicle j such that

$$U_j = \min_{V_i} \{U_i\}$$

Like the LIV rule, the LUV rule acts as a workload balancer among the vehicles in the shop.

Up to this point, attention has been focused on decision rules involving work-centre-initiated dispatching rules. As mentioned earlier, rule pairs are required to accomplish the vehicle dispatching requirements of an AGV-based material handling system. The second rule category is presented next. This is followed by a discussion of experimental results obtained to demonstrate the behaviour of these rules in a job shop setting.

Vehicle-initiated task assignment problems

From an operational point of view, the most desirable level of handling effectiveness is that which ensures that unit loads completed at a work centre are removed promptly and transported to their subsequent destinations with a minimum of delays. Actual operating conditions do, however, deviate from this scenario. The degree of deviation is a function of vehicle availability, shop loading, shop size, and the layout of the guidance network. Operating conditions do arise when requests for vehicles from work centres cannot be immediately satisfied (i.e. the request is uneventful). This is the case when all vehicles are engaged on missions. Such uneventful requests are therefore logged and considered for satisfaction when a vehicle becomes available. The procedures by which the work centres are dynamically prioritised for service by the released vehicles is the item of presentation in this section.

Like the work-centre-initiated task assignment problem, several heuristic rules are available for ranking work centres requesting unassigned vehicles. Possible assignment rules are presented as follows.

Random work centre (RW) rule

Under this rule, a list of all work centres requesting the service of vehicles is obtained. From this list, a work centre is randomly selected. The released vehicle is therefore dispatched to the selected work centre.

Shortest travel time/distance (STT/D) rule

The basis of this rule is to minimise the percentage of time vehicles travel empty. Under the STT/D rule, the decision rule is to dispatch the released vehicle to the work centre whose unit load pick-up point is closest to the vehicle. Closeness is measured in terms of travel time or distance along the shortest path and in the direction of traffic flow. (Again, the format for distance calculation is as shown in the Appendix.) Although the advantages of this rule seem obvious, it is worthy to note that it is very sensitive to the layout of facilities and location of equipment within the facility. Vehicles are available for reassignment when they are released from a previous assignment. Therefore vehicle release points generally correspond to delivery stations. If by chance (or by design) the pick-up point of some work centre turns out not to be the nearest to any vehicle release point, then according to the STT/D rule such a work centre may never qualify to receive a vehicle dispatch. This reduces material flow rate out of the affected work centre. Since new deliveries continue to take place in the centre, the output queue will grow to its maximum capacity, and thereafter machines will be blocked. As the machines continue to remain blocked, new deliveries accumulate in the input queue. Eventually, the input queue grows to its limit. Subsequent vehicles arriving at the station will be blocked from making deliveries. Thus a stalemate will result, and the possibility of locking exists. The locking phenomenon will be discussed in detail later.

Longest travel time/distance (LTT/D) rule

As an antithetical rule to SST/D, this rule assigns the highest priority to the work centre that is farthest away from the vehicle. Again, other than for system experimentation, there is no attractive quality to this rule.

Maximum outgoing queue size (MOQS) rule

Suppose q_k denotes the number of unit loads in the outgoing queue of work centre k awaiting pick-up and R_k is the number of unit loads in the same queue not yet assigned to any vehicle, where $1 \leq R_k \leq q_k$. The decision is to dispatch a vehicle to work centre j such that

$$q_j = \max \{q_k\} \text{ for all } k, R_k \geq 1$$

Note that $R_k = 0$ for all k implies there is no unit load in the facility awaiting pick-up that has not yet been assigned to a vehicle. In such a case, the released vehicle is set idle.

Minimum remaining outgoing queue space (MROQS) rule

The dispatching decision under MROQS is based on a criticality index that is a function of outgoing queue capacity and length at each work centre. If Q_k is the capacity of output queue at work centre k, S_k is the current length of output queue at work centre k, and R_k is the number of unassigned unit loads waiting in the output queue at work centre k, then for each centre $k(R_k \geq 1)$ compute the index C_k, defined as

$$C_k = Q_k - S_k$$

The decision is to dispatch the available vehicle to centre i such that

$$C_i = \min \{C_k\}, \text{ for all } k, R_k \leq 1$$

The basis of this rule is to reduce the possibility of work centre blocking occurring.

Modified first come–first serve (MFCFS) rule

This rule is a modification of the traditional first come–first serve rule. The rule attempts to assign vehicles to departments sequentially in chronological order as requests for empty vehicles are received from departments. When a department places a call (request) for an empty vehicle and the call cannot be immediately satisfied, the time the call was generated is saved. The saved call and time are used for future vehicle assignment decisions. If subsequent calls emanate from a department before an earlier saved call from that department is satisfied, the times of these subsequent calls are not saved. In other words, no department can have two or more outstanding saved calls simultaneously. When a vehicle becomes available, it is assigned to the department that has the earliest outstanding saved call and time. At the moment a saved call from a department is satisfied (i.e. a vehicle is dispatched to the department), the vehicle need of such a department is evaluated and updated in one of two ways:

- A zero outstanding call is recorded for the affected department if it needs exactly one vehicle. The number of vehicles a department needs at any time is equal to the number of unit loads awaiting vehicle assignment for pick-up from the department.
- If more than one vehicle is required at the department, a new call is saved immediately against the department and the corresponding time associated with the new call is set equal to the time the old saved call was satisfied.

Although this rule does not consider any impending blockages of departments due to imminent exhaustion of queue space, it does ensure that the elapsed time between the placing of a vehicle request by a department and the satisfaction of that request is reduced. The number of assignments made to a department is related to job traffic intensity in that department.

Unit load shop arrival time (ULSAT) rule

This rule is useful if the design objective is to reduce the time jobs spend in the shop. It accelerates jobs or unit loads through the shop in their order of entry into the shop. At the time a vehicle dispatch decision is to be made, the arrival time into the shop, T_k, of the unit load at the end of each queue that is not yet assigned is determined. The dispatching decision is to send a vehicle to department i such that

$$t_i = t^* = \max \{T_c - T_k\}, \text{ for all } k, R_k \geq 1$$

where T_c is the current time or the time the decision is required. The above decision is also consistent if a vehicle is to be dispatched to department i if

$$T_i = T^* = \min \{T_k\}, \text{ for all } k, R_k \geq 1$$

Performance evaluation

Several combinations of the above rules were tested on a facility by using a simulation technique. A simulation program, AGVSim, was developed specifically to simulate an AGV system[17]. In the simulation, the AGV system guide path is modelled as a collection of nodes and arcs, as shown in Fig. A1 in the Appendix. All arcs are considered unidirectional. Nodes represent guide wire intersections, merges, diverges, load pick-up stations, and load delivery stations. Safety zones, as shown in Fig. A2 in the Appendix, are constructed around every node to facilitate traffic control at node points. Vehicle movements between points in the network are modelled as series of transitions from one arc to another through interchange ramps (as shown in Fig. A2). Detailed discussion of the simulation modelling can be found elsewhere[1].

The shop whose layout is shown in Fig. 2 formed the test facility for the study. Fifteen rule combinations were included in the testing. The

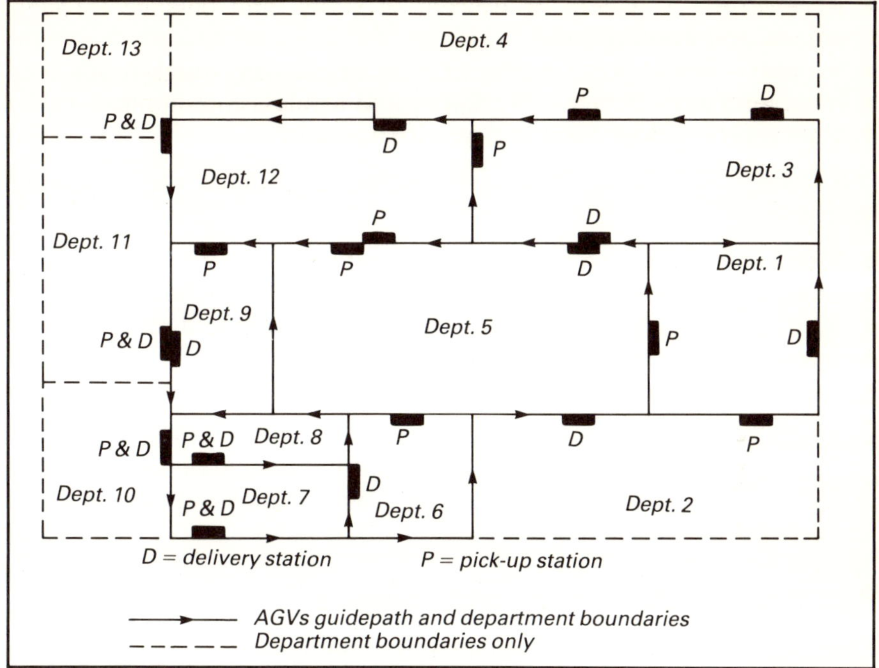

Fig. 2   Network layout of demonstrative facility

facility is a job shop with 12 departments, multiple identical machines per department, and six automatic vehicles. Departments 1 and 13 are the receiving and shipping areas respectively. There is a unique input queue and output queue in each department. With the exception of the input queue in the receiving department and the two queues of the shipping department all queues are full. Each job that arrives at the shop has a certain number of identical components. This lot is then broken down into equal size groups to form unit loads. A unit load, therefore, defines one or more components bound together and transported as a unit[18]. Unit loads belonging to the same job follow the same job route. A job is not completed until its last unit load is completed.

The rules included in the experiments were the following:

- *Vehicle-initiated task assignment rules:*
  - maximum outgoing queue size (MOQS).
  - shortest travel time/distance (STT/D).
  - longest travel time/distance (LTT/D).
  - minimum remaining outgoing queue space (MROQS).
  - modified first come–first serve (MFCFS).

- *Work-centre-initiated task assignment rules:*
  - nearest vehicle (NV).
  - farthest vehicle (FV).
  - longest idle vehicle (LIV).

Using load throughput as a measure of rule performance, Table 1 shows the results of 30 simulation trials, two trials per rule combination. All experiments were conducted under similar conditions. The number of asterisks in a cell indicates the number of times the rule combinations resulted in a locking of the shop. A shop is considered locked if the following conditions exist:

- The input and output queues are simultaneously full at some or all the departments and the machines in the affected departments are blocked.
- All vehicles transporting unit loads cannot make their deliveries because the input queues are full and there are no available vehicles to free some spaces from the output queues.
- All empty vehicles dispatched for load pick-up cannot get to their destinations due to interference from other vehicles.

Table 1   Shop throughput measured in unit loads (locking encountered by the shop)*

		Vehicle-initiated task assignment rules				
		MOQS	*STT/D*	*LTT/D*	*MROQS*	*MFCFS*
Work-centre-initiated	*NV*	*	*	*	766	775
task assignment rules		*	*	*	763	770
	FV	*	*	*	765	767
		*	*	*	770	777
	LIV	*	*	*	766	775
		*	*	*	763	770

When a shop locks, there is a complete seizure of further material flow throughout the shop. As indicated by Table 1, the first nine rule combinations produced shop locking. From the actual statistics there is a strong indication that locking occurred at very early stages of the experiments. Actual throughput for the locked cases ranged from 1 to 108 unit loads as compared with over 700 unit loads for all rule combinations that involved MROQS and MFCFS.

Explanation of the shop locking phenomenon

If one observes the data in Table 1, two questions are obvious. First, what are the conditions that can lead a shop to lock? Secondly, why is there little difference in shop performance amongst the rules within the work-centre-initiated task assignment rule?

With regard to the first question, the results above demonstrate the importance of a well-coordinated vehicle dispatching plan, especially when existing facility layout is involved and material flow volume is large. With large volumes of material flow, vehicles are rarely free to allow the invoking of a work-centre-initiated rule for dispatching vehicles. Instead, dispatching is governed completely by the vehicle-initiated rule. Since members of the vehicle-initiated rule subclass are generally invoked only when a vehicle is released, and furthermore, since vehicle releases occur at delivery points only, vehicle-initiated task assignment rules are very sensitive to the location of material delivery points. However, when STT/D and LTT/D are employed some departments almost never satisfy the dispatching criteria. Eventually, the output queues become full and subsequently block the machines in that centre. This is followed by a blocking of vehicles making new deliveries in that centre as the input queue itself grows to its maximum capacity. This chain reaction gradually propagates to all vehicles in the shop, causing a complete seizure of material flow. Unless pick-up points are well located relative to delivery points, shop locking or at least work centre locking is an inevitable phenomenon whenever rules that are derivative of distance measures are involved.

The answer to the second question follows from that of the first. The last paragraph asserts that, when material flow rate is high, work-centre-initiated task assignment rules are rarely invoked. This implies that the rules become virtually inactive after a few hours of shop operation. Therefore, if no significant performance difference between the rules is realised during the early phase of a shop operating session, no performance differences amongst the rules may ever be realised. In the example problem here the rules did not exhibit any significant performance differences during the early part of the shop operating session; thus the similarity in performance.

Operation with automatic intervention and a central buffer area

In practice, when a shop locks a measure is taken by the shop operators to unlock the shop. This involves dispatching some lift truck operator to release the queues. In order to avoid the necessity for such action, the intelligence can be built into the vehicles to recognise and initiate actions to diffuse conditions that could lead to shop locking. This scheme requires diverting loading and blocked vehicles to a central buffer area to have their loads delivered. Thereafter, they are directed to the shop floor to release blocked departments. The vehicles are subsequently returned to pick up the buffered items for delivery at their appropriate destinations. This is just one of a number of automatic intervention concepts.

With all shop conditions remaining unchanged (as previously described), and with the automatic intervention concept just presented, the shop output increased under the rule combinations that involved MOQS, STT/D and LTT/D. The new output levels ranged from 290 to 596 as against 1 to 108 under the no intervention case. The output levels for rule combinations under MROQS and MFCFS are the same before and after intervention. Table 2 shows statistics of the central buffer for the 15 rule combinations. Information of this type is useful for specifying the capacity of buffer areas. Incidentally,

MROQS and MFCFS do not show any need for intervention and subsequently a central buffer area is not required.

Table 2   Maximum[+] and average buffer queue length using system antilock mechanism*

		Vehicle-initiated task assignment rules								
		MOQS		*STT/D*		*LTT/D*		*MROQS*		*MFCFS*
Work-centre-	*NV*	*17[+]*	*6.22**	*5*	*1.06*	*9*	*1.31*	*0*	*0*	*0 0*
initiated task	*FV*	*15*	*5.80*	*6*	*1.04*	*8*	*1.26*	*0*	*0*	*0 0*
assignment rules	*LIV*	*17*	*6.22*	*5*	*1.06*	*9*	*1.81*	*0*	*0*	*0 0*

Shop performance based on infinite queues

Operating rules notwithstanding, the blocking of machining centres is the direct result of full queues. By relaxing the queue capacity constraints, one can eliminate locking effect. This provides the opportunity to assess the actual buffer space requirements under unconstrained conditions. To explore this postulate, the shop was simulated under the unconstrained (infinite) queue capacity. The simulation results are as shown in Table 3.

Looking at the table row-wise, there are virtually no performance differences among work-centre-initiated task assignment rules as compared with the differences among the vehicle-initiated task assignment rules. All rule combinations that have elements of queues in their derivation performed worse than all other cases. MFCFS outperformed all other rules. Ideally, MOQS and MROQS should yield the same throughput, since at the infinite queue condition they collapse to the same rule. The difference shown in the table is caused by the variation in tie-breaking rule employed under each policy.

On the other hand, while STT/D performed competitively under the system throughput criterion, it performed poorly when queue characteristics are considered. While queues grew to a maximum length of 13 unit loads under the MFCFS rule, output queues under STT/D rule grew as high as 231 unit loads in some departments. Actually no loads were picked up from these departments. Given the existing shop layout, STT/D did transform these departments to sink nodes. The enormous buffer space required by STT/D makes it an inoperable rule in facilities where the necessary layout condition is not met and modification can only be achieved at very high cost.

Table 3   Shop unit load throughput at infinite queue condition

		Vehicle-initiated task assignment rules				
		MOQS	*STT/D*	*LTT/D*	*MROQS*	*MFCFS*
Work-centre-	*NV*	*348*	*669*	*431*	*346*	*764*
initiated task	*FV*	*334*	*664*	*434*	*350*	*758*
assignment rules	*LIV*	*348*	*669*	*431*	*346*	*764*

Sequential vehicle dispatching rule

A common practice for vehicle dispatching strategy is to preprogram available vehicles to visit the load pick-up points in a sequence that forms a loop. If a pick-up station has a load when it is visited, the load is picked up and transported to its destination. Otherwise, the vehicle proceeds to the next pick-up station in the sequence. An inherent quality of this strategy is the elimination of any possibility for shop locking. In a new facility, this requires locating the departments along a loop to minimise vehicle travel time. However, in complex facilities with a large number of departments or ones with existing layouts, the flexibility to locate all pick-up points along a loop is highly constrained. In such cases, a sequence has to be selected through a heuristic procedure or by solving the corresponding travelling salesman problem.

The sequential dispatching strategy was demonstrated in this study and the resulting shop throughput compared with those obtained under two rule combinations. The tested departmental sequence of

$$1 \to 4 \to 13 \to 11 \to 10 \to 8 \to 6 \to 2 \to 12 \to 5 \to 9 \to 7 \to 3 \to 1$$

used for the demonstration was not chosen through any optimal producing algorithm. Fig. 3 shows the resulting vehicle fixed route for the sequence visitation. Vehicles conveying items were still routed through the shortest path between their origins and destinations.

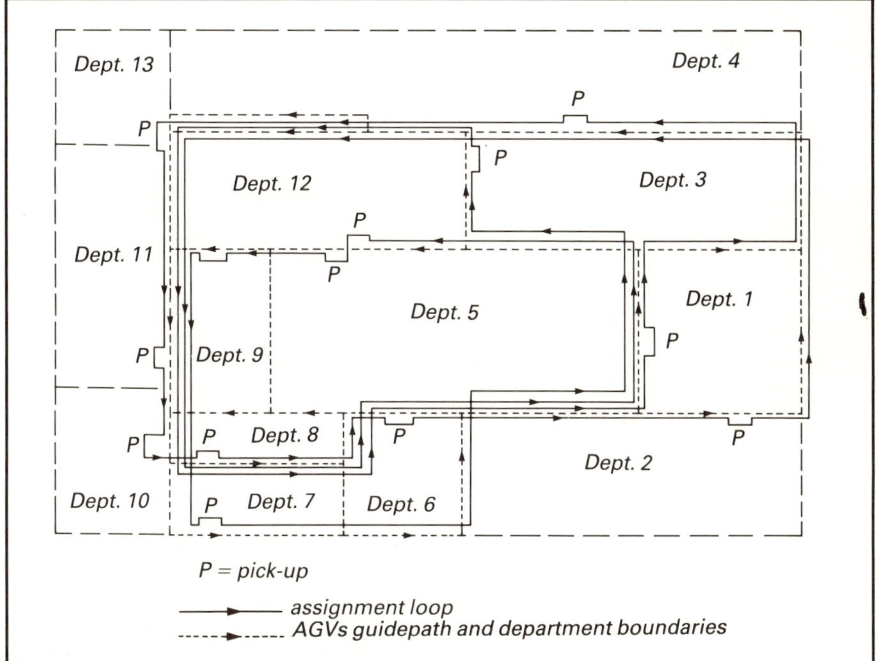

Fig. 3   A single-loop vehicle assignment system based on demonstrative facility

Table 4 Unit load throughput under sequential and nonsequential dispatching rules

Rule combinations		Sequential dispatching
LIV-MROQS	*LIV-MFCFS*	
740	770	406
754	783	499

Table 4 shows the shop output under three vehicle dispatching rule strategies, with two simulation runs per strategy. As can be seen, the sequential dispatching rule performed much lower than the other two rule combinations. This is partly due to the fact that the sequence selected is not optimal and partly because sequential dispatching strategy involves a large number of unproductive vehicle visits to departments at times when such departments have no need for vehicles. In an existing layout of the type given in this problem where to make a complete round of visits a vehicle may pass through some aisles more than once, sequential vehicle dispatching policy (if adopted) should be justified under different criteria, such as the simplicity of traffic control algorithms and the elimination of shop locking possibility, rather than on its ability to accelerate jobs through the shop.

Concluding remarks

Heuristic rules for dispatching automatic guided vehicles in a job shop with an existing layout have been presented. The characteristics of these rules were demonstrated under several shop operating conditions. The demonstrations indicated that rules which are derivatives of distance measures have several drawbacks if the appropriate layout conditions of a facility and equipment locations are not met. The results also demonstrated how the vehicle dispatching problem and its associated resolution techniques can affect shop design elements such as buffer space requirements in each department, central facility buffer requirements, shop throughput, and identification of poor layout designs. There are, of course, other design issues that are not addressed in this study. These factors include traffic control problems, load transfer mechanisms at load pick-up and delivery points, and other operation-related issues.

Appendix

The wire guide path of an AGV system (Fig. A1) can be modelled as a network consisting of nodes and arcs. The position of a node can be specified in terms of the Cartesian coordinate system where (x_α, y_α) represents the coordinate of node α.

For traffic control reasons, let a safety zone (i.e. check zone) be defined around every node as shown in Fig. A2 and let a check point be the intersection point between a check zone and an arc entering a node. The

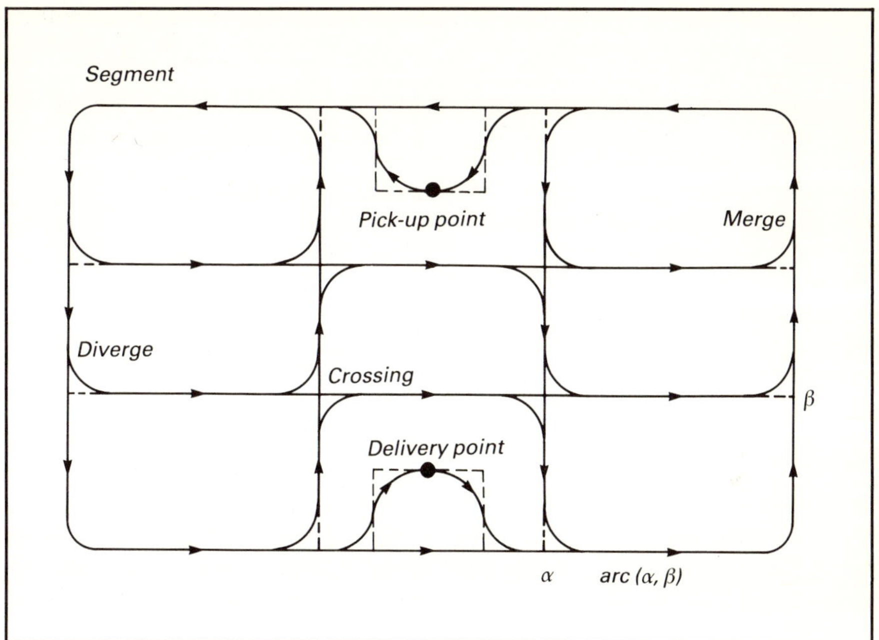

Fig. A.1   A guidepath system

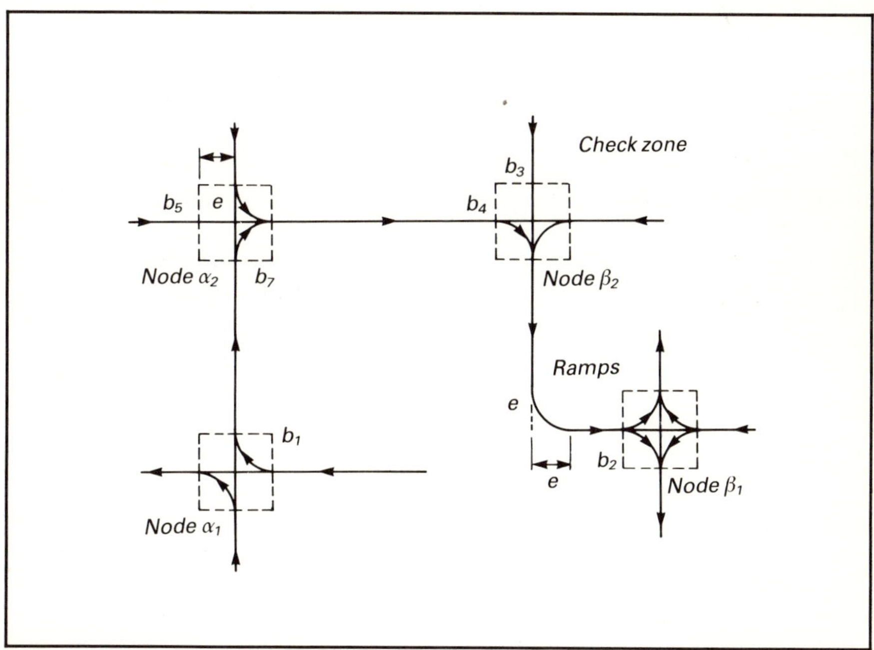

Fig. A.2   A section of a guidepath network

check points are denoted by b_i, $i = 1,2,3 \ldots, n$ in Fig. A2. check points are decision points for vehicles in the network. These decisions include intersection crossing, holding for blocking, and negotiating a turn. All vehicle turns are made through interchange ramps as shown in Fig. A2. Since decisions are made at check points, the flow of vehicles can be modelled as a series of transitions or discrete jumps between successive check points. The transition time required between the check points of two adjacent nodes α and β is a function of the vehicle travelling speed, the degree of traffic congestion between the nodes, and the physical separation of the nodes. Thus the distance between two check points of adjacent nodes can be represented as $d_{\alpha\beta}$, where

$$d_{\alpha\beta} = \begin{cases} \{(x_\alpha - x_\beta)^2 + (y_\alpha + y_\beta)^2\}^{1/2} + (f - 2e)\phi, \\ \quad \text{if the distance from node } \alpha \text{ to } \beta \text{ is Euclidean} \\ \\ |x_\alpha - x_\beta| + |y_\alpha - y_\beta| + (f - 2e)(\phi + 1), \\ \quad \text{if the distance from } \alpha \text{ to } \beta \text{ is rectilinear} \end{cases}$$

$$\phi = \begin{cases} 1, \text{ if a turn is required at the check point of node } \alpha \text{ to reach} \\ \quad \text{node } \beta \\ \\ 0, \text{ otherwise} \end{cases}$$

where e is the distance between a check point and the node it serves
and f is the length of the smoothed ramps required to negotiate turns. The value of f is a function of the turning radius of the vehicles that use the network and the maximum allowable vehicle speed. f can be approximated by $(\pi e/2)$ if the check zone approaches the shape of a square.

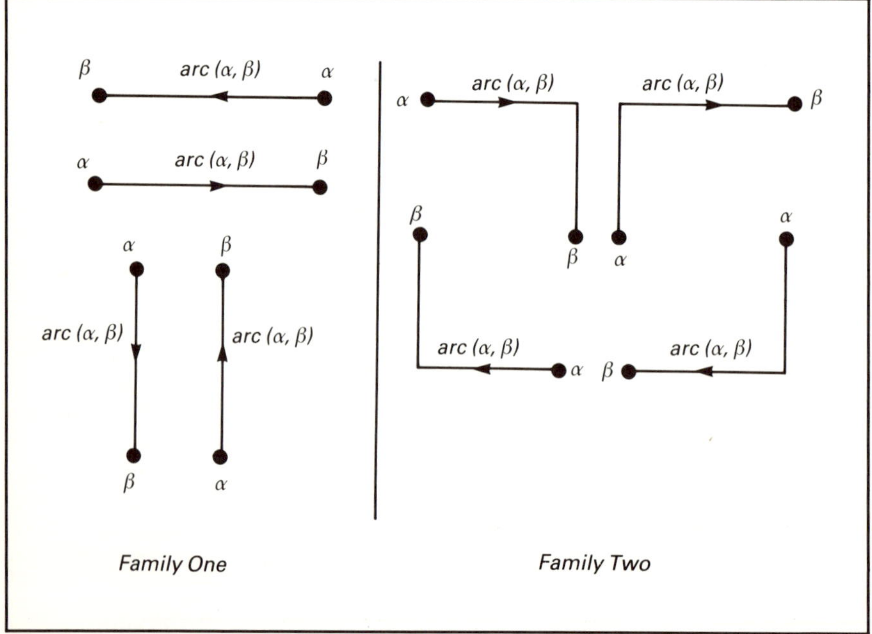

Fig. A.3   Characterisation of arc according to orientation

The distance between two adjacent nodes α and β is defined as Euclidean if and only if

$$x_\alpha = x_\beta \ \text{ or } \ y_\alpha = y_\beta$$

but not both. A distance is considered rectilinear if and only if

$$x_\alpha \neq x_\beta \text{ and } y_\alpha \neq y_\beta$$

That is, the node coordinates are completely distinct for the adjacent nodes. The distance equation above assumes arcs can enter or leave a node only along a coordinate axis. Also, arc shapes other than those classified in Fig. A3 are excluded.

References

[1] Egbelu, P. J. 1982. A design methodology for operational control elements for automatic guided vehicle based material handling system. Ph.D. Dissertation, Virginia Polytechnic Institute and State University, Blacksburg, VA, USA.

[2] Maxwell, W. L. and Muckstadt, J. A. 1982. Design of automatic guided vehicle systems. *IIE Transactions*, 14:114.

[3] *Modern Material Handling*, 1975. A musical driverless cart? Yes for two sound reasons, 30:44.

[4] *Material Handling Engineering*, 1979. Driverless load carriers make the long runs at Lennox. 34:86.

[5] *Modern Material Handling*, 1978. Driverless system makes old warehouse productive. 33: 62.

[6] *Material Handling Engineering*, 1977. Driverless tractors add shrink tunnels to their routes. 32:80.

[7] Dantzig, G. B. and Ramser, J. H. 1956. The truck despatching problem. *Management Science*, 6:80.

[8] Bodin, L. and Golden, B. 1981. Classification in vehicle routing and scheduling. *Networks*, 11:97.

[9] Irving, J. H. et al. *Fundamentals of Personal Rapid Transit*. Lexington Books, Lexington, MA, USA.

[10] McCloskey, J. R. and Hanssman, F. 1957. An analysis of stewardess requirements and scheduling for a major domestic airline. *Naval Research Logistics Quarterly*, 4:183.

[11] Richardson, R. 1976. An optimisation approach to routing aircraft. *Transportation Science*, 10:52.

[12] Wilkes, F. L. and Klee, H. 1976. Determining priority rules for underground rail-transport by computer simulator, *Automation in Mining and Metal Processing*. Proc. 2nd IFAC Symp., Johannesburg, South Africa, Sept. 13–17.

[13] Russell, R. S. and Tanchoco, J. M. A. An evaluation of vehicle dispatching rules and their effects on shop performance. *Material Flow*, (submitted).

[14] Maxwell, W. L. 1981. Solving material handling design problems with OR. *Industrial Engineering*, 13:58.

[15] Egbelu, P. J. and Tanchoco, J. M. A. 1982. Operational considerations for the Design of Automated Guided Vehicle Based Material Handling Systems, Technical Report No. 8201. IEOR Department, VPI and SU, Blacksburg, VA, USA.

[16] Blair, E. 1980. Vehicle routing and material handling subject to stochastic demand, *ORSA-TIMS Bulletin*, 72.

[17] Egbelu, P. J. and Tanchoco, J. M. A. 1982. AGVSim User's Manual, Technical Report No. 8204. IEOR Department, VPI and SU, Blacksburg, VA, USA.

[18] Tanchoco, J. M. A. and Agee, M. H. 1981. Plan unit loads to interact with all components of warehouse systems. *Industrial Engineering*, 13:36.

SIMULATION – AN AID TO THE DESIGN OF AGV NETWORKS

R. Hollingworth, R. B. Beadle and G. M. Hook
Istel Ltd, UK

Many designs of manufacturing facilities are now based around an automatic guided vehicle network. These AGV networks offer the designer flexibility of manufacturing and transportation but place more emphasis on scheduling and facility control. In many cases the scheduling and facility control can reduce the output of the production unit. This paper will illustrate by example how visual interactive simulation can be used to identify and overcome these restrictions.

The use of an AGV network, whether it be as a part of an FMS cell or merely as a replacement for a conveyor system, represents a significant step forward in the technology of in-factory logistics. There are many benefits that accrue from the use of AGVs, but to achieve these benefits many problems have to be resolved. These problems include:

- How many AGVs should be used?
- What kind of track layout to use?
- Should the track be unidirectional?
- Which routes should the AGVs use?
- Should alternative routes be considered?
- Should a taxi or merry-go-round system be used?
- When an AGV is needed, how is the choice of which AGV to use made?
- How should the task of AGVs be prioritised?
- How should the priorities at track junctions be allocated?

It should be noted that a majority of the problems in the above list are qualitative and not quantitative. This implies that mathematical analysis is insufficient to solve the overall problem of what is the best design for an AGV system. Furthermore, all of the above problems are interrelated, e.g. if the scope of the track layout is increased, this might lead to a reduction in the number of AGVs needed, and so cannot be resolved independently.

A failure to design an AGV system correctly in any one of the problem areas will lead to an under-performance of the system in terms of return on capital invested. The capital invested in AGV systems is never insignificant. The under-performance can occur in many different ways: failure to achieve design output, increased capital costs, high operating costs and extended commissioning periods.

The design of an AGV system is a complex task with many interrelated elements, each of which can have a major influence on the operation and performance of the whole system. Standard analytical techniques which can tackle an individual element are not sufficient to provide a comprehensive solution. For instance, optimising a number of different elements in isolation will almost certainly not provide an optimal solution overall.

In addition, these standard techniques would usually look at an average situation which often would not reflect a typical set of operating conditions. In fact, the nature of AGV systems is such that there may never be a typical situation. Simulating just the AGV system can lead to a similar misconception. The simulation model must encompass the whole factory system, including those machines and processes which actually generate the transport task which the AGV system is expected to fulfil. This scope of the simulation project is essential for success.

Simulation provides the opportunity to perform a comprehensive evaluation of the whole system, and allows the effect of small changes to particular elements in the system, e.g. a minor route modification, to be determined.

The extension to the use of visual interactive simulation is the process of designing the layout and operation of an AGV system can provide the designer and decision maker with a much greater sense of confidence that the installation will perform as required. This is because they have actually seen a representation of the end-product function on the computer screen.

How is simulation used

There are many different approaches for simulation and correspondingly many different packages to use. This section concentrates on the latest generation of simulation packages, which not only give a dynamic visual picture of the simulation, but are interactive and are quick and easy to build. The leading example of this type of package is the SEE-WHY/WITNESS system.

The process of building a simulation model consists of defining all the constituent elements of the AGV network to be studied (number of AGVs, speed section of track, the number of machines to be fed and the number of load/unload stations). The quantity of each element is normally parameter driven, such that the effects of alternative number of AGVs or alternative speeds of AGVs can be evaluated. Incorporated in this process is the design of a screen layout that closely mimics the proposed AGV network.

The next stage is to enter the control strategies to be used. These include rules for the selection of AGVs, routing criteria, scheduling algorithms and junction priorities. Such a control system will be a major cost element in the overall cost of the AGV network and can take many man-years of programming to implement. The simulation equivalent should take considerably less time, to be measured in hours, and should not require detailed knowledge of programming languages.

This whole process of building a simulation is extremely useful in itself, for it ensures that the purchasers of the AGV network fully understand the operation of the system and that suppliers have considered all the relevant points.

Once a model is built, inadequacies in the design are easily spotted by all interested parties. The screen will show idle machines, it will show blocked AGVs and it will show extended travel routes. Such inadequacies prompt the design team to ask many questions and subsequently to prompt examination of alternatives. This process leads very quickly to a robust design and to a general confidence in the design.

The simulation also provides detailed performance statistics on machine and AGV utilisation and can include information on the distance travelled by each AGV. This latter measure being a vital piece of information for assessing the battery size and recharging requirements of the AGVs.

The following three examples illustrate the benefits of simulating AGV networks. The first example shows how simulation was used to test the flexibility of using an AGV network in a car assembly operation, the second example is an FMS model, and the third is of a car rectification area.

An automatic welding facility

An advanced engineering study had proposed that a system of automatic welding gates fed by AGVs should replace the conventional assembly line welding systems. Management were concerned about this approach on two counts. Firstly, they could not easily size the facility requirements (how many welding gates in what kind of configuration, fed by how many AGVs?), and secondly, how variable would the output from such a system be? If the variability is high then storage would be required to keep production flowing to subsequent operations.

A schematic of the proposed layout is shown in Fig. 1. Before this layout could be simulated, there was a need to determine how to route the AGVs through the six stages of manufacture. Initial constraints

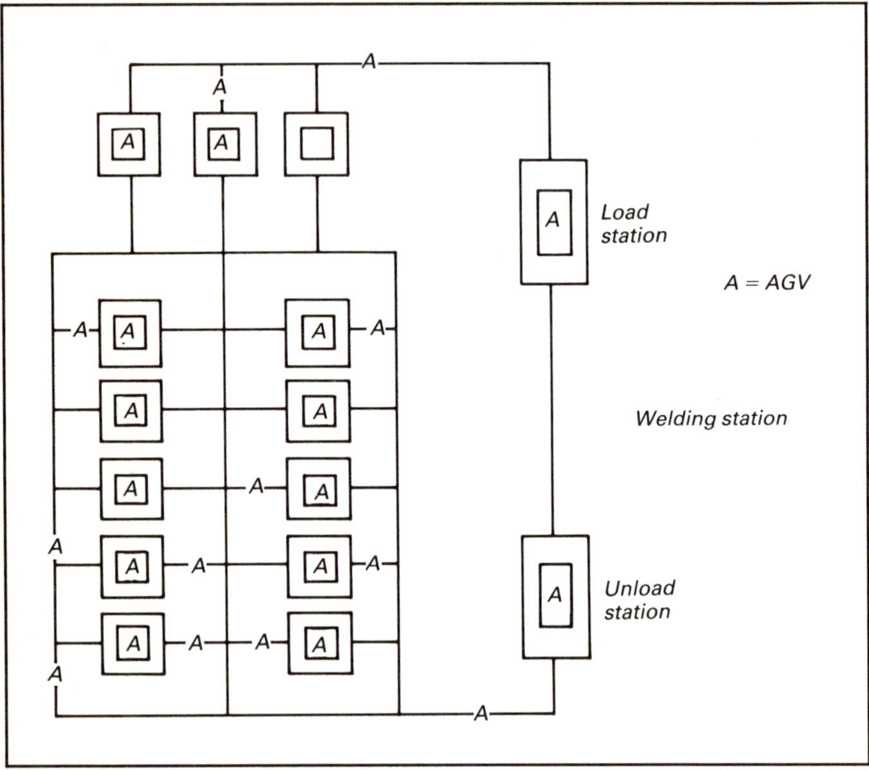

Fig. 1   AGV system for welding car bodies

were that the AGVs must visit the primary weld stations first and then they could visit the five secondary weld stations in order.

Step one in modelling was then to put an initial set of routes into the model and observe the flows, under both normal operating conditions and under simulations covering breakdown of weld stations. These experiments were carried out in the presence of design engineers and, therefore, the problem of communication of the results was eliminated. The results of this stage of the model were a set of control rules that depended upon the secondary weld stations being visited in sequence. This was the only way, given the proposed layout and all feasible alternatives, to maintain a seemingly steady output from under all types of operating conditions.

The imposition of sequential welds on the system meant that some of the expected flexibility of the system could not be achieved in practice. This would have been very difficult to explain to the design engineer if they had not been able to participate in the experiments and view the flows through the system.

The second stage of the modelling was to determine how many AGVs were needed and how variable the output would be under expected operating conditions. These experiments were again carried out with management and design engineers present and again the results were surprising to them. The initial estimate of the number of AGVs required gave rise to a very variable output, and the reason for the variability was obvious from watching the dynamic display of the model. Whenever a breakdown occurred this caused the AGVs to bunch up, which alternatively starved and then overloaded the output stations. The immediate reaction to this problem was to increase the number of AGVs, this being easily achieved in front of management with the interactive capabilities of the system. The effect of increasing the number of AGVs was to increase the variability of the output and the bunching became more severe. Reducing the number of AGVs did reduce the variability of the output but it also reduced the maximum achievable level.

By observing all of these effects on the model in the space of one or two hours both management and the design engineers became fully aware of the operational characteristics of such a system. Obviously many different configurations were tried of track and machines, but none of the configurations that were physically possible (there were space constraints on the layout) gave a suitable level and stability of output. Therefore, such an approach to manufacture for this situation was abandoned at a relatively early stage. Subsequent designs for this kind of manufacture have been much more realistic in their concepts, indicating that the design engineers have learnt a great deal from visual interactive simulation.

A flexible manufacturing system

Flexible manufacturing systems represent one of the largest changes in manufacturing procedures that are currently taking place. As with most radical changes many oversights occur that result in either underachievement or in redundant elements. Visual interactive simulation offers many benefits in this area which has the sizing and control elements more closely aligned than any other manufacturing example.

The case study from this area is an eight machine system plus a wash and load/reload area designed to produce 16 different components in vastly different volumes. Transportation with the system is provided by AGVs.

The questions that were asked of the simulation were:

- How many AGVs are required.
- What routing and priorities should the AGVs follow.
- What effect does the sequence of manufacture have on the efficiency of the system.

The base design of the system had the machines aligned in two banks of four with a simple loop track for five AGVs providing the link between the machines. This design was simulated and was proven to be adequate in terms of the design objectives. However, the design engineers in watching the simulation observed queuing of the AGVs taking place at several points in the system. Consequently they asked that the simulation model be amended to reduce the queuing. Their suggestion was that a feed loop on the AGV track be included for each machine, the contention being that this would eliminate queuing at each machine but they wanted to know if these loops would reduce the total requirement for AGVs. This second design was simulated and as anticipated the queuing in front of each machine was eliminated but the number of AGVs required to meet demand varied between four and five dependent upon the sequence in which the various parts were made.

It was also apparent that the AGVs were travelling relatively long distances between operations. A reduction in travel between operations would be achieved if bypasses were introduced across the centre of the AGV loop. Again, the effect on production was significant. Three AGVs could meet design production requirements independently of the sequence of production. The only cost being an increase in the complexity of the control of the AGVs which when compared to the cost of two AGVs was not significant. The final layout of this system is shown in Fig. 2.

A car rectification area

Car rectification tasks can be quite unpredictable in the short term, say over an hour or two, even if over a long period the time required to rectify different faults can be forecast. Consequently an AGV system is an ideal way of transporting the vehicles between the different rectification bays.

The case study described here contains five different types of rectification stations, linked by an AGV network, upon which AGVs can travel in either direction (see Fig. 3). One hundred percent of vehicle production passes through this area even if the majority only visit the polish/body inspection stations.

The visual interactive simulation model of this facility was designed to answer the following questions:

- How many AGVs are required?
- Are there sufficient stations?
- What are the optimum control rules for AGV activity in the area?

Although capacity calculations from average rectification requirements showed that sufficient stations of each type had been provided in the facility, the model showed that only 90% of the desired throughput was achieved. Floor space was a very strong constraint in this layout, as was of course cost. Consequently the client did not want to add more stations.

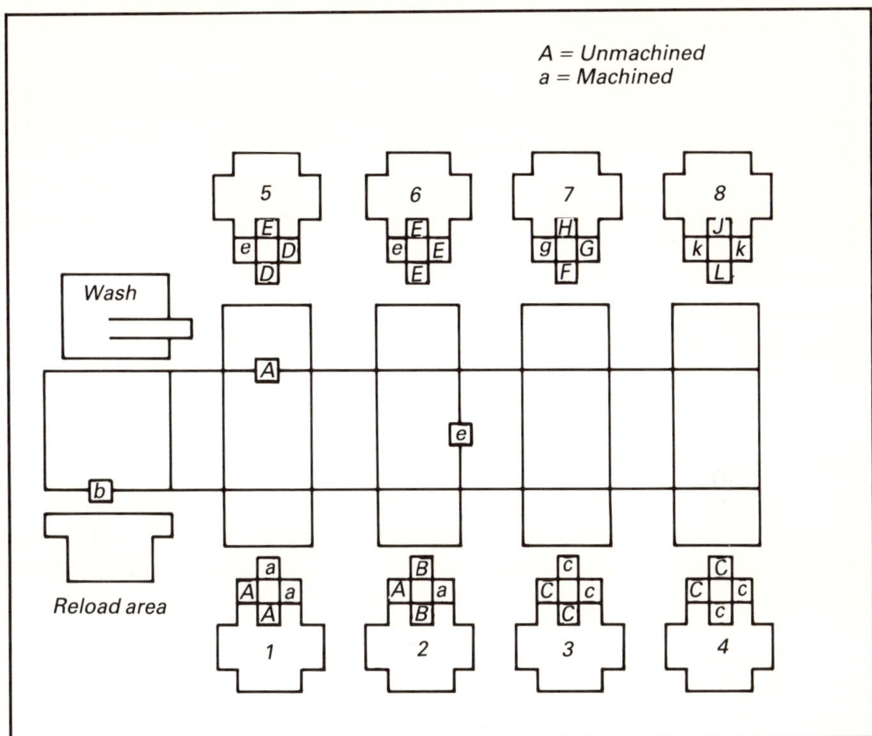

Fig. 2   FMS simulation

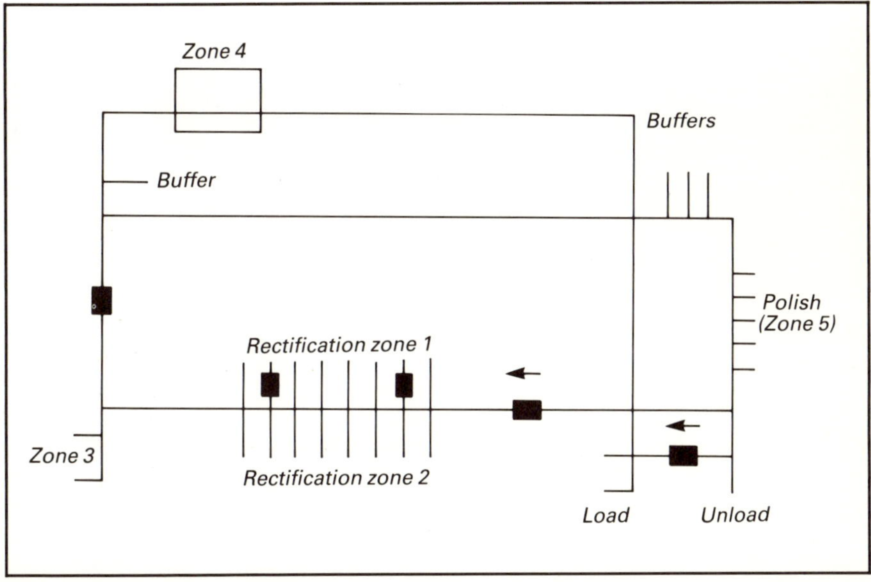

Fig. 3   Car rectification area

The control rules originally designed for the facility dictated that vehicles which could not be accommodated in the rectification station required had to circulate around the system on an AGV. The first solution tried was to provide more AGVs; this was not successful, it only caused more to circulate. The second solution which was investigated was to leave vehicles at the previous station, when the next station was busy, even though all the possible work which could be undertaken there had been completed. This solution was also found to be wanting, in that blockages occurred and the required station utilisation could not be achieved.

The basic problem was due to short-term fluctuations in the demand for particular types of rectification station. It became obvious from watching the visual simulation under a variety of operating rules that a buffer area was required. The problem that then required to be addressed was that because of space limitations where could this buffer be? The top right-hand area of the layout was redesigned to accommodate three buffer stations and a fourth was installed by the roller test area. These additions to the layout were put into the simulation model and fully investigated. AGV control rules were amended, and it was observed that the various rectification stations were able to operate at a higher utilisation and the required throughput achieved.

The customer was then confident that the proposed layout would provide the required rectification capability. At this point, an order for the facility installation was placed.

EVALUATING AGV CIRCUITS BY SIMULATION

B. Duffau and C. Bardin
Seri Renault Automation, France

The sizing and evaluation of automated guided vehicle (AGV) system circuits by analytical, statistical and simulation approaches is presented. Particular attention is focused on evaluation methods by simulation (after an analytical presizing), including the use of queue software and Petri networks. A detailed description is then given of the functions required by a specialised AGV simulator, in particular with regard to: the physical description of the network (division into sections, components, etc.), operation and management rules (calling modes for vehicles, conflict management, etc.), additional procedures (e.g. for statistics, breakdown modelling, batteries' autonomy computing, etc.), compatability (with Petri and queue networks, etc.) and software extension facilities (adding new functions or user algorithms), and dividing software into levels of increasing complexity, for progressive user training.

Faced with a problem concerning the design of a handling network, many designers possessing an automated system to aid design rush towards the display unit, first to draw the network and then simulate it, and only come to reflect on the problem afterwards. In such cases the drawing and the simulation are often inadequate and the designer runs the risk of bypassing an optimised solution. This article aims to show how modelling and simulation can be used to arrive at a balanced solution for the flow in these networks.

A simple handling system

Consider the simple AGV circuit shown in Fig. 1. The role of the AGVs is to transport loads from a workshop A to a workshop B. There are no traffic conflicts (no switches), nor is there any decision to be taken for the assignment of the AGVs to the stations.

If, with consideration of the length of time for transportation, of the operation times and of the flow of loads to be transported, a simple

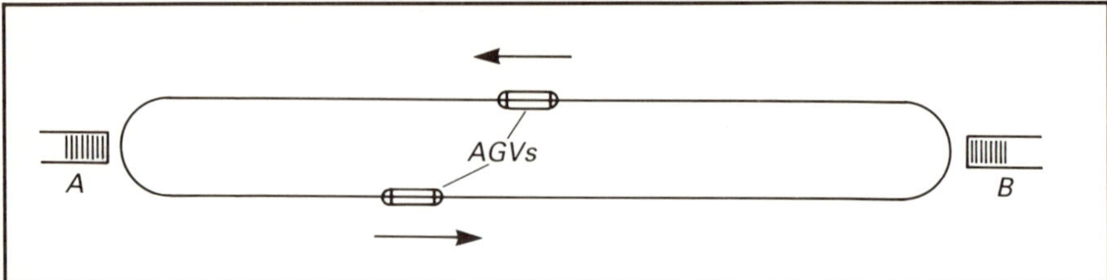

Fig. 1   A simple AGV circuit

'rule of three' leads to a need for 1.31 AGVs (that is in practice 2 AGVs), it is unlikely that simulation would lead to a different result, especially if there exist analytical methods with which the behaviour of the queue at A could be calculated if required. However, it will be of use to verify on the technico-economic plan if the AGV solution is, in this particular case, the most suitable.

An AGV network and analytical analysis

In the second example (Fig.2) an AGV solution is advantageous since it can be adjusted to the variations of the workshop load and, by adjustment of the network and its management, it allows the addition of new assembly lines in the workshop to be carried out with flexibility.

The workshop itself is an assembly of several types of mechanical sub-units. It is composed of ten parallel lines, connected together by a handling system. The average output rate is 120 sub-units an hour, with peaks of 160. The components fixed onto pallets at A, are transported 4 × 4 towards the first section of the assembly at B. At the entrance and exit of each line there are roller tables which serve as buffers. At each line exit the components are recovered by the AGVs and transported towards the following operation line.

A brief analysis of the workshop allows two nearly independent loops, on each side of the lines, to be isolated. Consequently a first quick calculation, of the stationary operation type, can be carried out from:

- Determination of the flow (number and length of transportations carried out on average each hour, empty and loaded).
- Length of time for handling (loading/unloading).
- Speed of the AGVs.

An addition and a 'rule of three' calculation allow the minimal theoretical number of AGVs to be determined, below which the workshop cannot function (in this example, there are 5 AGVs on the left loop and 4 AGVs on the right loop).

But this calculation is inaccurate since:

- The pallets do not arrive regularly at A, but in a random fashion, causing load variations in the AGV transport.

- The line operation times are subject to fluctuation.
- Certain AGVs may have to wait, in front of certain lines, for the last AGV to finish its handling (queue phenomenon).

In this way, the AGVs will have a dead time, which will not be caught up, and the estimated number above will be insufficient.

To improve the calculation a ratio of the type 'normal average utilisation factor of an AGV' could be utilised, but past experience has shown that stable ratios are difficult to guarantee, except in relatively simple cases. One therefore turns towards analytical methods of the type 'statistical calculations of queues'. In order to do this, several methods can be envisaged, such as the CAN-Q algorithm or algorithms of the MVA (mean value analysis) type, with Reiser or Hildebrant heuristics.

CAN-Q is a program written in FORTRAN which, with hypotheses of random circulation of components and of random service time of the AGVs, calculates:

- The mean production rates.
- The mean utilisation factor of the AGVs.
- The mean average waiting times in the workshop.

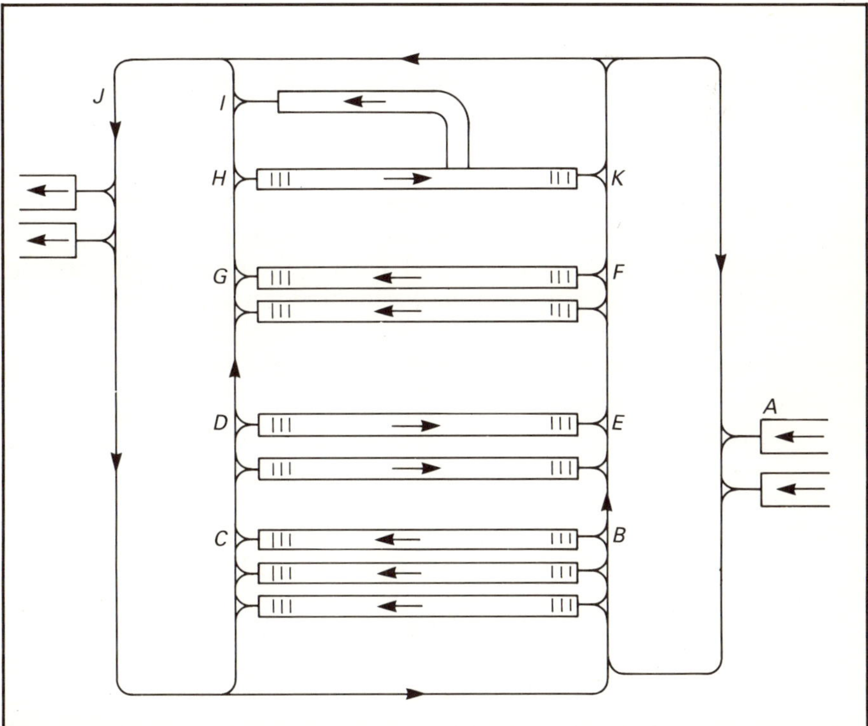

Fig. 2   AGV network: Cleon Renault Workshop

The main parameters required by CAN-Q are:

* The number of workstations.
* The number of components in the system.
* The number of servers at each station.
* The frequency of visits to each station.
* The average transportation which can be simultaneously transported (number of AGVs).
* The number of operations per component.

In order to carry out an analytical calculation, the program takes restrictive hypotheses, of which the principal ones are:

* The activity times are distributed in accordance with an exponential law: the average of each distribution corresponds to the average time of each operation, per machine.
* Each station possesses a queue with infinite capacity.
* There is no possible blocking phenomenon.
* The routing of components is carried out in a probabilistic manner.
* The number of components in the system remains constant: a finished component is immediately replaced by a raw component.
* There is only one single transportation time for the whole system (average journey times between two stations).

Applied to the AGV network in Fig. 2, this method has indicated a number of AGVs necessary per loop which is superior to that provided by the static calculation. This is explained by the need to consider the risks of load variation in the system. But as the restrictive hypotheses of the analytical model do not allow these results to be considered as certain, a simulation was necessary. After the requirements for battery recharging and AGV self-sufficiency had been considered, the simulation gave results fairly close to those of the analytical method.

The analytical method thus acted as a 'forecaster' and avoided trial-and-error efforts in the formulation of simulation hypotheses.

Further examples and general approach for sizing AGV systems

In the following example (Fig. 3) the role of the AGV is to transport loads between a warehouse A, a coach station B, and a railway station C. A quick calculation can show that, in regard to the flow, traffic conflicts will not be numerous, and that in this way, a first approximation of the transit time between A and B, B and C, etc., is available.

The essential problem for the sizing of the quantity of AGVs is to successfully control the flow of AGVs, notably the organisation of empty returns and the distribution of the empty AGVs as well as the manner of assigning the AGVs for a load call. The actual drawing of the network will be of importance only in a second stage. In order to

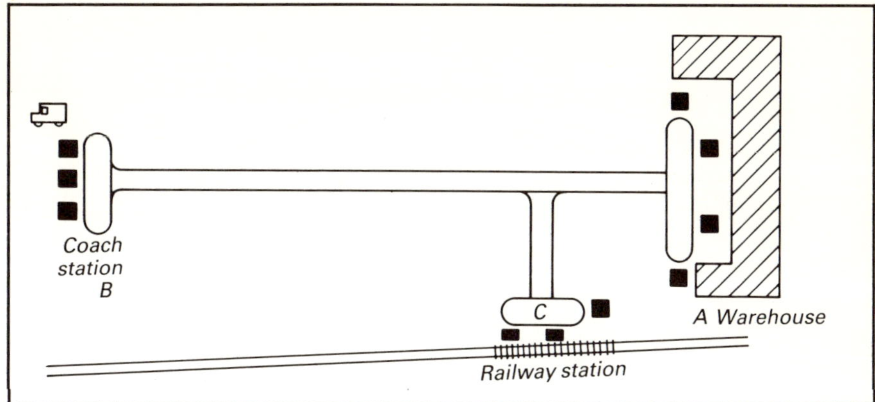

Fig. 3   AGV system

study the strategies of return and assignment of the AGVs, a program which will consider the stations, the flows, and the strategies of return and assignment, and which will evaluate and simulate the behaviour of the system studied will be preferred to a program for aiding the drawing.

The basic elements of such software are presented here. They form an integral part of the simulator presented below.

Before all else the designer must determine the workstations (loading/unloading) which will constitute the junction points of the flows. Each flow from one station to another is described by specifying its capacity (in maximum number of AGVs at a given time) and the transit time for each AGV (Fig. 4). Once entered into the flow, an AGV meets no blocking problem and exits after the given time. Certain flows could be determined for the empty return of the AGVs, or for the arrival at an assignment station.

In order to use the software in this quick form, it is necessary to be able to anticipate the types of problems the circuit can present. With a slightly more complicated circuit than the one in the example, estimation errors are easily committed. It is possible for important points of conflict within the system to go undiscovered.

As a general rule, flow methods are all the more efficient as the designer has already implemented networks of the same type. The next example shows how analytical methods can be inefficient.

Fig. 5 represents a planned AGV network intended for the handling of packets of finished products in a warehouse (storage 16,000 places). The AGVs convey loads between the road transport loading/ unloading platform shown at the bottom of the diagram and the automated warehouse, both in the direction from the platform to the warehouse and from the warehouse to the platform (unstocking). A part of the handling is from platform to platform (transit) and certain loads pass through administrative or technical stations (e.g. for customs clearance). The average flow is 400 loads per hour.

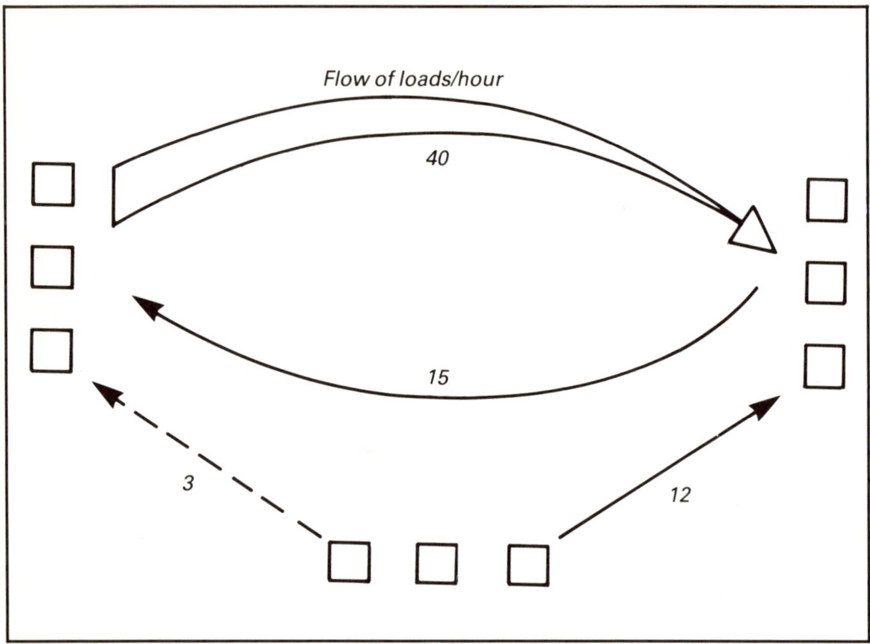

Fig. 4   Flow method

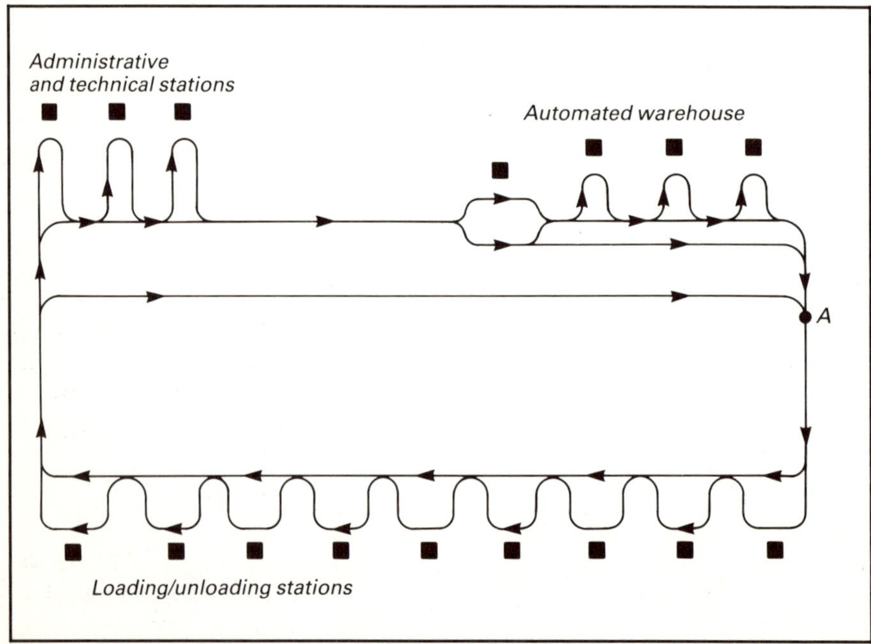

Fig. 5   Original plan of AGV network and automated warehouse

Fortunately, Fig. 5 does not correspond to the actual construction; this is explained below.

With the preceding analytic methods it has been estimated that the necessary quantity of AGVs is 35. A simulation, however, has shown that with this estimate only 65% of loads can be handled. Therefore, a network built according to this diagram would have a bottleneck between the road transport unloading platform and the automated warehouse, since 35% of the packets could not circulate. The critical point is point A on Fig. 5. All the empty or loaded AGVs come together there and finally form a serious jam. If in an attempt to remedy the situation the number of AGVs were to be increased, not only would the cost go up appreciably but the results would be worse because the flow would decrease.

The curves shown in Fig. 6 enable the results obtained with the different methods to be compared. The deviation between the analytical and simulation methods is easily explained by the fact that analytical methods do not take into account all the blocking or saturating phenomena in the queues.

Thanks to the simulation results, the mistake which could have been made has been avoided. Fig. 7 shows the final drawing of the network.

In conclusion, it is possible to draw a general approach for sizing such systems:

(1) Research the main difficulties and characteristics of the behaviour of the network, by analogy with other experienced cases (designer's know-how).

(2) If it is felt necessary, balance the network and implement a management strategy. The efficiency and the duration of this phase of the study, which is based on the use of flow simulators, will chiefly depend on the experience of the designer or engineer.

(3) If there are risks of conflicts within the system or other particularities of the network which must be analysed in more detail, use detailed simulation of the network, as described below.

In this approach there are two main elements, each of equal importance:

- Flexibility: If the network is supposed to undergo future development and if an idea of this evolution is known, it is useful to verify that the choices made will keep their advantages in the future. Simulation can test rapidly the different possible evolutions of the network, its workload and its management.
- Reliability: It is necessary that the studied network will be as resistant as possible to breakdowns of its component parts. A combined approach using simulation and Markov graphs is to be detailed in a future report.

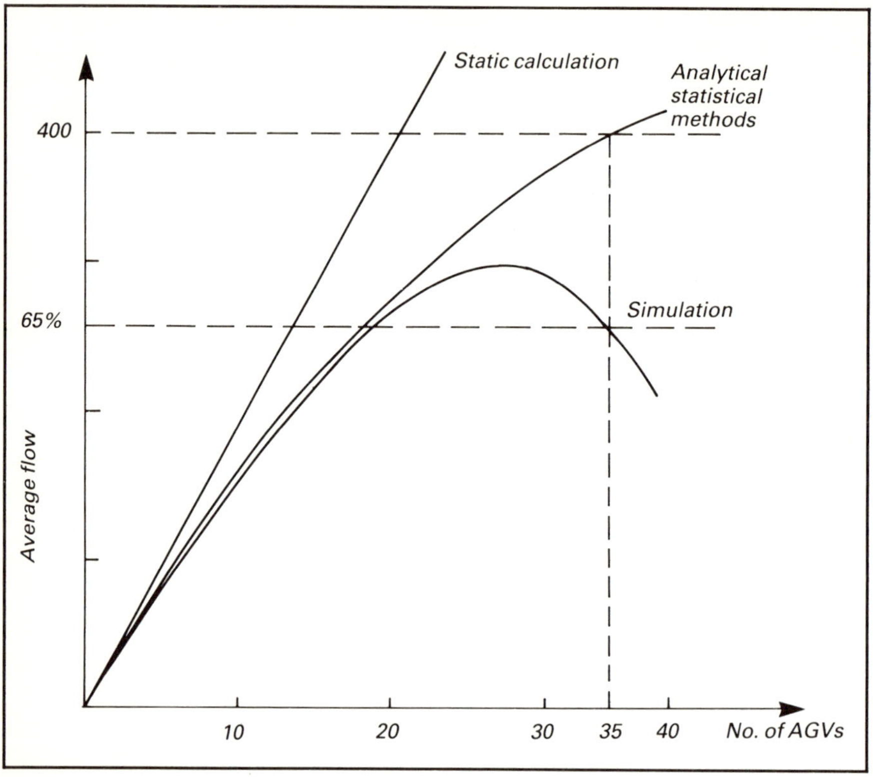

Fig. 6   *Comparison of evaluation methods*

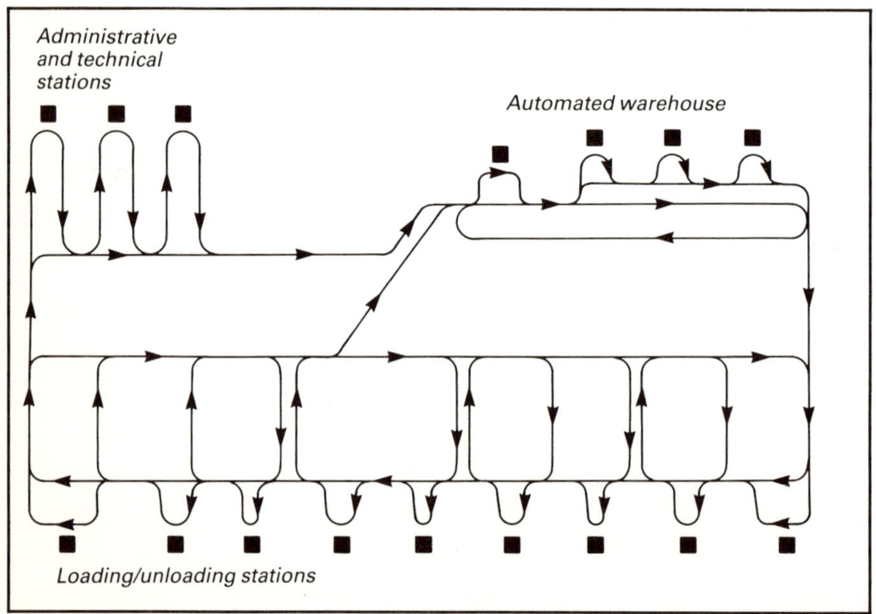

Fig. 7   **Final layout of AGV network and automated warehouse**

Simulation tools

A network may of course be modelled with general simulators, whether they are, for example, queue simulators of the SLAM type or state and transition simulators like Petri networks.

Fig. 8a shows a divergent switch modelled in SLAM. The loaded AGV arrives at a queue H, where it awaits the release of the divergent switch. After an unblocking period, it releases the component (COMP). At this stage a FORTRAN program decides which exit will be taken with regard to the AGV data: assignment, load, etc. The AGV is then allowed, after a certain period of time (transit timer), to position itself in queue I of the left exit (LE), or in J, that of the right exit (RE). Once the exit is free, it will be possible, after the unblocking period, to release the divergent switch for subsequent traffic.

Fib. 8b shows the switch modelled by a Petri network. Here, the AGV will advance only when the tokens mobilised by the path of the preceding AGV are returned to it, thereby crossing the Petri transition.

It can readily be seen that for this type of switch an AGV specialist-orientated language (AGV/SL) capable of expressing the switch's function directly will be easier to use. Fig. 8c shows the switch in such a language.

However, in order to represent, in a network, more unusual behaviour (e.g. concertina effects between AGVs in a picking aisle with a large number of racks), it will be useful to call upon formalisms such as SLAM and the Petri networks. That is why it is desirable that a specialised simulator (with components like those in Fig. 9) should function together with general simulators (adapted to special cases).

In 1980-81 Séri Renault Automation created an AGV simulator like the one in Fig. 9 for the design of several complex networks. This simulator was tested on other different networks, its advantages and defects being analysed. A working team, bringing together AGV users, a project-study group and a manufacturer, was then formed to produce an improved version, as discussed below. All of the software is not presented here, since that would be rather tedious, but for the purpose of illustration several of the simulator's functions are considered.

'Cruising loops' and assignment centres

The importance of the strategy used for the consideration of the empty AGVs was discussed earlier in the article. In the end the proper functioning of an AGV system will depend on its correct distribution.

One of the distribution strategies currently used is to have, for each empty AGV, a 'cruising loop'. An AGV, as soon as it is empty, will try to make its way back to this destination, by passing through calling stations.

Another distribution possibility is the assignment station. This is in

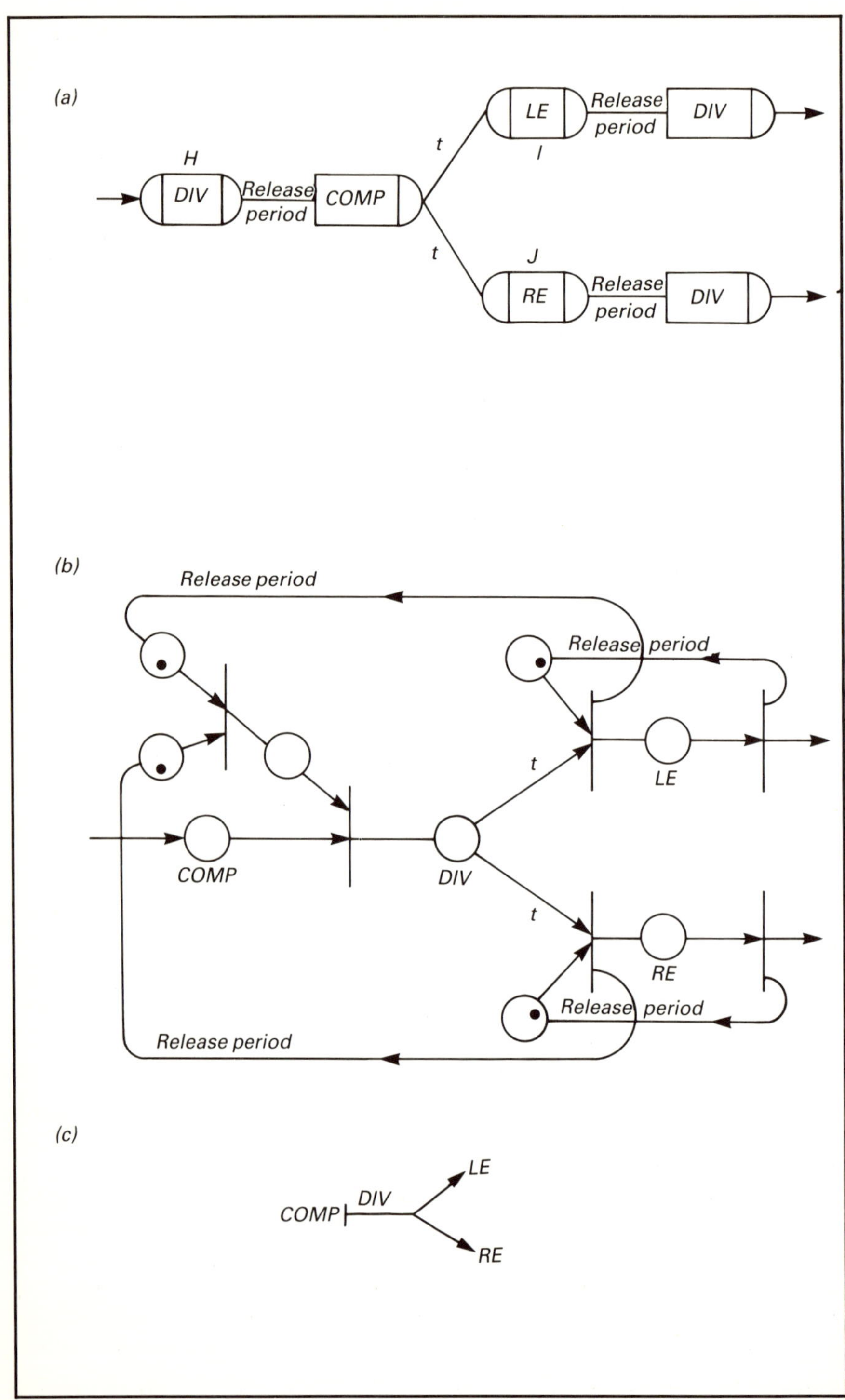

Fig. 8   Modelling of a divergent switch: (a) SLAM; (b) Petri network; (c) AGV/SL

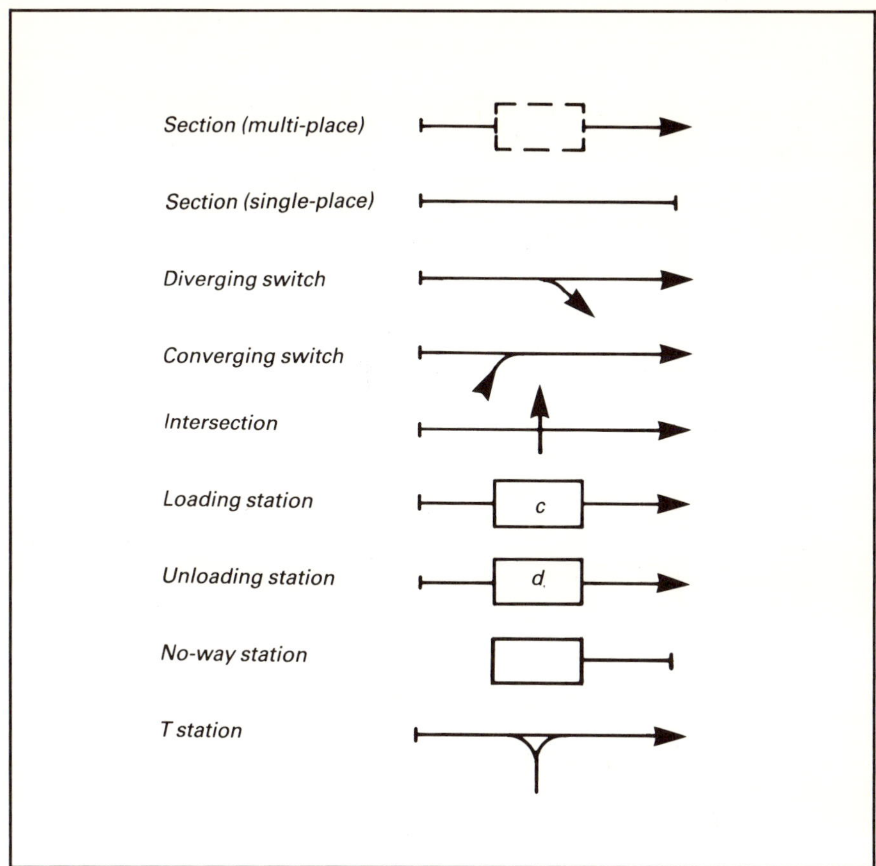

Fig. 9   Components of an AGV specialist-orientated language

fact a dialogue point which allows the empty AGVs to be guided towards certain loading stations where the loads are waiting. This station can, if none of the loading stations concerned request an AGV, choose between blocking the empty AGVs or letting them pass towards their cruising loop. This solution enables only those AGVs which are strictly necessary to be sent to certain zones which have a difficult access.

Re-orientation of AGVs during assignment

It is possible that the AGVs may have to be re-orientated during their assignment. This redistribution of the AGVs can be cyclical, random, etc. towards a certain number of newly determined destinations. For example, Fig. 10 shows a section of a network where the 10 identical work lines should be served cyclically wherever the AGV may be, so as to avoid waiting.

The AGV/SL simulator considers up to four types of distribution, according to the problem to be dealt with.

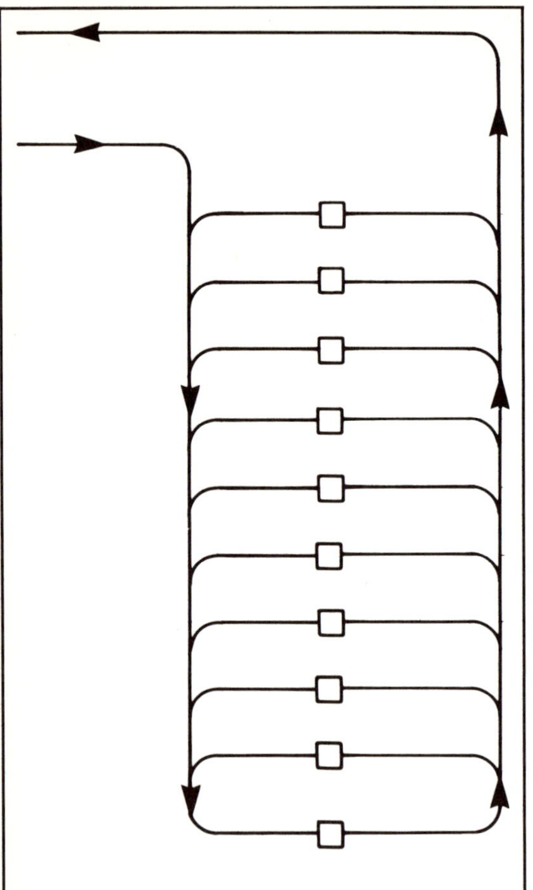

Fig. 10   Cyclical routing of AGVs around 10 work lines

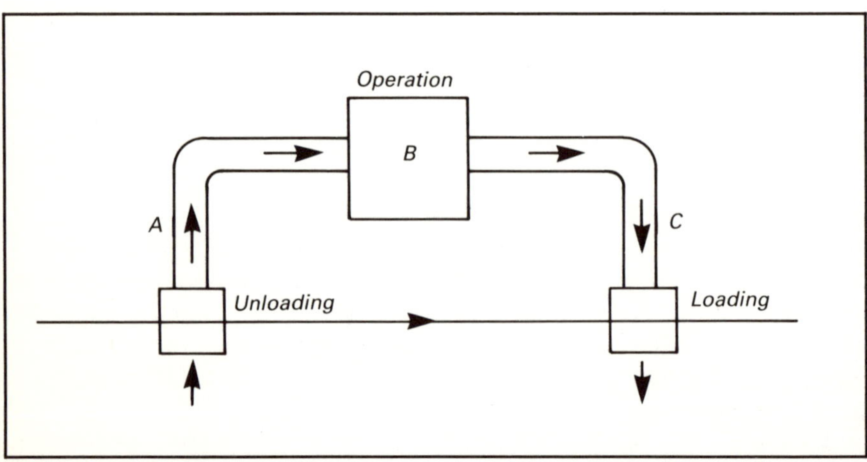

Fig.11   Integrated workstation

Interface with workstations with queues (FMS type)

An AGV simulator must not be totally disconnected from the different operations that the components can undergo between two handling processes. In fact, the loading stations, with arrival of components in accordance with statistical laws, do not always allow a system to be modelled accurately. It is possible for the arrival of components to be conditioned by the actual transportation of another component carried previously. Workstations integrated into the AGV model (Fig. 11) give this possibility. The components, after they have obtained a position on leaving the unloading station (A) and after the operator's availability has been verified, begin to be processed (B). When this is finished, and if they obtain a position upstream of a loading station (C), they can then call an AGV.

Thus, thanks to the management and sizing of the queues, the interface with a workshop of FMS type can be represented, with due account of all the saturation and blocking problems (e.g. when a queue is full after unloading).

Analysis of battery self-sufficiency and consideration of battery charging zones

In a number of cases the evaluation of a circuit without consideration of the problem of the AGV batteries proves to be incomplete. It must be possible to describe the zones where the recharging of batteries is carried out, as well as to obtain, at the end of the simulation, the statistics on this recharging for each AGV. It could then be decided if these zones are sufficient, or if special management of this problem is necessary, with a slight increase in the number of AGVs.

Sequencing AGV types and speed control

In certain projects that have been processed, Séri Renault Automation has been confronted with the problem of the AGV characteristics. Certain AGVs meant specifically for a determined job had a speed which differed considerably from the others. In other cases the AGVs had speeds which differed according to whether they were loaded or empty. Each AGV must therefore be individually defined by specifying its empty and loaded speed.

Extension of the software to other types of components

It is difficult to represent certain network sections, such as very complex intersections, with classic components, such as sections, switches, etc. But it is possible to determine certain sections of the network in the form of a more flexible yet more abstract description: description with arches.

Each arch, by definition, can contain only one AGV at a time, and can be declared to be in mutual exclusion with other arches or basic components (sections, divergent and convergent switches, stations, intersections, etc.). Therefore each route from one point to another is modelled simply. The various exclusions prohibit the presence of an AGV in certain zones.

Compatibilities

In rare cases where description with arches is not possible, the structure of the program allows for the connection of the network to components modelled by waiting queues, Petri network simulator or other specialised simulators. Indeed, Séri Renault Automation has developed a family of specialised simulators which are compatible with each other and correspond to workshop structures and to the equipment most often used in automated production. The AGV/SL simulator belongs to this family, and was developed by Séri at the request of the Technical and Scientific Affairs Division within Renault.

Levels of increasing complexity of the simulator

For training purposes, and in order to facilitate the research and development of new algorithms for management, it was decided, at the specifications stage, to break down the simulator into three levels:

- *Level 1:* This simulator is already operational, but has only a small number of system configurations. Nevertheless, it can be used after a very rapid training period.
- *Level 2:* This is the complete simulator, possessing all the system configurations.
- *Level 3:* This is the interface of all the points of decision-making (e.g. progression of an AGV, calling of an AGV by a station) with an environment which uses FORTRAN. This interface enables the user or the searcher to build new management algorithms without having to build the physical structure of the network in FORTRAN.

Drawings and graphic animation of the network

Graphic animation is necessary for the simulation of an AGV network in order to study in particular:

- The transitory states.
- Occurrence and dispersal of jams.
- The network behaviour the moment after breakdowns and repairs have occurred.
- Control and management of abnormalities.

Also, working meetings of project design teams have shown the utility of presenting at any time, either during a meeting of designers or on site, the results in the form of animated graphics of the transitory states behaviour in addition to the usual graphical or statistical presentation.

Thus the simulator itself (without graphics) was developed in FORTRAN, and it is portable so as to adapt to most workstations or minicomputers. Colour graphic elements which can be personalised have been developed on a 'home computer' which can be easily carried in a suitcase. The 'home computer', connected to the RS 232 port, receives the results of simulation from the 'mainframe', and then, after transportation, gives back information on a VDU.

Concluding remarks

This study has tried, through the presentation of several AGVs networks, to explain the problems encountered during AGV system design and simulation as well as the special functions required of an AGVs simulator.

The simulator developed by Séri Renault Automation, which is not orientated towards specialists of data processing but towards users whose function is to design AGV systems, and which is compatible with other simulators of automated production, facilitates the integration into the simulation of a complete production plan.

At present this simulator is being interfaced with a software of artificial intelligence, 'Prolog', in order to enable study, at the research and development stage, of new powerful management rules which are easier to modify than if they had been written with an algorithm.

4

Applications

This chapter is divided into two sections, dealing firstly with AGVs in flexible manufacturing systems and flexible assembly systems, and secondly with their use in warehousing.

Of the six papers in the first section, the later part of the first and the next four are concerned with various examples of FMS and the way in which an AGV system is integrated with the other subsystems. The fourth paper describes an installation in which AGVs are used to transport workpieces, fixtures and tools within the same manufacturing system and experience with maintenance of the systems. The interesting concept of mounting a robot on an AGV to improve mobility is the subject of the following paper. The earlier part of the first paper and the sixth paper both discuss examples of flexible assembly systems pointing out their advantages over conventional methods.

The second section contains three papers, the first of which includes an outline of a warehouse using an AGV system linked to automated narrow aisle trucks within the bulk store. Another highly automated distribution centre using an AGV system is described in the next paper and the chapter finishes with an examination of load transfer techniques between AGVs and storage racks, drawing on a number of application examples.

4.1
FMS and FAS

AUTOMATED MATERIAL HANDLING IN FMS AND FAS

F. Schneider
Schindler-Digitron AG, Switzerland

The present economic climate demands continual improvement of productivity. This in turn demands that in production systems due consideration should be given to flexibility with reference to the present product range and to future products, and to maximisation of the efficiency of production factors. Flexible manufacturing and assembly systems are discussed and examples given to illustrate these productivity-related concerns.

Pressure of competition is forcing continual reduction of unit prices. This demands maximum exploitation of production factors. For example, by means of shorter non-productive waiting periods, unmanned shifts, improvement of manual workplaces from the ergonomic aspect, and reduction of the fault rate.

Zero growth and/or declining markets compel enterprises to exploit even difficult market sectors. The result of this is a wide range of products and/or a range that can be adapted to the rapidly changing market conditions. This demands production plant that allows either a greater variety of alternatives within large production runs or economic production of even medium-sized and small runs.

Hence production plant must be flexible with reference to the current range of products. This means, for example, multi-purpose machines that are suitable for a variety of workpieces and rapidly convertible.

The speed of innovation is leading to progressively shorter product life cycles. Production plant must therefore be flexible with reference to subsequent and replacement products. This means, for example, capacity for economical conversion and for reuse of components.

Thus the present economic situation demands constant improvement of productivity, which is determined by technical performance and the flexibility of production plant.

Flexibility

Flexibility with reference to the product range

The aim of flexibility with reference to the product range makes it essential to turn away from the organisational modes conventionally used in production, such as workshop and assembly line production. These are being replaced by flexible assembly line production as a new mode of organisation. The definition of this given by D. Elbracht is as follows:

"Optional external linkage of work stations with their situation determined according to various organisational aspects, preferably on the principle of performance production. Flexible assembly line systems allow different lots to be processed at work stations that can be linked together in any way required."

A flexible assembly line system can include one or several of the following subsystems:

- Automated store for unmachined parts.
- Flexible manufacturing system (FMS).
- Automated store for assembly units.
- Flexible assembly system (FAS).
- Automated dispatch store.

In the case of workshop production the lots of workpieces are transported from workshop to workshop in batches, with relatively little temporal relationship to the manufacturing process. In the case of assembly line production the workpiece is often rigidly tied into the conveyor system. The flow of material is therefore relatively simple. The flexible assembly line system, in contrast, makes much heavier demands on the material flow system:

- Logically, on the availability and on the capacity for integration into the production control system.
- Physically, on the layout design, the adaptability of the load suspension device and the capacity of the conveyor system.

Maximum efficiency of production factors

Whereas for the operatives an improvement in the ergonomic features of the workstations can increase efficiency in the widest sense, for the production factors, machines and materials the main object that must be aimed at is an improvement of the relative lengths of productive and non-productive periods and downtimes of machines on the one hand, and conveyance times and lay-days of materials on the other; at present these relationships are often unsatisfactory. Flexible systems that react quickly are required for flow of materials, they must reduce the times machines are out of action to a minimum without the necessity for large buffer stores, which tie up capital to a fairly large extent and occupy expensive space that could be used for production.

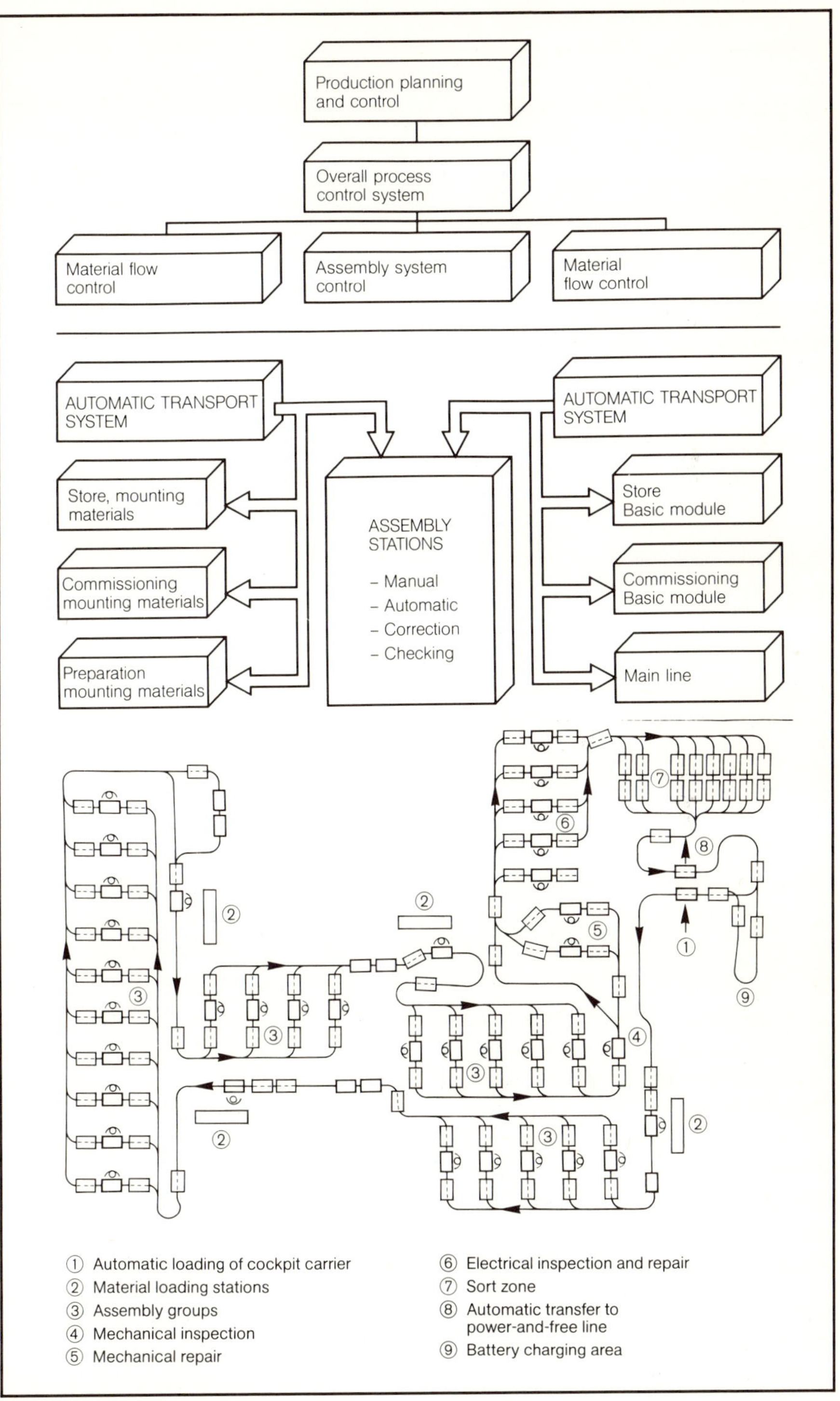

Fig. 1   *Possible FAS subsystems*

Fulfilment of these demands necessitates the highest possible degree of automation in conveyor, storage and transfer systems, computer-assisted distribution with an on-line link to the production control systems and high availability of all components of the material flow systems.

Flexibility with reference to subsequent products

Flexibility with reference to subsequent or replacement products means being able to use as large a part as possible of the production plant again even in the case of drastically changing production processes. Besides a long service life, this requires the greatest possible versatility. This can be achieved by shifting as many processes as possible into the logic level (e.g. control and measurement processes) and by using physical equipment that can be handled and moved in many different ways (e.g. robots). In the case of material flow systems this leads to an additional requirement for installations whose location can be varied as widely as possible.

Flexible assembly

A flexible assembly system may be made up of the subsystems shown in Fig.1. The most important components include:

- Automatic or automated warehouses.
- Flexible transportation systems not tied to fixed cycles.
- Handling and assembly robots.
- Manual assembly stations.
- A very efficient control system for individual system parts as well as for overall control and coordination purposes.

In contrast to sequentially cycled production, flexible assembly offers a multitude of possibilities. In addition, rapid technological development continually offers even further opportunities. Only installations for which all important points such as:

- exact analysis of the work content,
- establishment of the required degree of flexibility:
  - variation in the work content,
  - product mix,
  - ability to change to new products,
- design of the automatic and manual assembly station,
- quality controls/repairs,
- material flow,
- overall control and information system

have been exactly analysed and defined will result in a system that can meet all requirements over a number of years, can survive even a generation change, and, despite higher basic investments, will yield a profitable return-on-investment.

In order to deal with the main points of this multi-faceted problem, this presentation is divided into the following subjects:

- Important planning points.
- Main components of the FAS:
  - Overall control system.
  - Transportation of the assembly unit.
  - Workplace.
  - Material supply.

FAS planning

The FAS can be characterised in the following manner, whereby it differs radically from the conventional assembly line:

- It is able to combine serial and parallel workplaces in any sequence and in any combination.
- The assembly work at the individual stations can be performed by persons or by assembly robots.
- The individual assembly stations are not coupled to any basic cycle.
- The dwell time and the work performed at the individual assembly station may vary:
  - Plus/minus tolerance for a specified assembly operation.
  - Different assembly work at the same workplace.

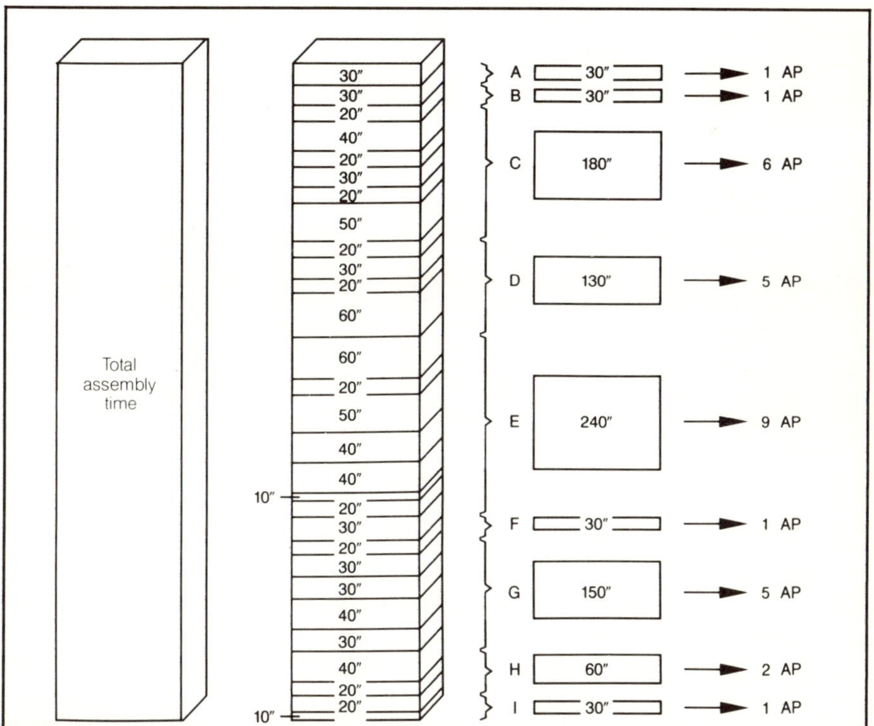

Fig. 2   Breakdown of work steps in flexible assembly

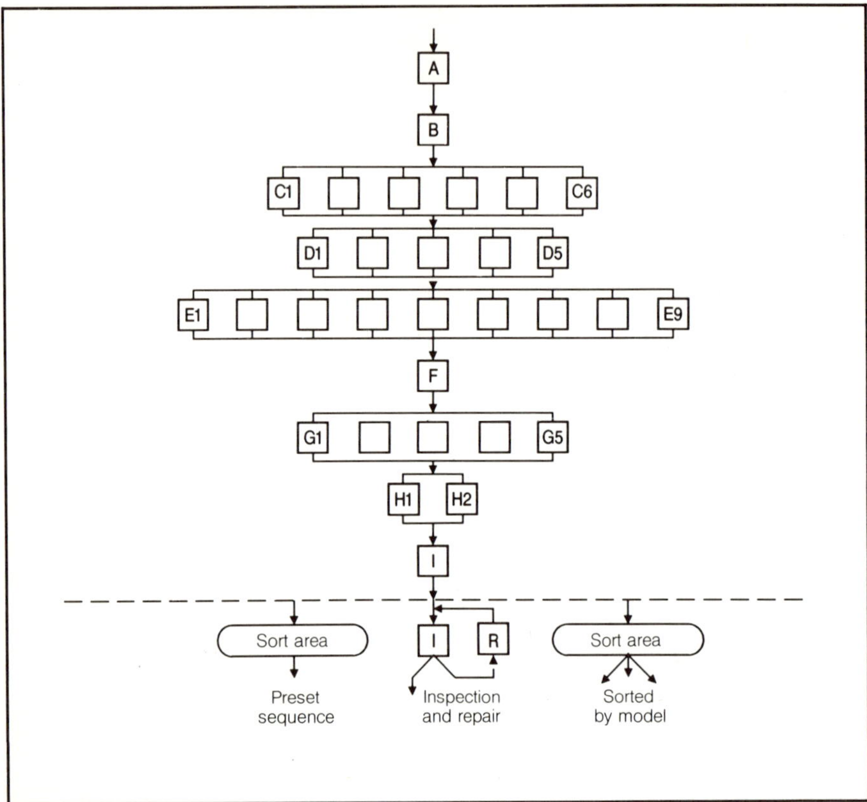

Fig. 3   Organisation of work flow

- The material supply problem can and to a certain extent must be solved in different ways.
- The breakdown of a system element should not stop the entire assembly facility.
- Widely varying quality control and rapid feedback to the initiators is practicable.

In order to utilise fully the potential offered by flexible assembly, the planner must thoroughly familiarise himself with the assembly procedure. The entire assembly operation is first broken down into the smallest possible individual steps (Fig. 2, column I). The individual assembly steps are then combined for an operation best performed by an assembly operative or by an assembly robot (Fig. 2, column II). Neither with the first fine division of the work steps nor with the following combination into logical assembly sequences is any allowance made for basic cycle time.

A basic time span is now estimated for each individual assembly step and this is divided by cycle time (number of units to be produced per hour). This procedure allows the compilation of the organisation chart which determines the number of parallel and sequential workstations.

In addition, the organisation chart can also indicate:

- Whether one or more quality control stations and perhaps repair stations should be included.
- Whether at the end the units should be arranged in the original or in some other specified sequence (the sequence can be changed during assembly as a result of parallel assembly stations not being bound to cycle time).
- Whether specific cycle times must be maintained when the assembled item is delivered to the assembly process (Fig. 3).

After the procedure has been determined according to the work content, the next points in the planning have to be established:

- Transportation of the assembly unit.
- Layout of the workstations.
- Supply of the materials to be assembled.
- Overall control procedure to guarantee assembly procedure reliability and flexibility.

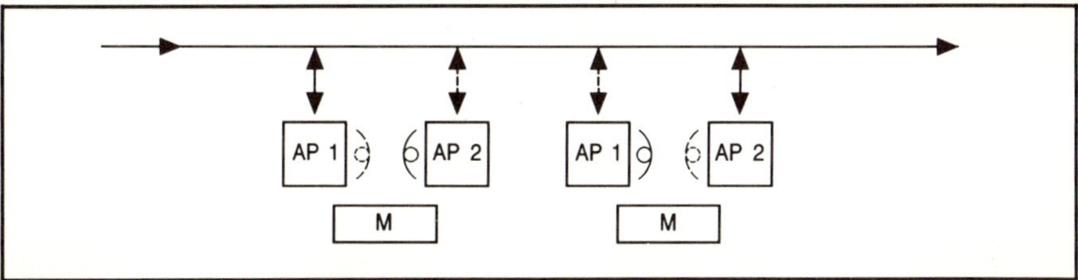

Fig. 4   Taxi system with ROBO-vehicles

Components of the FAS

Transportation of assembly units

Transportation systems with independent drives for each transport (assembly) unit are needed. Among the transportation systems available today, automated guided vehicle (AGV) systems are particularly suitable, as are overhead conveyors in some cases.

The vehicles may convey the unit to be assembled as a taxi system between fixed workplaces, or they may act as travelling assembly platforms.

- *Use as taxi system* – With the taxi system (Fig. 4), one or more workers normally work alternately at two neighbouring locations. While work proceeds on one assembly station, the one on the second station is exchanged. In this way there is no waiting time when changing to the next assembly job after completing the one in hand.

- *Use as assembly platform* – Three stops for vehicles are installed within the workplace area (Fig. 5). At the middle stop the assembly work is done. A fresh vehicle moves up to the first stop. The third (optional) stop enables all vehicles to leave the workplace at once after completing the work, without being obstructed by other vehicles in the main line. Minimum changing (i.e. waiting) times are assured by this.

Both types of transport afford full freedom of travel to serial or parallel workstations in addition to the temporal decoupling of the individual workstations from the overall procedure.

Layout of workstations

Both transport systems (taxi systems and assembly platforms) are eminently suitable for inclusion in the layout of the workstations.

Manual assembly stations. The assembly stations with a taxi system as well as those on assembly vehicles can ideally be laid out in ergonomic terms:

- All-round accessibility.
- Adjustable height.
- Rotatable worktable.
- Tools in the immediate vicinity over or near the workpiece.
- Optimal material availability.
- Individual determination of completion of work.

Automatic assembly stations. The assembly units (workpieces) can be accurately positioned by both transport systems for presentation in the automatic assembly stations. The central control system can, if necessary, provide the assembly robots with the type of designation or with the program to be executed and at the same time undertake quality control functions.

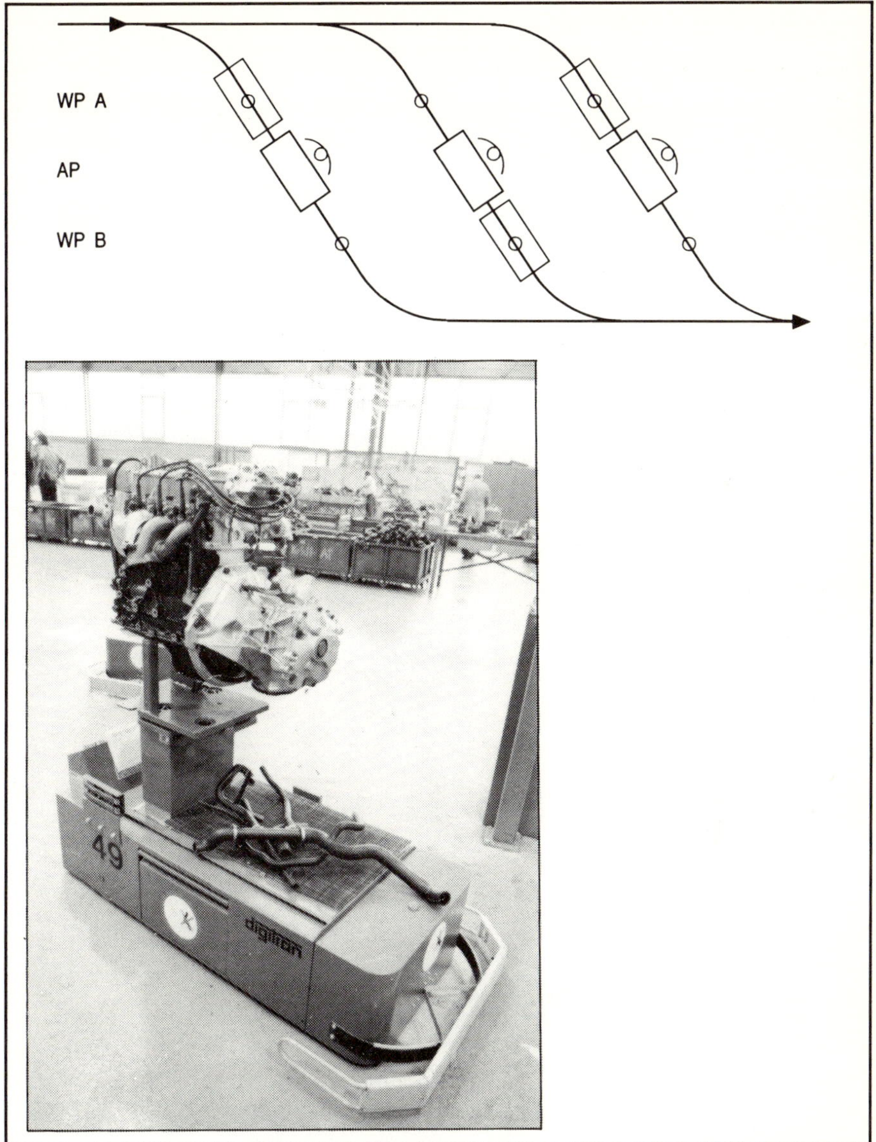

Fig. 5 Assembly platform with ROBO-vehicles

Material flow

Besides the transport of the assembly units, special attention must be given to bringing up the parts to be assembled.

With flexible assembly, it is not certain in advance which workstation will receive a given assembly job. Consequently, parallel workplaces are not supplied directly with material (except small parts). Normally the transport vehicle is taken to a material dispatching station. Here the assembly materials for the next

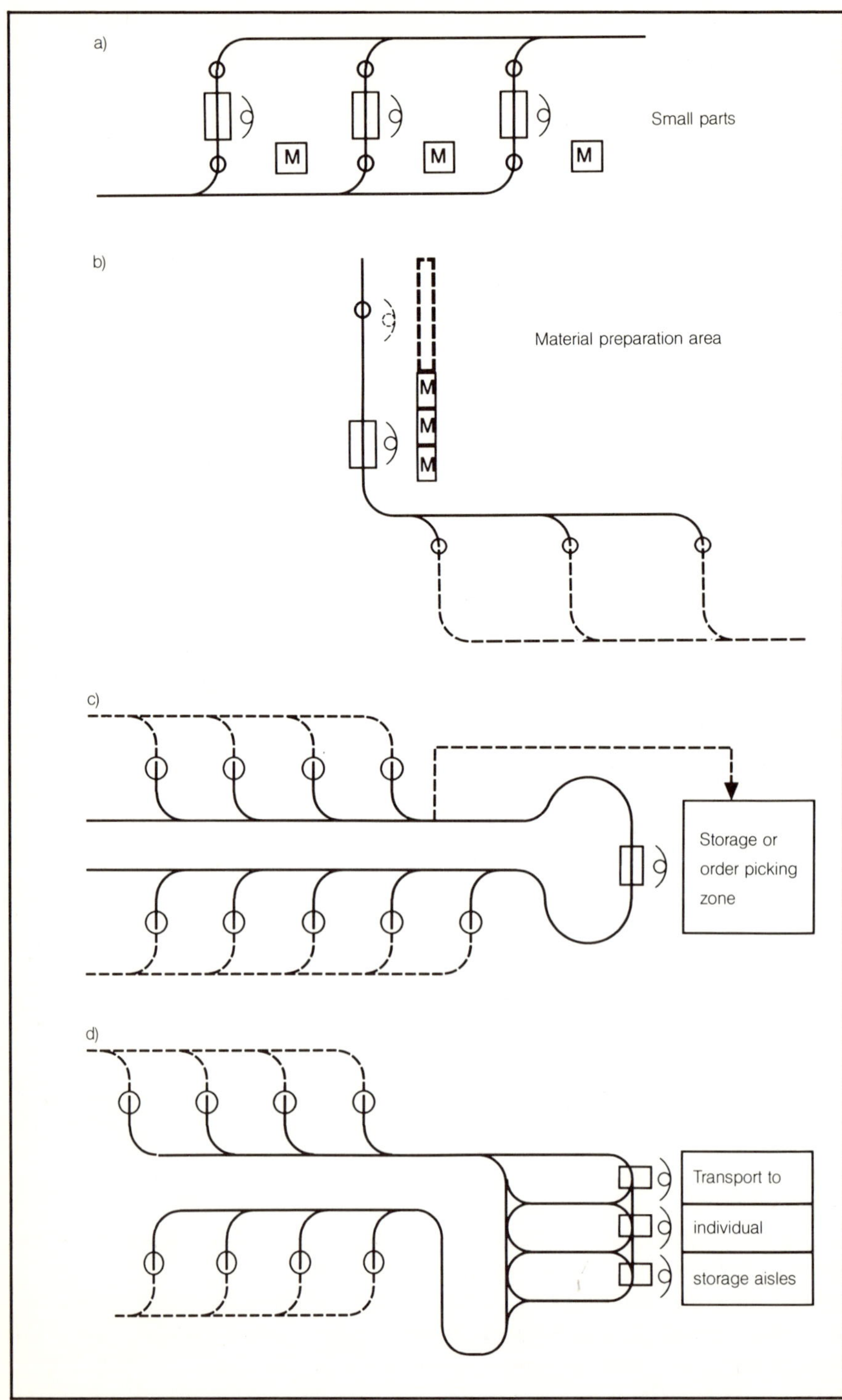

Fig. 6   Opportunities for supplying materials in flexible assembly

workplace(s) are loaded onto the vehicle (or pallet) according to instructions from either the central control system or type and material lists accompanying the vehicle. When the vehicle reaches the next workplace(s) with the material to be assembled, the worker does not need to worry whether all the necessary parts are there. Everything is ready beside the workpiece.

Flexible assembly generally offers the following opportunities for supplying the materials (Fig. 6):

- Materials at the workplace (small parts).
- Central material preparation station(s).
- Travelling to a store or to an order picking area.
- Selective travelling to storage aisles, picking places, etc.

Overall control system of the FAS

Data flow

Above all else, the operation of a flexible assembly system involves a substantially increased data flow, which is very complex in its interrelations.

Flexible assembly is intended to allow a wider product mix which can be altered at short notice according to market demands or the availability or non-availability of materials. This necessitates constant access to the production and stocks data.

The non-synchronous assembly sequence with parallel workstations and possible changes in the sequence call for more follow-up and direct intervention in the sequence of operations. At the automatic workstations the appropriate production data must be transferred with the proper timing. The introduction of selective quality controls with feedback to the originator also requires and gives rise to increased communication.

The voluminous data flow in a flexible assembly system is due primarily to the following circumstances:

- Every workpiece must be followed up directly, because the sequence may be altered on its way.
- For manual workstations with widely variable job content, the particular work may have to be prescribed by the central control system.
- Assembly robots with variable work content must be programmed and monitored for every operation.
- The sequences of the workpieces must be monitored and possibly assured.
- Material stocks must be matched to the production programme. Rush orders must be initiated if shortages appear, or the production programme must be altered.
- Certain materials must be supplied to the consumer stations at the right time and in the correct sequence.

- Standards may have to be laid down for quality control inspections.
- Defective workpieces must be directed to the reworking stations.
- Defective assembly must be reported back to the station responsible.
- Preparation of performance statistics for the individual workstations.
- Preparation of production statistics with specific data on variations in assembly times, quality shortcomings, etc., for tracing possible correlations.

This escalating data requirement must be coordinated in the production sequence with the proper timing. Despite the volume involved, no waiting must occur anywhere with data requiring real-time processing. Plant start-up and shutdown (beginning and end of shift) must be assured without problems. Despite the complex flows of materials and data, attendance and maintenance of the system must be simple.

One of the most important requirements is that, after any breakdown resulting from defective plant components or faulty handling, it must be possible to restart the plant quickly and without problems. Here there is an enormous difference compared with line assembly, because with asynchronous assembly the exact status must be known of every individual workstation as well as of the workpieces, if the plant is to start up again without trouble after a breakdown.

Control system concept

As seen above, the control system of a flexible assembly plant must satisfy exacting demands. The system employed largely determines whether a plant can attain the specified performance and availability.

Control technique development

Already years ago, rapid progress in data processing techniques made it possible to control plants with similar duties by means of a very efficient process computer. Nevertheless, this approach – largely involving direct control of an entire plant by a central process computer – is attended by manifold problems, such as:

- Once the computer type has been selected, it may have to be replaced if its storage capacity, reaction time, etc., prove inadequate.
- The same may happen if the plant is enlarged without allowance being made for expansions from the outset.
- When testing out subsystems, all stations must report to the central computer.
- Because different computers must be used depending on plant size, no generally applicable standard software packages can be provided.
- The same holds true of efficient test-and-verify software.
- Some of the demands are too strict to be satisfied by a computer.

State of the art

Further developments during recent years, above all in the domain of microprocessors, make it possible today to employ hierarchic control systems such as the Schindler-Digitron system, which has performed successfully since 1980.

The structure of a hierarchic control system (Fig. 7) is based on the following principles:

- Every system is conceived hierarchically from bottom to top. The smallest system operates only through the bottommost or the two lowest hierarchic levels.

- If a small plant is expanded or various plant parts are amalgamated, this is accomplished via the next hierarchic level or by extending an existing level.

- The size of the system thus no longer dictates the computer size or type, but only how many standardised units are to be employed on what level. This applies without exception to the microprocessor, local control and area control levels.

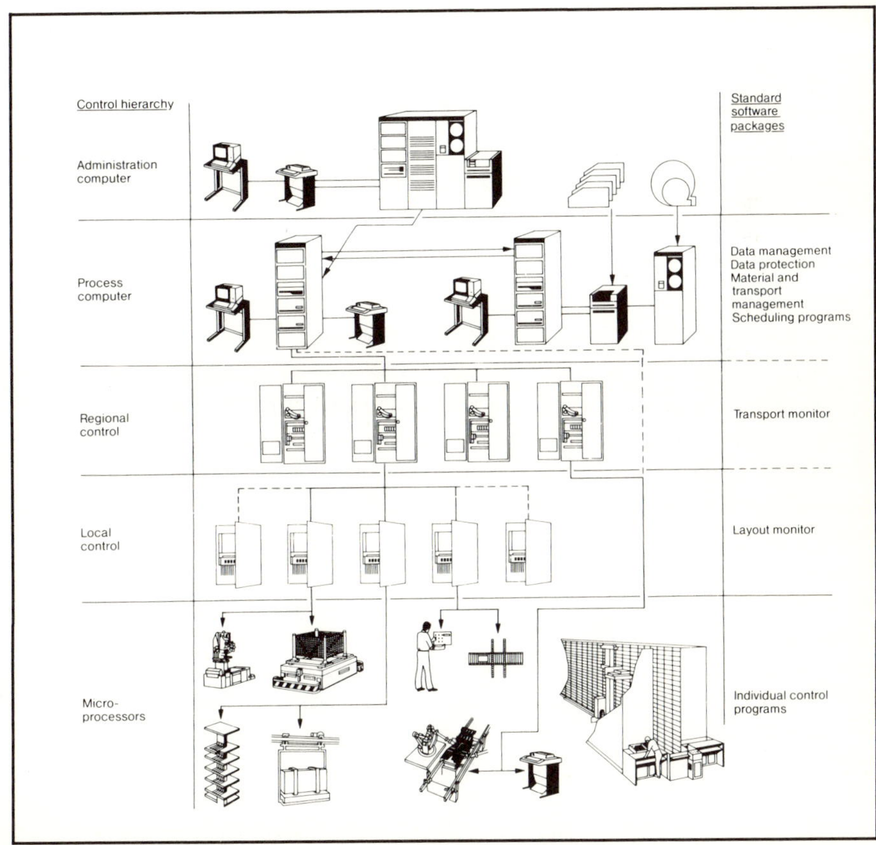

Fig. 7   Schindler-Digitron hierarchical control concept

At the superordinate computer level there is normally a customer mainframe capable of communicating with the control system.

- At every level the same standardised control units are employed for all applications.

- Matched to the products and control units, the hierarchic control concept includes a wide choice of standard software packages for duties like:
  - Controlling individual units like AGVs, conveyors, stacker cranes, etc.
  - Data communication via process computer grid.
  - Terminal attendance.
  - Traffic control for AGVs and overhead conveyors.
  - Traffic control for conventional conveying systems.
  - Standardised fault signals for all possible fault causes.
  - Test-and-verify programs for individual plant parts through group test to overall control.
  - Automatic program loading from master computer into lower hierarchic levels.
  - Constant data assurance through the entire process.
  - Automatic restart after breakdown with data and file reconstruction.
  - Remote software servicing via telephone line and modem.

- The software programs are structured so that they can be used on different hierarchic levels according to the control configuration of the particular plant.

Advantages of the hierarchic control concept

A hierarchic control system with the properties described is eminently suited for controlling flexible production plants.

Customisation. The ability to configure the control system freely, horizontally and vertically, allows a tailor-made solution to be achieved for every plant. For extensions or problems with reaction time, the system may be expanded in both dimensions at any time without problems. For this the hardware elements and software packages employed are provided for all levels.

Universal application. Through its structured nature the system is equally suited (i.e. is appropriately configured) for the processing of large data volumes as for the monitoring of widely dispersed installations in different buildings. Moreover, with the same elements the control system is capable of covering the entire logistics of a plant, ranging from the control of all manner of conveying systems, through automatic storages, data inputs and (automatic) data acquisition, monitoring and initialising robots, all necessary coordination functions to the various areas of superordinated control, administration and data evaluation (statistics, quality report, etc.).

Commissioning. With its consistent standardisation, distribution of intelligence and available test-and-verify software, this control system is ideally suited for speedy commissioning without problems. This is true also of phased commissioning (e.g. conversions in existing buildings or extensions not planned from the outset).

Fault identification. Quick and selective fault identification is extremely important with flexible production plants which are sometimes widely dispersed and operate with different conveying, handling and storage elements, corresponding to man-machine stations and super-ordinated data processing systems. The control system described has the advantage that fault detection and diagnosis have been consistently engineered and adapted from the bottom up, so that a maximum of selective information is available at the control station as at the individual elements for speedy rectification. The principle is:

1. Annunciation at control station: where and what trouble has occurred?
2. The serviceman, already orientated regarding the cause of the trouble, goes to the spot, where if necessary, he can obtain additional selective information about the source and cause of the fault.

Data assurance, restarting. After the fault that brought the plant or parts of it to a stop has been remedied, the ability to restart the installation as quickly as possible is critically important. With bigger and widely dispersed plants, where the control concept for restarting calls for manual collection and input of data or the status of individual elements, this procedure can be very time-consuming and may lead to difficulties of such magnitude that the profitability of the plant is called into question (i.e. the required performance is never achieved in the long run).

The hierarchic control system described here overcomes these problems as follows:

- During the working process all data relevant to it are stored continuously on two different media.
- After any system breakdown (whether due to an attendance error, the control system or storage media), and after repair or switchcover to a standby computer, the exact state of the entire system (over all levels) is restored automatically with the help of the stored data. Then the plant resumes operation exactly from the state it occupied at the time of the breakdown. This restart procedure takes a minimum of time and is absolutely free of error. Any system demanding manual data collection and data inputs claims far more time for restarting; moreover, it always involves the risk of grave errors.

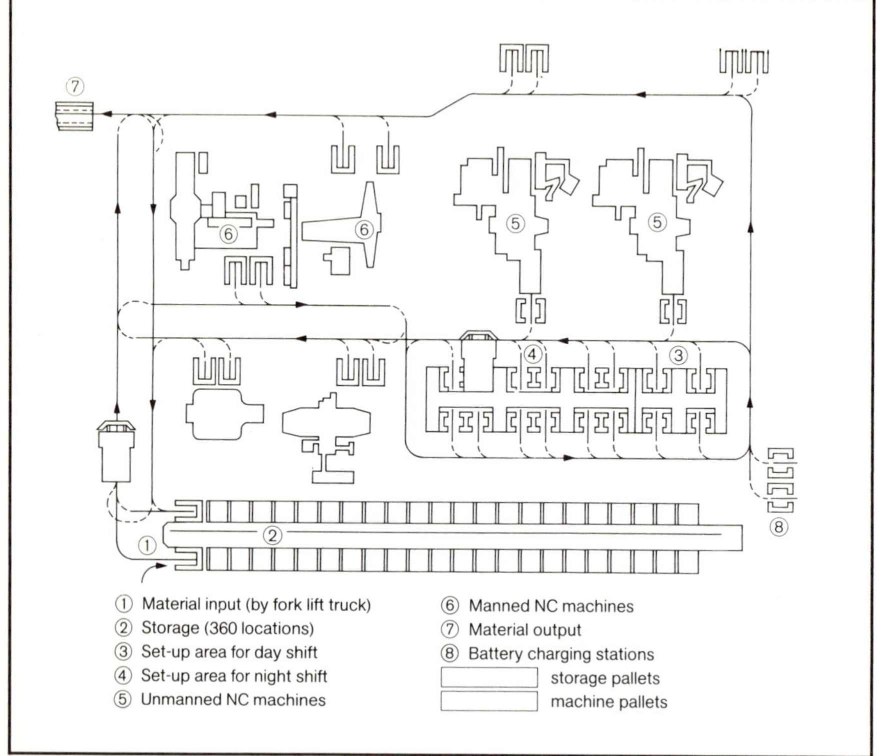

Fig. 8   Example layout of an FMS (Schindler-Digitron)

Example of an FAS

The example described here is taken from the automobile industry. It concerns an FAS in which doors are assembled parallel to the main assembly line (Fig. 8). The doors are fitted to the vehicle on the main line and the final fitting out of the car is then defined and identified at an information point. After painting the doors are detached from the bodywork and conveyed to the FAS area, for the doors to be fitted out, by means of an overhead conveyor plant that also acts as a buffer.

In this area the two doors that make up a pair are loaded onto a ROBO-vehicle. This goes to an initial fitting station, where material is added to the trolley, and then to the initial workstation, where the material is assembled. This is repeated four times. The different work periods mean that the vehicles have a mixed sequence. For this reason, synchronisation is carried out at the end of the FAS.

Once the assembly of the doors is complete the two doors are again suspended from an overhead conveyor system and returned at precisely the right moment to the main assembly line and fitted to the original vehicle. The order document accompanies the doors throughout and defines the materials to be fitted. This makes any configuration of door fittings possible.

Doors for 90 cars are assembled every hour in the plant. In a 1700m layout 140 ROBO-vehicles are in operation. The plant is controlled by a duplex process computer (in common with a similar FAS for cockpits), which is linked on-line to the works control computer.

Let us examine this example of flexible assembly with reference to how far it fulfils the demands formulated earlier in the paper on the subject of increased production.

Flexibility with reference to the product range

- Production of a large range of variants in small piece numbers is chaotic – possible with different assembly times and material requirements.
- No space problems within the assembly system for a wide variety of materials, since the supply is effected synchronously from outside.
- Output can be tailored to demand by hook-up and cutoff or workstations.
- Cutoff of individual workstations makes it possible to convert them without adverse effect on the overall assembly process.

Maximum efficiency of production factors

- Ergonomic workstations, with no clock constraint, lead to increased output and a reduction in the fault rate.
- Automation of the workstations is simpler.
- As production is not organised in lots the flow time for material is reduced.
- Testing, checking and correction of preassembled parts away from the main line ultimately reduces processing time of the final product.

Flexibility with reference to subsequent products

- Flexible assembly systems are almost immune to change of model.
- The modifications are restricted to the fixing units on ROBO-vehicles and possibly to the workstations and the software.
- When AGVs are used layout changes are simple.
- Capacity changes can be carried out easily by using more vehicles and inserting additional workstations.

Flexible manufacturing

In an FMS an automatic multi-stage production process with a variety of end products is possible. It includes several automatic machine tools and possibly also manual or semi-automatic workstations. The workstations are linked together by an automatic material flow system in such a way that simultaneous processing of different workpieces is possible, which go through the system by different routes.

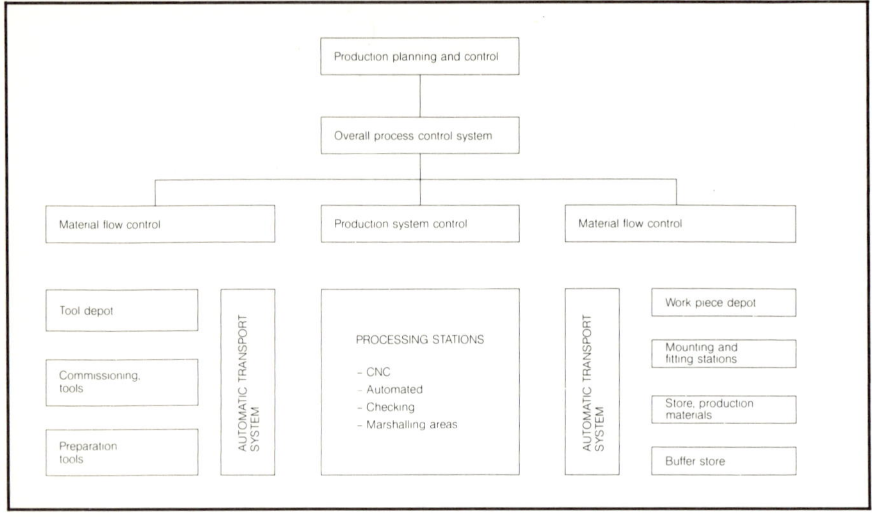

Fig. 9   FMS subsystems

An FMS is made up of the following subsystems (Fig. 9):

- Control system.
- Processing system.
- Storage systems for workpieces and production material.
- Storage system for tools.
- Conveyor system.

A differentiation must be made between prismatic and rotating workpieces for the conveyor system for workpieces and production materials:

- Prismatic workpieces are mounted on uniform machine pallets in an external mounting and fitting zone, conveyed from there to the workstations, and, in the case of automatic tools, transferred automatically to changing stations.

  If different machine tools make it impossible to use uniform machine pallets the workpieces are moved on standard pallets that are deposited automatically in marshalling areas immediately adjacent to the machines. The mechanical pallets are then fitted out separately at the individual machines.

- Rotating workpieces are usually conveyed to the marshalling areas in the processing centres in magazines. Between magazine and work fixture they are 'handled' by robot or by loading portal.

Owing to the lack of restrictions and the variability of the line direction selection and to the high availability, for FMS the AGV system often offers more economical alternatives to the conventional systems, such as roller conveyors and overhead conveyors. Problems with positioning at the interface points are often mentioned, but these can be solved with no great difficulty if the mechanical engineers concentrate their usual tolerances on the machine area and do not transfer them to the overall conveyor system.

Automation of the supply of tools from a central tool depot is still rather rarely achieved. Long idle periods and redundancies in use seem to work against the trend toward automation in this case. The tools are conveyed between tool store and machines automatically in magazines that are also conveyed by AGVs or by overhead conveyors.

At least in the case of fairly large systems, the linkage of the subsystems on the information side for distribution purposes is now integrated into the production planning and control system by way of various stages. A communication link and/or centralisation of the actual processing programs of the DNC machines with the option of supplying the appropriate processing programs on-line as required in the form of a down-line loading is being promoted by the Japanese with the CAM systems, but this has hardly been achieved in Europe.

Among other reasons this is due to the low suitability of the PC control systems for communications and the lack of standardisation with respect to programming languages.

Example of an FMS

An example of an FMS currently being realised is described here. Its degree of automation and the technical risks it involves can be borne even by a medium-sized enterprise. In the material flow sector the automation is restricted to the conveyor system for workpieces and manufacturing material and to storage of the workpieces.

The plant is designed for processing prismatic and rotating workpieces with chip removal. It includes two fully automatic DNC machines with automatic conveyor devices and six semi-automatic processing machines with automatic conveyance up to the transfer areas and a marshalling depot for workpieces and remounting and/or buffer areas (Fig. 10).

The unprocessed workpieces are on wooden pallets and are deposited in the input area by fork-lift trucks, recorded for the information system, and then deposited automatically in the production store. Larger prismatic workpieces (up to 1000kg) are conveyed to the remounting areas by AGVs, where they are transferred and mounted manually from the wooden pallets on the machine pallet. The same ROBO-vehicles then convey the machine pallets to the two fully automatic DNC machines, where they deposit them on the transfer areas with an accuracy of about 5/10mm with the aid of appropriate centring devices. From this point they are taken up by changers and fed to the machine fully automatically. For the unmanned nightshift the appropriate workpieces are prepared on the machine pallets, kept on waiting places, and transferred automatically by the AGVs to the appropriate machines from these during the night.

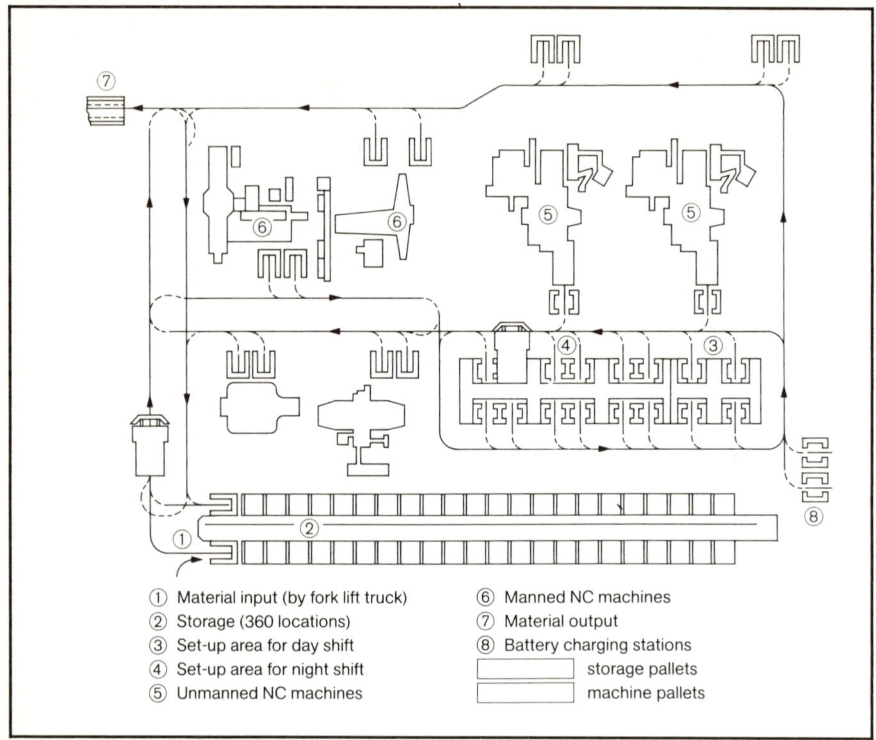

Fig. 10   Example layout of an FMS (Schindler-Digitron)

Tools are supplied and taken away from the remaining six semi-automatic machines directly from the tool depot; the AGVs convey the appropriate wooden pallets to the transfer areas. The workpieces are mounted by hand. The workpieces and/or pallets may be stored temporarily between the individual processing stages in the areas provided or in the production store.

Once processing is complete the workpieces leave the FMS by way of the outlet points.

The plant consists of an automatic one-aisle warehouse with 368 pallet places, two ROBO-vehicles, each with a loading capacity of 2t, approx. 270m layout and 44 transfer areas, and of one central control system for the processing and material flow.

Let us now examine this example of flexible manufacturing with reference to how far it fulfils the requirement for increased production formulated earlier in the paper.

Flexibility with reference to the product range

- Production of a large number of variants is chaotic with small production runs, possibly with different processing times.
- It is simple to tailor output to meet demand by activating and deactivating workstations.

- It is possible to combine the automatic process with manual conveyance on the same route.
- Fully automatic and semi-automatic machines can be combined in the same system.

Maximum efficiency of production factors

- The system allows the introduction of one unmanned shift.
- The flow time of material is minimised and there are no unnecessary buffers.
- The centre-zero relay devices and the good distribution system eliminate waiting times of the machines.

Flexibility with reference to subsequent products

- Changes of layout are simple in the machines and transfer area sector.
- There is no problem in inserting additional machines by extension of the layout and/or increasing capacity by an increased number of vehicles.
- Stepwise conversion from semi-automatic processing machines to fully automatic processing centres is possible.

AGV SYSTEMS IN FMS

J.-O. Myhrman
MTS-Systems Norden AB, Sweden

Today, the trend in flexible manufacturing systems (FMS) is toward fully automated production cells comprising integrated units, such as machine tool systems and handling systems. To integrate these systems effectively and to complete the automation, an AGV system is necessary to transport materials and work-in-process among the various modules. The economic and technical aspects of FMS and, in particular, AGV systems are discussed.

Between manual single-piece manufacturing processes and mass-production machine lines there exists a growing need for facilities in which individual machines can be combined into groups which, together, form a flexible automatic and fully combined production unit. Such a production unit can be divided up into a number of integrated working function modules:

- Tool system – easily exchangeable tools complete with feed hopper or magazine.
- Machine system – a number of interacting machines for tools and blanks.
- Handling system – robots, manipulating units, computerised part-changers and automated guided vehicle (AGV) system.
- System for control and supervision of machines and peripheral equipment.
- Transportation and warehousing system complete with control equipment.

This paper highlights the coupling together and integration of individual modules by means of an AGV system with the aim of achieving an efficiently functioning unit.

Economic background: Capital rationalisation

It is widely held that growth is necessary for industry to survive. And growth must, at least partly, be financed by higher equity, unless industry is willing to accept falling solidity, as has been the case in many branches in the last 15 years. Thus capital-restricting issues play a central role in any expansion.

An analysis of restricted capital in many companies would most certainly prove to be worrying. This is because turnover and total assets in many companies have developed at the same rate and sometimes in a ratio of one to one. In many cases, liabilities have grown faster than shareholders' equity; in other words, expansion has been gained at the cost of solidity.

The endeavours to increase turnover during the last 15 years have also made their mark on balance sheets, and today the capital tied up in inventories forms a dominating proportion of restricted equity. A look at the distribution of physical capital within the engineering industries shows that materials in warehousing, stocks and distribution account for about 55%. In Sweden, the materials flow investment is estimated to amount in total to approximately 160 billion Swedish Crowns.

Understandably, companies are faced with a troublesome situation due to this tying up of capital, while at the same time capital becomes more expensive and difficult to procure. Therefore it must be asked: can a contribution to capital rationalisation be achieved by applying more efficient material administration methods, including better transportation? The answer is a resounding yes.

Let us take a look at how the 160 billion is distributed and how much of it can be influenced in some way. Since information concerning inventories in Sweden is remarkably inadequate, the figures discussed here are only estimates. The hope is that there will be access to better and more reliable material in this area in the future.

According to the present author's rough estimates, approximately 90 billion can be attributed to an acceptable inventory level based on budgeted service levels and to prevention of production disturbances, etc. This figure includes various forms of inventory: variant, seasonal and transit stocks, as well as stocks of a more strategic nature and even stocks which it can be beneficial to hold for reasons of taxation.

On the assumption that this is fundamentally correct, there remains 70 billion which can be influenced. Through better internal handling, which is to say better material planning, a higher level of automation, use of AGV systems etc., a total of about 25 billion can be influenced. The figure is up to 35 billion with better information systems, such as administrative systems and computers. Through inter-organisational systems, etc., it is estimated that about 17 billion can be influenced.

Flexible manufacturing systems

Sweden is already among the leading countries in the world when it comes to the number of industrial robots in use. In future such robots must be combined with automated material handling of both the

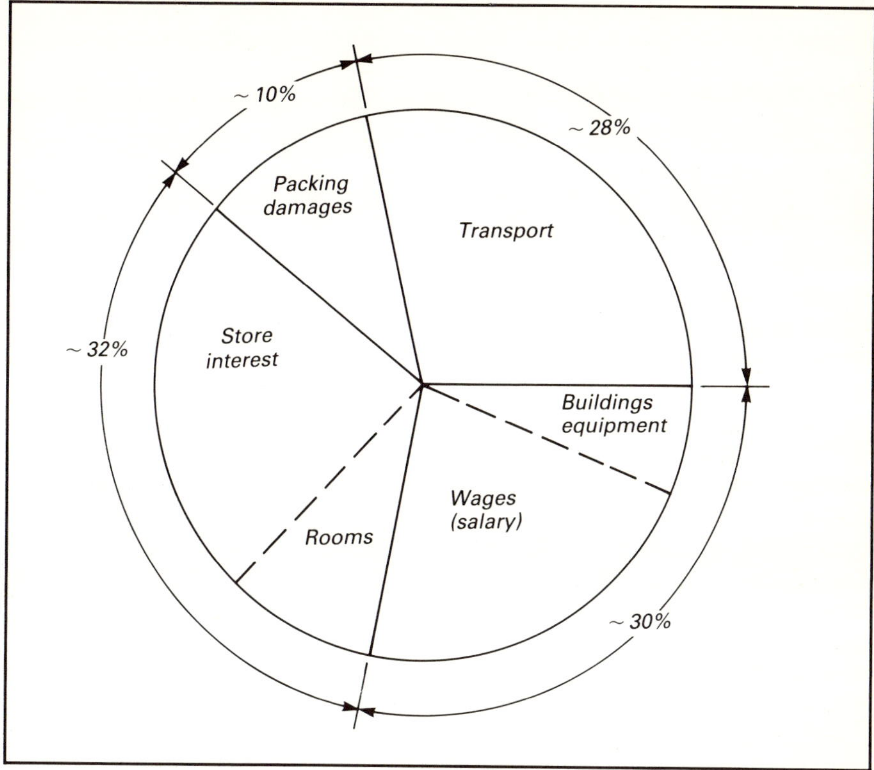

Fig. 1   Material handling components and their costs

incoming and outgoing flows. Against the background of previously mentioned equity which is tied up in inventories, a great deal of effort and money must be invested in new systems and new equipment. The objective is to automate the entire throughput from delivery, through production, machining and assembly, to the finished product.

Material administration represents between 30 and 40% of a product's processed value, which in Sweden amounted to a total of 254 billion Swedish Crowns in 1980. It consists of a number of components. Fig. 1 gives a breakdown of the costs presently involved in material handling. Through the introduction of an FMS-AGV system, restricted equity will be reduced in the form of:

- Materials.
- Work-in-process.
- Finished stocks.

In addition, through the use of three-shifts it will be possible to reduce production costs by making better utilisation of:

- Buildings.
- Machines.
- Transport systems.

The system will also make for a better working environment by reducing:

- Handling damage.
- Tool damage.
- Workforce injuries.

Construction of the system

The transport and handling of materials and tools in a typical modern factory is approximately as shown in Fig. 2.

Experience shows that large amounts of the incoming flow of material often gather together in the buffer store(s). From there, the materials are transported to the various production lines for machining or assembly. When a workpiece reaches the machine group, it is not always the case that the necessary tool is available. In fact, very often workpieces have to be placed in buffer storage between production stages. This makes for quite a lot of unnecessary handling by fork-lift truck and difficulty in keeping a check on the flow. A typical breakdown of throughput time is 1.5% machining, 3.5% transportation and handling, and 95% storage. Something must be done about this.

The introduction of FMS-AGV systems can provide the answer for future production systems. In Sweden, the concept of Production and Limited Manning (PLM) has been discussed for many years, its objectives being the following:

- Automated production, either unmanned or with low manning supervision for periods of 1 to 16 hours.
- Automation of often and regularly repeated work tasks.

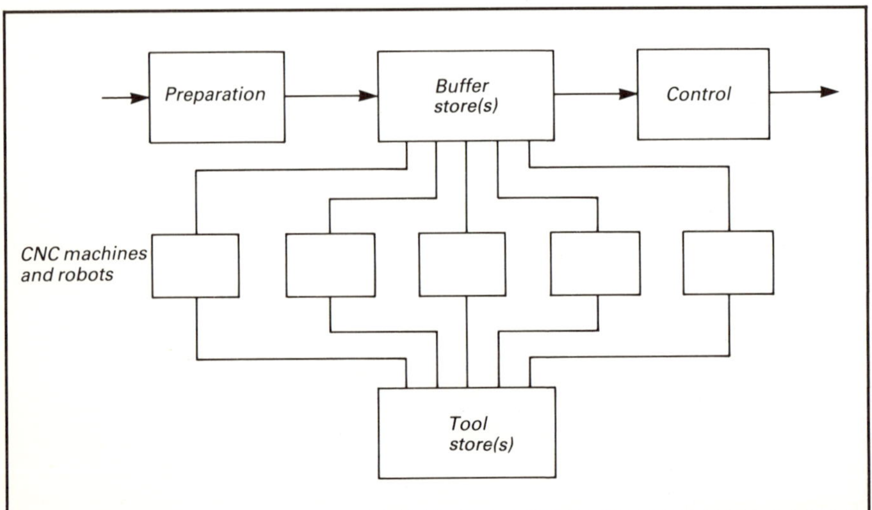

Fig. 2   Normal way of material transport handling in a typical modern factory

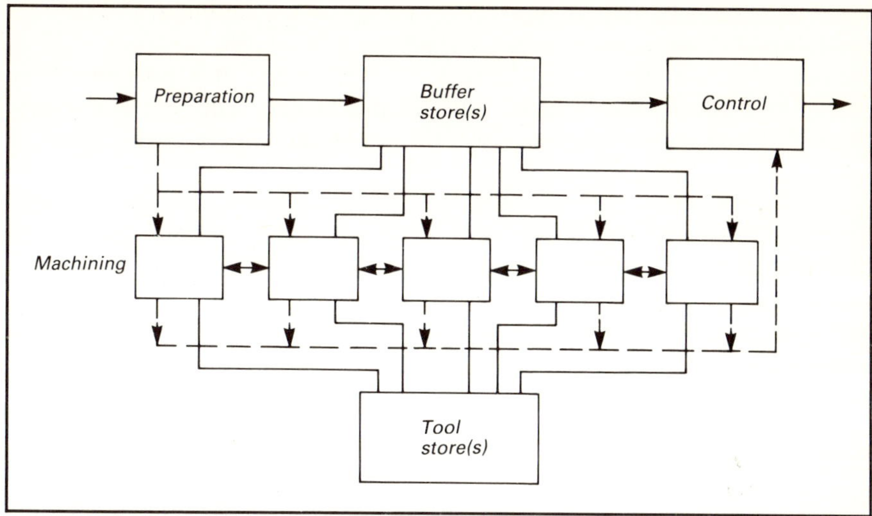

Fig. 3   Flow pattern in an FMS factory with an AGV system

- Storage of manual work by the preparation and buffering of workpieces, tools, programs, etc.
- Manual work to be confined to dayshift hours.
- Relatively short production runs.

Scale of operations

The smallest production system can consist of a self-contained and powered machine with simple, automatic handling equipment. A larger system can consist, for example, of a number of machines or machine groups linked together by a common automatic transport system and a computer for control and supervision.

Irrespective of the size of the system, its functions can be divided into the modules mentioned in the introduction:

- Tool system.
- Machine system.
- Handling system.
- Control and supervision system.
- Transport system.

All of these functions must be tied together to form an integral unit that can collaborate with other functions within the company.

Linking the various functions

With Tellus/Carrago transport and handling equipment and the related intelligent control system, a flow pattern as shown in Fig. 3 can be built up. Material is taken straight from the entry point to the first group of machines, from where it travels on for further machining or assembly and then to quality control and dispatch. At the same time, tools are also controlled and manipulated. That is to say, tools are ordered forward and transported to the correct machine at the correct time. As the tool system can collaborate with the approximate or spot-on positioning system of production planning, it is possible at an early stage to see whether or not the machining schedule is correct. Should this not be the case, then the material can be buffered at an earlier stage when its process value is lower.

Setting-up times must be minimised. A well-functioning transport and handling system is an important aid in this work. An FMS installation must be able to adapt to existing production, for which reason the automated machine groups should be able to collaborate with manual work systems also incorporated in the system. Careful study of any manual stations is necessary before they are linked into the layout.

Example of the function of a transport system in FMS

The AGV system is made up of two main components: the AGVs for the physical transportation of components and materials, and a related control system. The latter can be of the manual type, where addressing is done by push-buttons on the AGV or by push-buttons on a (separate) panel. However, in an automated environment where CNC machines and/or robots are used, a computerised control system is necessary.

Salient features of the FMS-AGV system are as follows:

- Automatic data reception and transmission.
- Automatic machining equipment.
- Robotic capabilities.
- Flexibility.
- Automatic transportation of workpieces by means of the AGV system.
- Up-to-date information concerning each individual component.

To function well, the manufacturing units need a high degree of mechanisation and automation, but above all they need better information administration. A troublesome matter today is the virtual explosion in indirect costs, which is to say costs for paperwork. Such costs are minimised in an automated factory because the data flow is automated using a network which runs between the machines and manufacturing units to the factory units and throughout the entire company, enabling current information to be retrieved as required by authorised personnel anywhere in the chain (Fig. 4).

To make FMS work, knowledge is required concerning plant layout, manufacturing technology, material and production control, CAD/ CAM, robot technology and handling technology. The latter is perhaps the most important consideration. In fact, an automated factory is, in one sense, really a material handling system with interspersed buffer and machining stations. There are two major supplier groups within the total system: suppliers of manufacturing

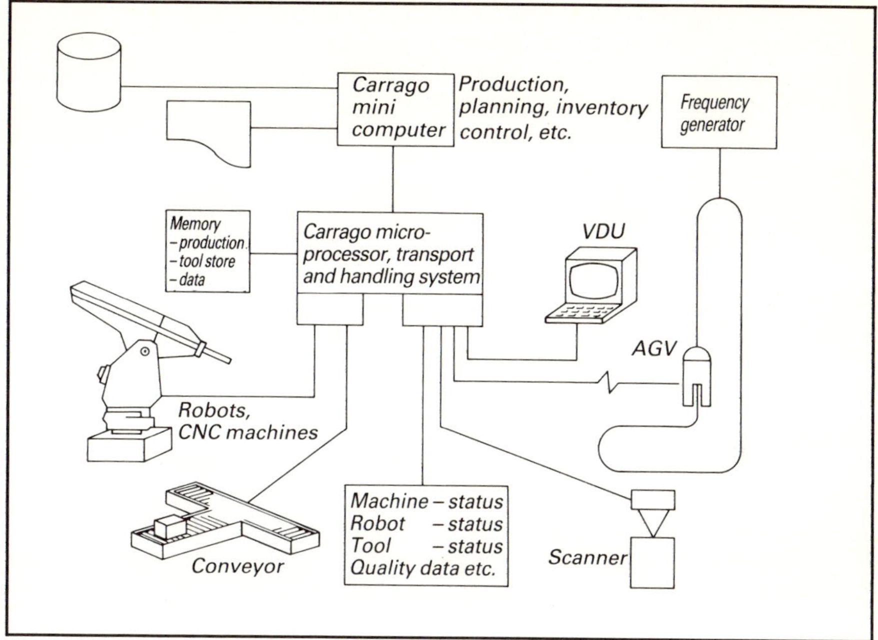

Fig. 4   Information system for FMS with AGV system

machines, and suppliers of material handling equipment. The demarcation lines between these groups are becoming more and more indistinct, which means that the control system must work equally well with both.

The physical and economic service life of new equipment must be of such duration that machines and equipment can be used not only for current needs, but also for those of the future. Production arrangements must also be flexible enough to accommodate fluctuations in the market as they occur. In addition, the philosophy of 'continuous' material flow through every stage of production and distribution leads to new starting points in the design of factories, warehouses and distribution terminals. The field of electronics is of course extremely interesting in this context.

The company management developments today are probably more akin to revolution than evolution. Sophisticated material administration methods allow capital to be released, while at the same time the control of material flow forms an instrument by which productivity can be measured on a wider scale and enables the manufacturing process to be looked at from a 'new' viewpoint. This approach results in shorter lead-times, with a related higher service level (and higher income level!), less capital tied up in the business, and less risk of system obsolescence.

Compared with these overriding economic benefits, the better cost efficiency in the actual handling process is of lesser importance. The latest material handling techniques offer the opportunity of higher manufacturing efficiency and the additional benefit of lower material handling costs – both of which are of course welcome.

Productivity is the key to reindustrialisation, and ultra-efficient material handling is the key to productivity.

The mechanical interface

Depending on the specific AGV system, the transportation of material can take place in various ways. For example, the workpiece, or a number of workpieces, can be mounted on a table or fixture for machining. In this case, the AGV system will often work together with some form of handling equipment such as robot or computerised part changer and conveyor. In turn, this handling unit feeds a CNC machine or a washing plant, for example. Here, it is important that the table or fixture can be positioned within tight tolerances.

The system must also be capable of transporting pallets for machine tools. The demands involved in such transportation are considerable since the pallets often vary both in weight and size. In addition, the AGV must be capable of off-loading pallets at different heights within the same system. The load weights can also vary.

What, then, are the benefits offered by an FMS-AGV system? They are as follows:

- Precise positioning of pallets.
- No extra demands on guiding and stopping vehicles.
- Chips do not affect positioning of pallets.
- Interfaced to different machine heights in the same system.
- Drip-proof vehicles.
- Auto, semi-auto and manual modes of operation.
- Pallet locked to vehicle during transport.
- Load weights from 300 to 15,000kg.

Signal exchange

The exchange of signals can take place at various levels depending on the complexity of the system. In the system represented in Fig. 4 the signals from the robots are connected to the microcomputer via a potential-free contact input.

As an example of signal exchange, consider the case in which a robot has machined all of the workpieces on a pallet. The robot then calls an AGV by causing the microcomputer to interrupt the signal 'pallet in position', after which the vehicle picks up the pallet and takes it to the next machining station in the chain or to the buffer store, where it can be stacked in racks up to 5m high. The AGV then fetches a new pallet-load of blanks, and when it is positioned at the robot, the microcomputer again sends the signal 'pallet in position' to the robot. Should the pallet move out of position at the robot, the signal 'pallet in position' is interrupted. The signal is recontinued when the pallet is once more in position.

For more complex systems (numerous robots, machines) there is also an exchange of signals between, for example, the machines and the overriding microprocessor. The microprocessor is used for the following:

- I/O signals from robots, machines, tools, etc.
- Quality data.
- Tool statistics.

Restarting

Every system, irrespective of how sophisticated it is, will sometime or other come to a standstill. For such an occurrence there must be clear routines on how restarting is to take place.

It is essential that all categories of personnel involved in a stoppage in the system are trained in their part of the restarting routine. The important thing is for the supplier to provide an easily understood documentation of the restarting process. However, the user must ensure that this batch of documents and instructions is not left to lie as a guarantee in itself for restarting. The involved personnel must study the routines described and train themselves in their use on special exercises so that everyone involved (current and new employees) is informed as to exactly what to do.

Emergency stop network

In an automated factory using AGVs, robots, CNC machines, etc., each unit must of course be equipped with emergency stop functions. In addition to these component-related emergency stop functions, there must be others designed to stop the entire system or parts thereof. Examples of emergency stop components include pressure mats, gates and photocell set-ups.

It is important to ensure that personal safety is ensured by all functions, but the way in which the emergency stop network is used is also important. The various emergency stop devices must operate in such a manner that machines and moving units are stopped in a sensible and acceptable manner, which does not necessarily mean that everything stops instantaneously. Parts or components which are not directly dangerous or which incorporate their own labour safety devices should be permitted to complete the ongoing operation. In this way, breakdowns are avoided which could jeopardise labour safety.

In order to design an effective emergency stop network (without it being an obstacle that will result in it being bypassed), careful study must be given to its function in the system and its physical locations, and must be introduced in a manner which is acceptable to all.

Back-up systems

An automated material handling system can also offer a number of back-up systems. If, for example, it has an overriding computer system for total control, this can be doubled to give an extremely high standard of reliability. For protection against failure of power supplies, microprocessors and peripheral equipment can be provided with battery back-up supplies.

For the local control of AGVs (Autotrucks) in the system shown in Fig. 4, a Carrago microprocessor is used. Should the main computer fall out, transport assignments can be fed into the microprocessor by means of a CRT keyboard. Should one Autotruck in the system break down, some capacity is lost but the function of the system is maintained. The lowest level of back-up system for the Autotrucks is to control them individually by means of a manual control box.

Concerning back-up systems, there is often a desire to have these modes at various levels. Unfortunately, many of these are only back-up systems in theory and cannot be applied in practice. Here, like in many other cases, it is necessary to decide on a sensible level of back-up and so achieve an effective back-up system which is well documented and with which the personnel are familiar.

Concluding remarks

This paper has referred to many of the benefits offered by FMS with automated material handling by AGV system. The following lists the more important of these, together with a few others worthy of note:

- Ability to fetch/off-load at various levels.
- Freedom to plan machine groups, buffers, stores, entry points and quality control without restrictions.
- Production rates can be varied.
- No obstruction of floor area.
- Machines can be run without staffing, for example when building up smaller buffer stocks.
- Intersecting material supply routes and production flows.
- Shifting between parallel and series-arranged machine groups.
- Flexible buffering possibilities between machine groups.
- Better product quality since each individual object or machine can be controlled.
- Any scrapping is reported directly.
- Environmentally dangerous jobs can be done in special enclosures although still being a part of the production flow.
- Individual interruptions do not disturb the total production system.
- AGVs can be serviced and maintained during the dayshift.
- Unobstructed floors facilitate cleaning and maintenance, etc.

- Transport also takes place during breaks and after production hours.
- Intelligent control systems that can combine and compare transport demand with machine capacity and production control and can communicate with the overriding administrative system.

With these benefits, the introduction of FMS-AGV systems within engineering industry will enable very significant capital rationalisation in the next decade.

FLEXIBLE MANUFACTURING SYSTEMS WITH AUTOMATED MATERIAL HANDLING

M. Annborn
AB Bygg-och Transportekonomi (BT), Sweden

To optimise production and/or to realise unmanned production in
flexible manufacturing systems, automated material handling is
required. BT plans and implements such implementations. Two of
BT's systems currently in operation, one using a stacker crane system
the other an AGV system, are described and compared.

Automatic machines and robots are becoming common within industry. Material handling to and from these machine systems is still
generally accomplished using manual aids, such as overhead cranes
and manual trucks. With these types of aids it is difficult to optimise
the production apparatus and/or to have unmanned production.

To reduce WIP (work-in-progress), investments in more or less
sophisticated planning systems are often made. These systems plan
production based on theoretical assumptions, which often means that
in practice the calculated savings do not occur.

BT markets cranes and AGVs (automated guided vehicles) for
production-orientated material handling. It implements the automation and reports the *true* state of production.

Flexible structure for material handling

To build a handling system for a complete workshop at one time often
appears difficult from the customer's point of view. To simplify the
problem, in the first phase of the development BT looks at material
handling in each separate production group. A production group can
contain anything from advanced FMS machines to manual work areas.
The size of the production group is decided by the customer's layout
and the capacity of the AGV/crane. The basic requirement for each
separate production group is the possibility of automatically handling
platens, robot magazines and Euro-pallets. Racking associated with

the production group area is used for storage of WIP. The racking is placed on both sides of an automatic crane aisle or in any chosen position for an AGV.

Each group has an input conveyor for unworked parts and an output conveyor for finished parts. The material handling systems in each separate production group is able to function during development phases II and III even if the computer is non-functional.

Each separate production group is able to function independently of the other systems. Each system is equipped with a microcomputer for production reporting and material control. Pallets with associated data may, if the layout allows, be transported among the various systems.

The system can be developed in two directions:

- Horizontally – installation of further independent production groups.
- Vertically – increase of computer capacity.

In the second phase of the automation the computer system is developed by placing a minicomputer over the production group microcomputers. This allows increased possibilities for production reporting, supervision and control of production – all with the aim to minimise WIP.

To transfer production planning and to feed back the actual production situation in the third phase the minicomputer is connected to the existing planning computer.

Production group with stacker crane

An example of a stacker crane installation can be seen in BT's own workshop in Mjölby (Fig. 1). An automatic stacker crane BT ASC 1000 is placed between two sets of racking (Fig. 2). On each side of this racking are placed 23 NC machines. Three of these machines are LMP machines. On a signal from the machine, the crane executes automatically a platen transfer on the machine's rotary platen transfer table. The number of platens in storage for machining is dependent on the machining time per platen and the length of time the machine operates unmanned.

The other NC machines have human operators who instruct the crane by means of a computer terminal. The crane then delivers a pallet to the input roller conveyor of the machine in question (Fig. 3). After the machining operations are complete, the crane is instructed to collect the pallet from the machine's output roller conveyor. This system makes the handling of Euro pallets, machine platens and robot magazines possible.

The machining of parts alongside the racking and their storing in the racking means that optimum use is made of shopfloor space. The total width of the crane and two sets of racking is about 4.3m. With larger

Fig. 1   FMS installation at BT, Mjölby

pallets than the Euro variety, this measurement will be increased. The length and height of the racking is mainly dependent on the needs and dimensions of the workshop.

The crane installation layout is very flexible. From a handling point of view, it is relatively unimportant in which order the machines are placed around the racking. There is a central input conveyor on which all pallets of unmachined parts are placed independent of which

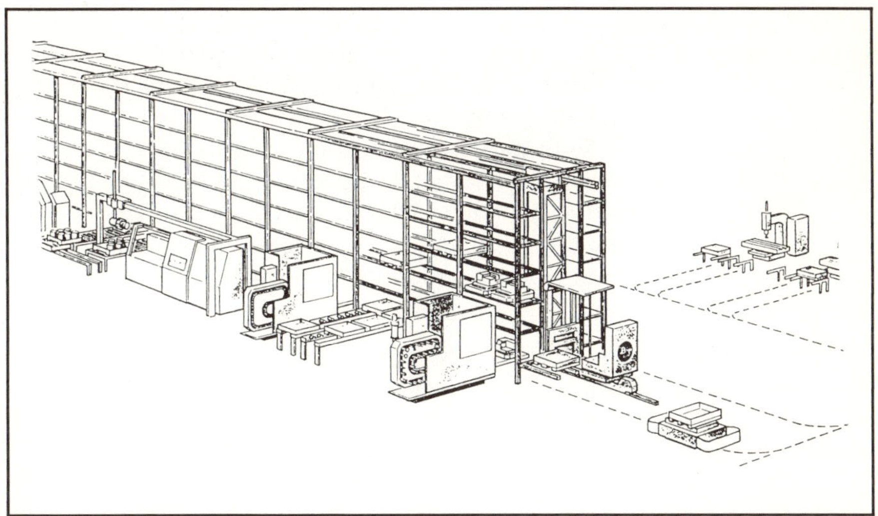

Fig. 2   BT stacker crane between two sets of racking

machine they will be transported to for machining, and on which parts are stored in order to be mounted on platens for LMP machines at a later stage. The output station is the same for all machined parts and is connected with a final quality control station. The storage of parts always takes place near the machine in question. This gives each operator and other related personnel a direct picture of the queue situation at each machine.

Automatic stacker crane

The automatic stacker crane BT ASC 1000 consists of three main parts: chassis, mast and lift platform. The crane has a single triangulated framework mast. Movement, lift and fork operation are driven by thyristor-controlled dc motors.

The crane's lift apparatus is equipped with a standing platform for manual operation for service purposes and as a back-up in the event of computer disturbance. With automatic operation, the crane's movement, lift, fine positioning and fork operation are controlled by minicomputer. Fine positioning is controlled by reflective tape on the racking: infrared photocells use the tape to locate the crane's lift apparatus.

Steering/control system

The control system for the installation is built around the minicomputer. Apart from the handling of pallets and platens, management and the planning department receive facilities for efficient and correct follow-up to production. All the monitored information reflects the real production situation. The unmachined parts on the pallets/platens at the system input point are identified on a VDU network. Among other things, article number, order/batch number, number of units, machining time per unit and machine number are recorded.

When a machine operator requires a new pallet for machining, he/she goes to a VDU terminal and calls for a pallet. A pallet is then automatically taken from the racking by the crane and transported to the machining station in question (Fig. 3).

The computer deals with pallets/platens on the basis of 'first in, first out' (FIFO). With a high-priority batch, however, a machine operator can, in cooperation with management, override this principle.

The network together with associated printers, assist management with planning and analyses. Valuable information can be listed in this way. For example:

- Print-outs of all pallets/platens which have the same machine number. This is the actual queue situation of the chosen machine.
- As above, but for a group of machines.
- Print-outs of all machined parts which have left the system after the specified time. Analyses are received concerning the number of approved and discarded parts as well as the total throughput time.

Fig. 3 *BT stacker crane feeding a machining centre with pallets*

Details are given on all print-outs of the date and time when the pallet was checked in at the input point.

BT's LMP system is a way of guiding production based on the actual situation. One difficulty in industry is to guide production within a department handling many articles in small batch sizes and with a complex flow. BT LMP works according to FIFO. The system constitutes a production point and can therefore be regarded as a planning point. The time of input of parts into the system guides their production order automatically according to FIFO. Through visual observation of the queue situation at each machine, or when needed terminal print-outs showing the queue situation, a controlled and minimised queue of parts awaiting machining at each workstation in the system is achieved. This leads to:

- Shorter throughput time.
- Maximum machine usage.
- High delivery reliability.

For a smaller system without production follow-up or one for all automatic handling and storage of platens to LMP machines a microcomputer is sufficient.

System advantages

The system's main advantages can be summarised as follows:

- Flexibility in flow of parts and layout alternatives, so that various functional group or line layouts are possible. It also includes the

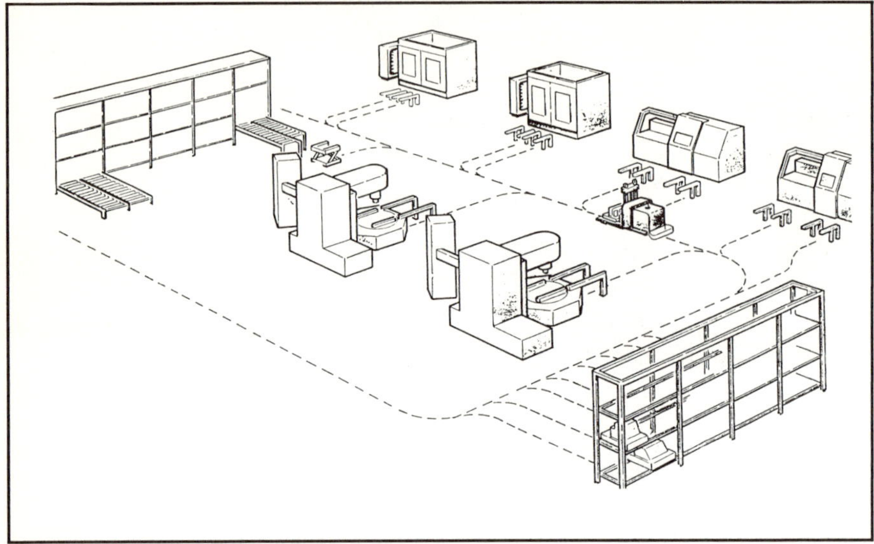

Fig. 4   FMS installation (ESAB, Laxå)

facility for simple layout alterations. When necessary, temporary flow-groups can be combined.
- Reduced throughput time and a reduction in the amount of work-in-process achieved by FIFO as well as availability of production follow-up.
- The possibility to machine high-priority parts quickly on instructions from the planning department.
- Saving on floor space by use of storage racking.
- The possibility to handle pallets, platens and robot magazines.
- Independent of the number of pallets/platens, manned or unmanned machines.
- The centralisation of loading areas for LMP machines means that it is relatively simple to arrange built-in compressed-air cleaners, washing units, lift units and other equipment to improve the working environment and reduce the risks of accidents.
- Transport passage for all materials is isolated from machine operators and other personnel, thus reducing the risk of accidents.

Production group with AGV

The example taken here is an FMS installation at ESAB's factory in Laxå (Fig. 4). The following main components are included in the system:

- AGV BT AHL 1000 (high-lifting fork-lift truck).
- Control system.
- Control loop layout.
- Racking and loading areas.

The AGV follows a buried guidewire. The vehicle is fitted with a mast which allows the stacking of pallets and platens up to a height of 3m in the buffer store (Fig. 5).

The object of the control system is to give the AGV assignments and to remember the contents of the buffer store: pallets/platens with raw materials and fully machined parts. The control system has the capacity to remember about 60 platens and 100 pallets with machined and unmachined parts.

The BT AHL 1000 allows freedom of layout. Machines, racking and loading areas are placed in positions best suited to the customer. The racking is conventional. I-beams are fitted to those positions where pallets/platens are to be stored. This allows the AGV's forks to go under the pallet/platen. Special guidance of the pallets/platens into the racking is not normally necessary.

Registration and ordering of pallets/platens

Pallets/platens are transported to the system by manual truck or AGV. The pallet/platen is identified at the input conveyor by an operator entering the machine number, article number, operation number, etc., via a VDU terminal.

The pallets/platens are stored in the racking. The AGV picks up the pallet/platen from the input conveyor and places it in position in the racking. Its position and data are automatically registered by the computer.

Fig. 5   FMS installation using BT high-lifting AGV

Automatic ordering of platens is as follows:

1. A machine tool signals the control system to instruct the AGV to pick up a finished platen.
2. The AGV stores the finished platen in the racking.
3. The AGV retrieves an unmachined platen from the racking (following the FIFO principle).
4. The AGV deposits the unmachined platen on the machine tool's platen changer.

Robot magazines are handled in a similar way.

In the case of machines with human operators, the operator signals for pallets via a VDU terminal. The strategy for each machine is via a VDU terminal. The required machine via the VDU terminal gives a presentation of all the pallets in the queue. The strategy is made up accordingly. Of course it is possible, when necessary, to change priorities. With no strategy the FIFO rule applies.

System advantages of the high-lifting AGV

- Flexible platen/robot magazine handling to one or more machines.
- Pallet handling to manual machines.
- One AGV for platen handling to machines of different manufacture and type.
- Work at different levels at the pick-up/deposit points.
- Storage of platens and pallets in racking.
- Flexible layout solutions, with machines and racking placed independently of each other.
- Automatic material handling makes possible the control and supervision of production on automatic machines and in unmanned work areas.

DESIGN AND USE OF AUTOMATED MATERIAL FLOW SYSTEMS

G. Handke
Messerschmitt-Bölkow-Blohm GmbH, West Germany

Based on the task and the premises related to planning and
realisation, the structures of three integrated material flow systems
are shown. With reference to the facilities management model,
maintenance of such systems and experience gained in this context
are described.

When automating production areas, the flow of materials must be
considered more important than in the past, since, in many instances,
the overall integration is the only way to achieve optimum production
results with machines and facilities. This involves a rethink process not
only with respect to production but also for service areas such as
planning and maintenance[1].

The task

Systematics

The material flow systems described and discussed here have been
designed to supply the individual tools or machine tool systems with all
the necessary manufacturing components, i.e. materials or workpiece
cutters, fixtures, work instructions, and program data.

They are systems of the centralised production control and, in
addition to their basic material flow function, the lowest level of
information flow. This provides high transparancy of the entire
manufacturing process, which is the prerequisite for production
optimisation.

The material flow system interlinks the functions necessary for the
supply and removal of tools at the manufacturing equipment; such as
storage, dispositioning, delivery, distribution, and buffering. These
functions result in a basically identical structure of these systems, each
having a circular characteristic.

Planning premises

The basic premises for all material flow systems are that:

- Material flow is realised by supply and removal circuits, i.e. from the storage area to the workshop and back.
- Supply and removal operations for the machines must not be critical in time.
- Reservation and provisioning of all manufacturing components required in the production area are done with adequate lead-time.
- Production control is effected from the production control room.
- The facilities must obtain a high degree of availability in three-shift operation.

These premises are certainly valid for all such systems that are integrated into the production of highly complex conponents.

Structure of the material flow systems

The material flow systems provide supply and removal of tools for machine tool systems, i.e. they are basically secondary peripheral facilities having fundamental significance with respect to the productivity of the production equipment. They are components of the computer integrated and automated manufacturing system (CIAM).

The set-up of this manufacturing system[2] is primarily designed for the manufacture of highly integrated aircraft components, however, it may also be generally applied. Fig. 1 shows the hierarchical structure, and the tie-in supply and removal systems.

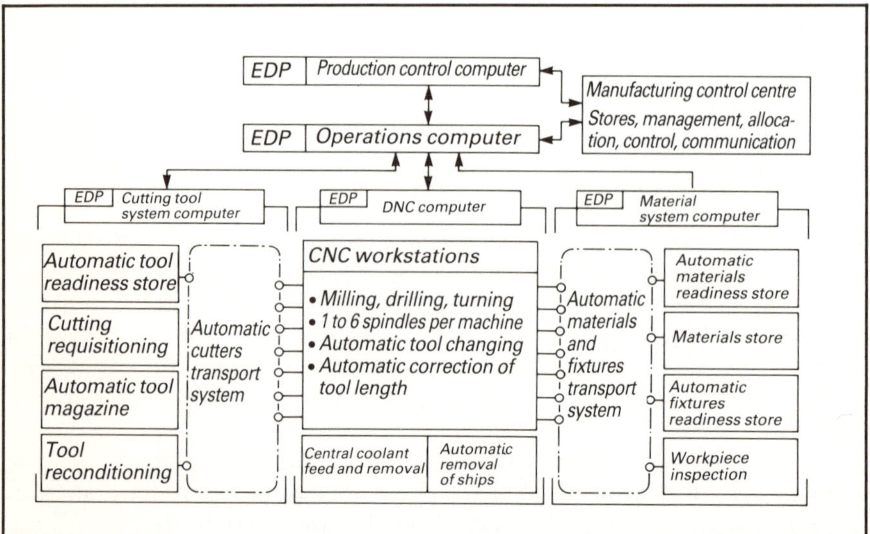

Fig. 1   Structure of the CIAM machining system for the machining of integral parts

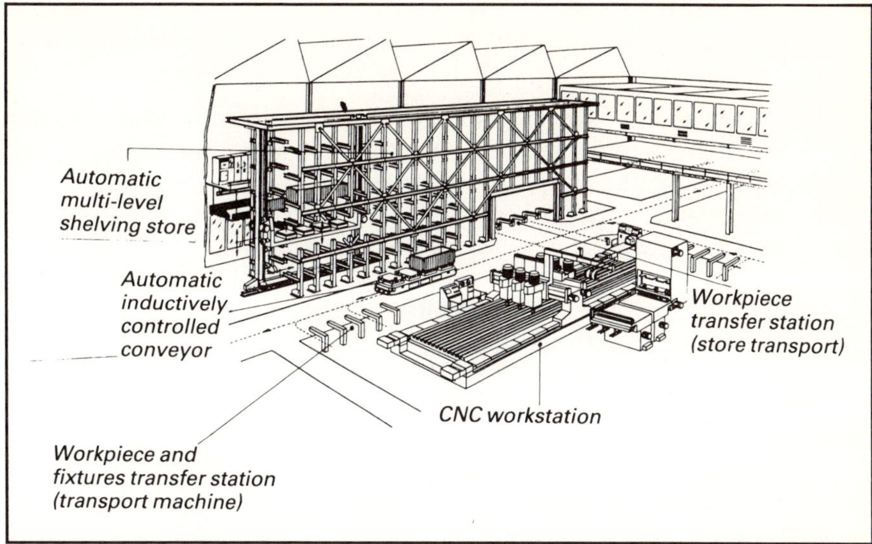

Fig. 2   Automatic transport and storage system for the feed and removal of workpieces and appliances in NC fabrication

Materials and fixtures system for the cutting of major components

The overall system is composed of three subsystems:

- Buffer store, configured as multi-level store.
- Conveyor system with inductively guided microcomputer controlled vehicles, moving lengthways and sideways. The overall transportation system is controlled by a process computer. Loading is by means of lift tables.
- Machine buffer configured as cantilever racks.

The system is shown in Fig. 2.

Workflow starts at the cutting area of the central store. Here, the raw components are dispositioned on special-to-component pallets as controlled by the control room and placed on the support fixtures in the cutting area.

The inductively guided vehicle picks up these special pallets according to disposition order from the central control room and, via the identification point, supplies them to the buffer store. In due time, before machining on a machine tool system commences, the job order (one pallet) is removed from the buffer store and moved to the machine by means of the inductive transportation system. At any time, there is only the job under work plus the following job located at the machine. This results in minimisation of the machine buffer and maximum disposition flexibility of production control. Subsequent to the work sequence, the job is moved back to the buffer store by means of the inductive transportation system. It is thus ready for new assignment and closes the supply and removal cycle.

Tool circuit (cutting tools)

The overall system of the tool circuit includes the following subsystems:

- Tool store, where cutting tools are stored by quantities and sorted according to identity numbers. Configured as a multi-level store, where the individual pallet is pulled and pushed in lateral direction. The mechanical grab technique used substantially improves the usable storage space – especially with low pallets.
- Dispositioning area for the arrangement of special-to-job tool sets using roller and belt-type conveyors.
- Buffer store for time-uncritical provisioning of the required tool sets configured as an automatic multi-level store.
- Transport system configured as a single-rail overhead conveyor system with individual vehicles. The system is arranged so that all acceptance and delivery points are positioned at the secondary tracks accessible via switches. This permits operation of all vehicles during periods of high demand.
- Machine buffers for two tool sets in paternoster configuration with direct supply and removal from the vehicles of the overhead system.
- Tool conditioning by means of an automatic tool rinsing facility followed by disposal of the tool sets.

The system is shown in Fig. 3.

All processes are controlled by the control room. The overall system receives all instructions and commands from the higher level operations computer.

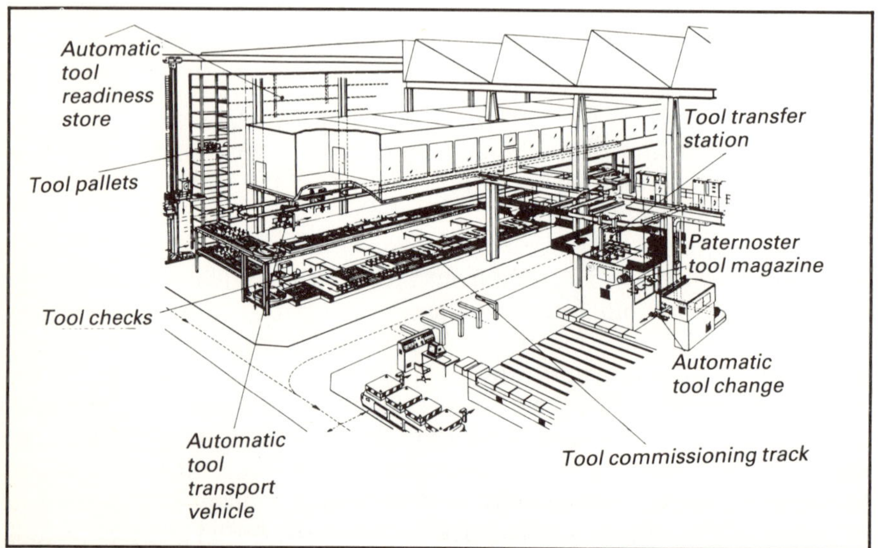

Fig. 3 *Automatic transport and storage system for the feed and removal of tools in NC fabrication*

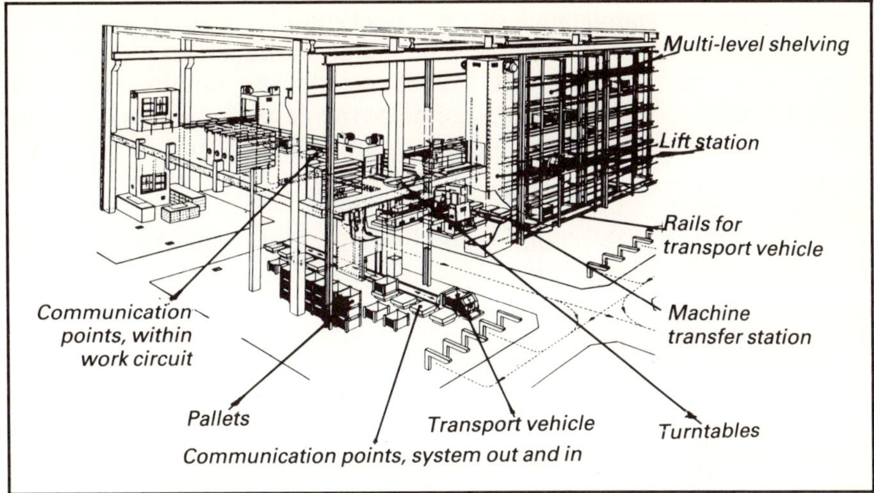

Fig. 4 Automatic feed and removal system for conventional machining

The identification point is arranged as a check point and is located between dispositioning place and buffer store.

Material flow systems for small components machining

With respect to the sizes of these components it was possible to use a special box pallet measuring 800 × 1200mm, where all job components are assembled – as compared to major component manufacturing.

The solution consists of a storage servicing cart system combining the functions of dispositioning, storage and distribution, i.e. the servicing carts assume the functions of a shelf conveyor and, in addition, the distribution functions of an automated guided vehicle system[3]. The system is shown in Fig. 4.

The processes compare to the material circuit associated with major component machining procedures. The service carts are controlled via radio by a process computer. Propulsion is effected by linear motors permitting high acceleration and deceleration rates. The vehicles are controlled by means of on-board microcomputers and move freely about the whole system.

Maintenance

Significance of maintenance of highly automated flexible manufacturing systems

The significance of maintenance has as yet not been fully assessed in production engineering. Whereas with process engineering and primary industries it is a matter of course to give all necessary consideration to maintenance, the maintenance of production machinery is in many instances still considered a necessary evil. However, maintenance must provide the prerequisites to enable maximum utilisation of highly automated systems.

The economic significance of the direct maintenance effort must of course be compared to the indirect maintenance effort, i.e. loss of production costs, loss of profit, loss of cost-sharing amounts, etc. This aspect makes it necessary to connect the layout for maintenance (direct and indirect costs) to the rate of utilisation of productivity of a company, which, however, seems to be hardly assessable by calculation and planning.

In conjunction with planning and procurement of highly automated manufacturing systems, the subsequent costs or cost curves, respectively, must not be neglected. This is true especially for highly integrated, extensive material flow systems, the economy of which is obtained by rationalisation.

Looking at the development of factors such as materials, production personnel, energy, and maintenance as the degree of automation increases, the superproportional decrease in the number of operators due to the strong rationalisation effect is evident. At the same time, a probably superproportional increase with respect to maintenance effort must be anticipated due to strongly increasing complexity of the systems. It is mandatory that this increase be kept within reasonable limits.

Facilities management/plant facilities

Prior to planning and realisation, due consideration had been given to the high degree of integration and complexity of all production systems and material flow systems. By extending the organisational unit 'plant facilities', the economic function had been considered already at the time of procurement and realisation. The 'plant facilities' area represents the functional circuit 'facilities management'.

Functional circuit 'facilities management'. It is the objective of 'facilities management' to optimally interlink planning, realisation and support of all facilities of a plant. Special attention is directed to the feedback of experience gained from maintenance activities as a basis for new procurements.

The overall support of investments is optimally taken care of and ensured by this area.

Linking the planning and realisation levels with the operational level, i.e. on-line interlinking of planning, realising and maintaining functional units to form an overall unit, optimal support is achieved. This functional circuit is depicted in Fig. 5.

Based on this type of set-up, the entire 'facilities management' is an in-house service company in a plant. The idea is that a central activity handles planning, realisation and maintenance of all investments. The decisive advantages are:

- Tie-in maintenance already at the planning stage.
- Rapid feedback of experience gained during maintenance operations as an aid for new procurements.

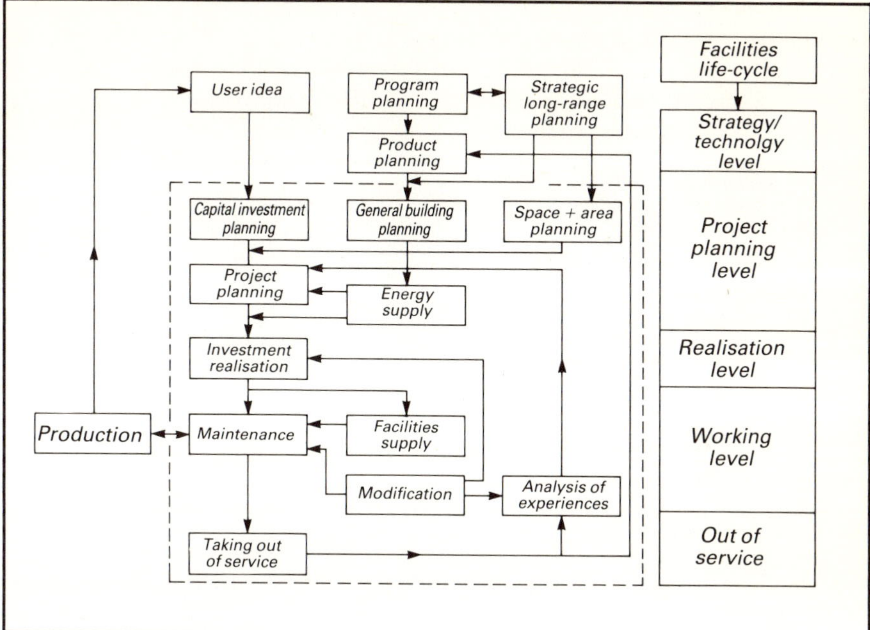

Fig. 5   Functional circuit facilities management

- Optimum coordination and investment when procuring replacements.
- Cost transparency in the investment and maintenance areas.
- Optimum possibility to improve facilities and to carry out follow-on development.

High transparency of the maintenance area is indispensable for effective operation of the functional circuit 'facilities management', only achievable by strict systematic structure within the overall maintenance complex.

Maintenance as part of facilities management. It is the objective of maintenance to ensure operability of all equipment required for production.

Since each facility, each energy net, each machine, etc. is subject to erosion and since a certain amount of stock level of erosion is required to offset wear – in order to accomplish the job – it is the responsibility of maintenance to retain or replace the stock level of erosion.

Reaching the damage limit may be avoided and reduced to equipment stoppages for short periods of time by specific, planned, on-condition maintenance. In order to enable the application of this approach to complex material flow systems, the maintenance structure – to include 'weak spot elimination' – equal to facility improvement. Facility improvement may be necessary for operational or maintenance reasons. Fig. 6 shows this type of maintenance structure.

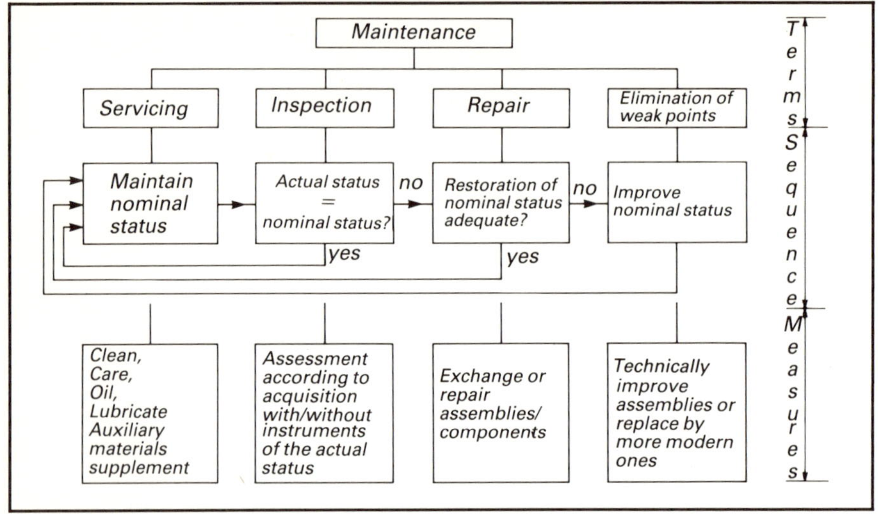

Fig. 6   Maintenance – terms, sequence, measures

It is the objective of any maintenance activity to achieve the cost optimum between:

- Direct maintenance costs corresponding to maintenance effort.
- Indirect maintenance costs, i.e. costs arising from damages, loss of profit, loss of cost sharing, schedule overrun, additional costs for rework, etc.

The optimum span for a plant, for a production area or similar activity, cannot be calculated in actual practice since:

- The impacts of system changes occur years later.
- Short-term changes to the structure of the capital investment prevent comparable figures.

Maintenance system at MBB Augsburg

Maintenance system. The overall system of maintenance at MBB Aircraft Division, Augsburg Plant, reflects a long-term basic philosophy. It comprises the 'planned maintenance' and is based on the fundamental production structure and long-term strategic planning in the plant. The overall system is dependent only on the fundamental, long-term production concept of the division. It is not subject to short-term peripheral conditions, such as shop-load, program planning, etc.

Maintenance strategy. The maintenance strategy is a medium-term concept. It is dependent on the medium-term program planning and shop-load planning.

All facilities are assigned priority groups. These priority groups indicate the medium-term significance of a facility in terms of importance for production in case of failure and the consequences of such failure. In addition, the priority groups assign the level of maintenance intensity to each facility.

Maintenance concept. The maintenance concept includes the short-term adaptation of all maintenance activities to suit at any time the situation prevailing in the production subareas.

The essential elements of the maintenance concept are:

- Activation of inspection work in order to learn on the basis of inspection reports about trends enabling timely corrective action.
- Utilisation of skilled workers trained also in related or alien fields to boost qualification.
- Detailed report for technical and cost control.
- Weak-spot analysis on the basis of individual jobs.

At Augsburg, machine maintenance is, of course, focussed on the highly automated production areas with their integrated material flow systems. This means, with respect to maintenance:

- Support of high-precision, complex machine tool systems.
- Support of the extended or hard-to-get-at storage and transport systems.

Organisational set-up. Technical machine maintenance is broken down in two separate groups, i.e. mechanics group and electrical/electronics group as specialists groups; and maintenance centre for management and cooordinating functions.

For the purpose of cost optimisation, inspection activities are given high priority. This inspection work carried out on the machines and facilities, all of which are practically prototypes, meets with special difficulties, since adequate inspection or maintenance instructions are in most cases not available from the equipment supplier. A production equipment instruction sheet has been worked out based on the experience gained in connection with these important major facilities. This 'procurement order' specifies also the structure, breakdown and contents of the overall documentation.

Work-flow organisation. Maintenance activities are carried out upon written job orders only. All pertinent data and actions are recorded in this work order. The maintenance work order is the basis for technical failure evaluation and weak-spot analysis.

With NC machines operating in DNC mode, all data collection is accomplished by means of the integrated machine data acquisition system in order to catalogue malfunctions and corrective actions taken and to enter them in the service record of a facility as a basis for systematic monitoring and weak-spot analysis.

Maintenance costs. Maintenance costs are, to any possible extent, assigned to the individual facility or to the respective cost account.

Cost control is accomplished by the maintenance department jointly with operator and finance department.

Maintenance of automated material flow systems

Technical results

Downtimes/availability. The availability of the material flow system may substantially affect production flow. In order to ensure emergency operation even in case of failure of dispositive and controlling computers, great emphasis has been placed on maintenance operation facilities during the planning and realisation phases.

The failure rate is used as the evaluation criterion for a facility and, especially with transport system, is assessed according to the following formula:

$$A_{\mathrm{T}} = A_{\mathrm{T\ facility}} \times A_{\mathrm{T\ vehicles}}$$

$$A_{\mathrm{t}} = \frac{T_{\mathrm{AA}}}{T_{\mathrm{B}}} \times \frac{\sum\limits_{i=1}^{n} T_{\mathrm{AF}i}}{n \times T_{\mathrm{B}}} \times 100\ (\%)$$

where A_{T} is the technical failure rate, $A_{\mathrm{T\ facility}}$ is the A_{T} of permanently installed component groups, A_{T} is the A_{T} of vehicles, T_{AA} is the downtime of the facilities for technical reasons, $T_{\mathrm{AF}i}$ is the downtime of the *i*th vehicle for technical reasons, *n* is the number of vehicles, and T_{B} is the operating time.

Even though this formula equalises the failure of all vehicles, thus distorting the failure relevancy of the vehicles with respect to the availability of a transport system in actual practice, the effort regarding data acquisition and evaluation according to this formula is already high. Yet this formula is sufficiently precise in actual practice if associated with a systematic failure analysis.

Thus, with highly automated material flow systems, technical failure rates are between 3% and 8%.

Experience shows that downtime distribution curves already permit a preliminary assessment with respect to reliability and also the current technical status of a specific facility.

Fig. 7 shows a typical example of a frequency scale of downtimes for transport and storage systems. It is evident that normally the downtimes for storage systems are shorter as compared to transport systems, the reason being that transport systems cover a large amount of floor space.

Fig. 8 shows general downtime distribution curves for transport and storage systems. It is clearly shown that with transport systems, there is a marked tendency towards longer downtimes. This tendency is based on the fact that with vehicle systems, the failure of the first vehicle does not incur high priority and, in many instances, the failure

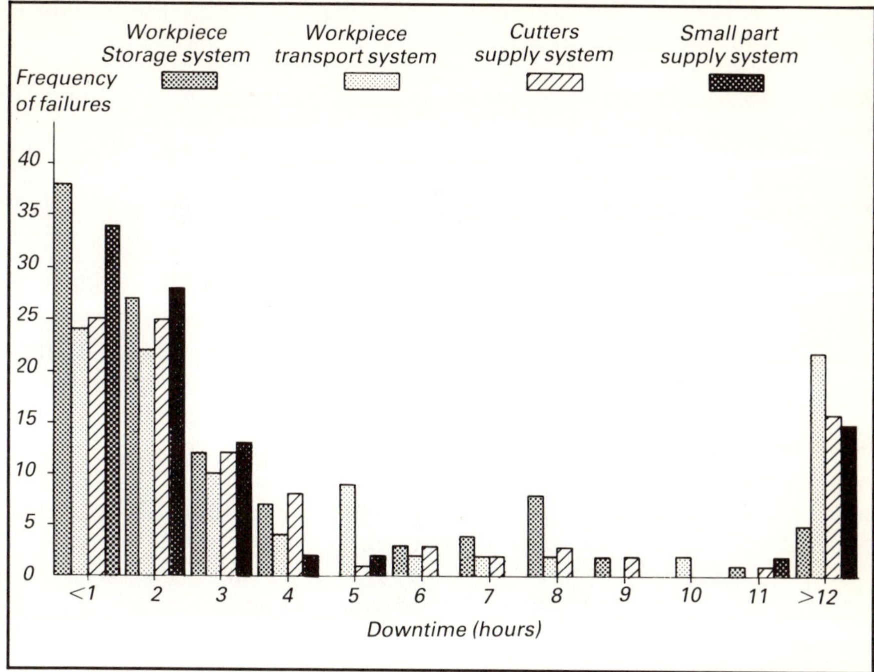

Fig. 7   Frequency scale of downtimes for transport and storage systems

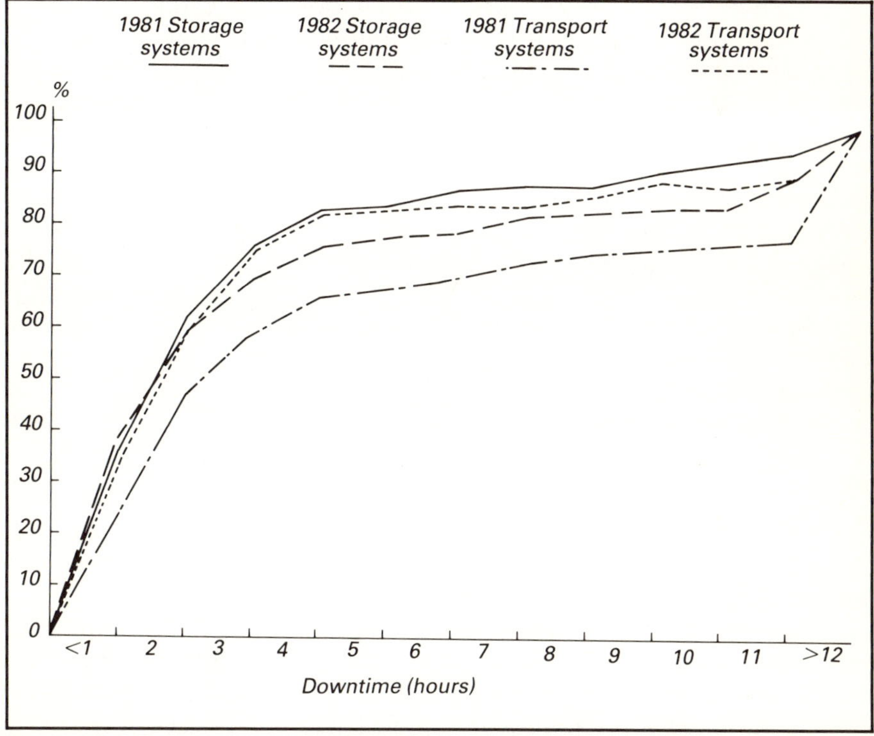

Fig. 8   Downtime distribution for transport and storage systems

of a single vehicle does not at all or just slightly affects the transport capacity of the systems.

Failure analysis. Weak-spot analysis is an essential part of planned maintenance, primarily designed to:

- Indicate to maintenance personnel the technical weak points.
- Provide investment planning with essential knowledge on technical approaches with respect to systems and subsystems in order to enable optimisation of new procurements.

Comparing the failures occuring with a highly integrated material flow system to an automated guided vehicle system shows that failure concentrations are almost identical.

The comparison shows that in the first place electrical circuit components and the mechanical components of a vehicle, i.e. the dynamically loaded units which exist in large numbers, reflect high failure relevancy with respect to equipment downtimes. This is not necessarily equal to plain failure frequency.

Maintenance intensity with these systems cannot be compared because of the different significance of vehicle failures on the grounds that the capacities of such material flow systems are designed for high utilisation rates.

Documentation. The evaluation of a facility can be only based on established facts and forecasts. To establish such facts with regard to maintenance effort, technical status, etc., today we need very detailed statistical material. It is only the service record of a facility that contains the summarised material permitting the establishment of facts while, at the same time, providing short-term historical information for the specialist on the job. The preparation of this type of documentation requires maximum effort, especially in the case of non-series facilities and is not obtainable from the supplier of the equipment.

Based on experience gained the maintenance instruction can be broken down into:

- System overview.
- Description of equipment components.
- Description of equipment installations.
- Job instructions.
- Records.

This breakdown also includes design documents, job instructions, inspection reports, or supervision records.

It is just the job instructions, acceptance procedures and error checklists for material flow systems which, in most instances, are system solutions adapted to the specific approach and which are rarely obtainable from the equipment supplier. However, such documents are indispensable for optimal equipment support.

Maintenance efforts

Direct maintenance costs. The most important item with respect to maintenance, in addition to technical availability of the facility, are the maintenance costs. These costs can be broken down into two blocks:

- Direct maintenance cost, i.e. salaries and wages, materials and outside services.
- Indirect maintenance costs, i.e. costs arising from downtimes, loss of profit, schedule overruns, outside production, etc.

Precise figures regarding indirect maintenance costs are established in very rare cases. In most instances there is no assessment. However, the two categories of costs must be added up to be meaningful with respect to facility economy.

Considering only the direct costs, these figures must be related to a definite value. With machines, the investment value at the time of procurement or the replacement value is taken as a basis. Fig. 9 shows the cumulative curve of direct maintenance costs for transport and storage systems.

It is obvious that with transport systems the gradient of the curves is steeper as compared to the gradient for storage systems, i.e. the cost rates for maintenance of transport systems are substantially higher than those for storage systems.

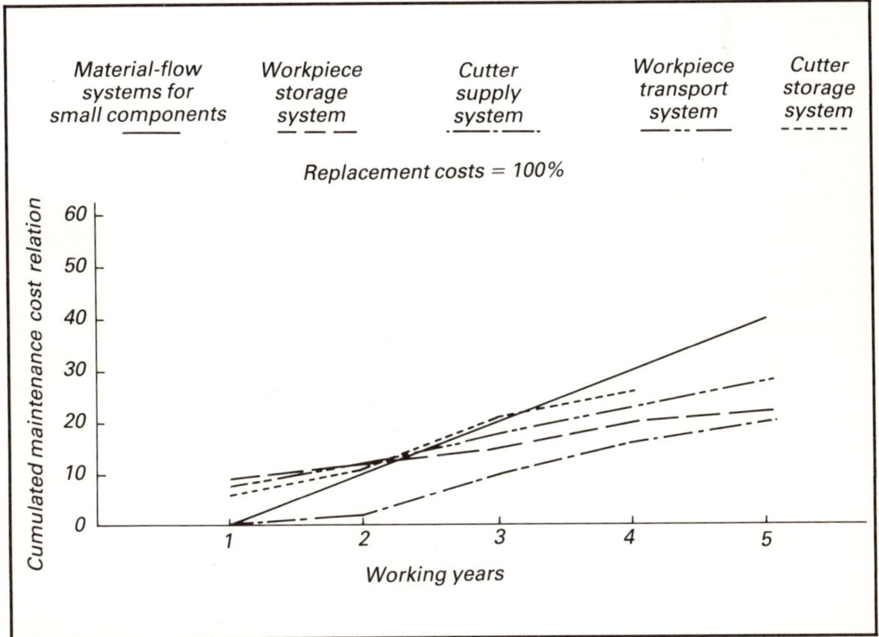

Fig. 9 Cumulated direct maintenance costs for transport and storage systems based on replacement costs

Personnel. It is very difficult to establish staff requirements for the maintenance of prototype facilities, since it is almost impossible to evaluate factors such as number of shifts, on-call working teams, weekend work, complexity of facilities, etc.

With regard to the qualification of staff members, two aspects are clearly identified:

* With increasing complexity and automation of production facilities the qualification of personnel must be higher in order to achieve reasonable failure rates.
* To the same extent that the number of personnel engaged in production on automated facilities decreases, the training of personnel must be broadened.

Highly qualified personnel carrying out maintenance work is indispensable for the efficient operation of highly automated systems. Since trained maintenance personnel are not available from the labour market even today, system operators must intensify their own in-house training and the results of MBB's efforts to effect this have been very positive.

Approaches to rationalisation

Requirements for material flow systems to be procured in the future

Based on the experience with respect to highly automated flow systems, the following requirements must be specified (in MBB's point of view).

Mechanics

* Conversion of theoretical possibilities of designs into practicable evaluation and acceptance criteria, such as vibration behaviour, stiffness, self-vibration resonance, etc.
* Integration of modern measuring and inspection procedures with respect to function, measuring chains, etc.
* Persistent use of self-diagnosis systems as is normal today with CNC machines, and, as far as material flow systems are concerned, not only within the control area but extended to all critical components and areas of a facility.
* Installation of redundant functional, monitoring, and control units if failures could result in substantial damage (limit switches, etc.).

Control technique

* Error self-diagnosis systems must be advanced and made compatible with decentralised evaluating units.
* The calculation of tendencies on the basis of cyclic interrogations of equipment status is an essential prerequisite in order to increase the portion of planned actions regarding maintenance and for the

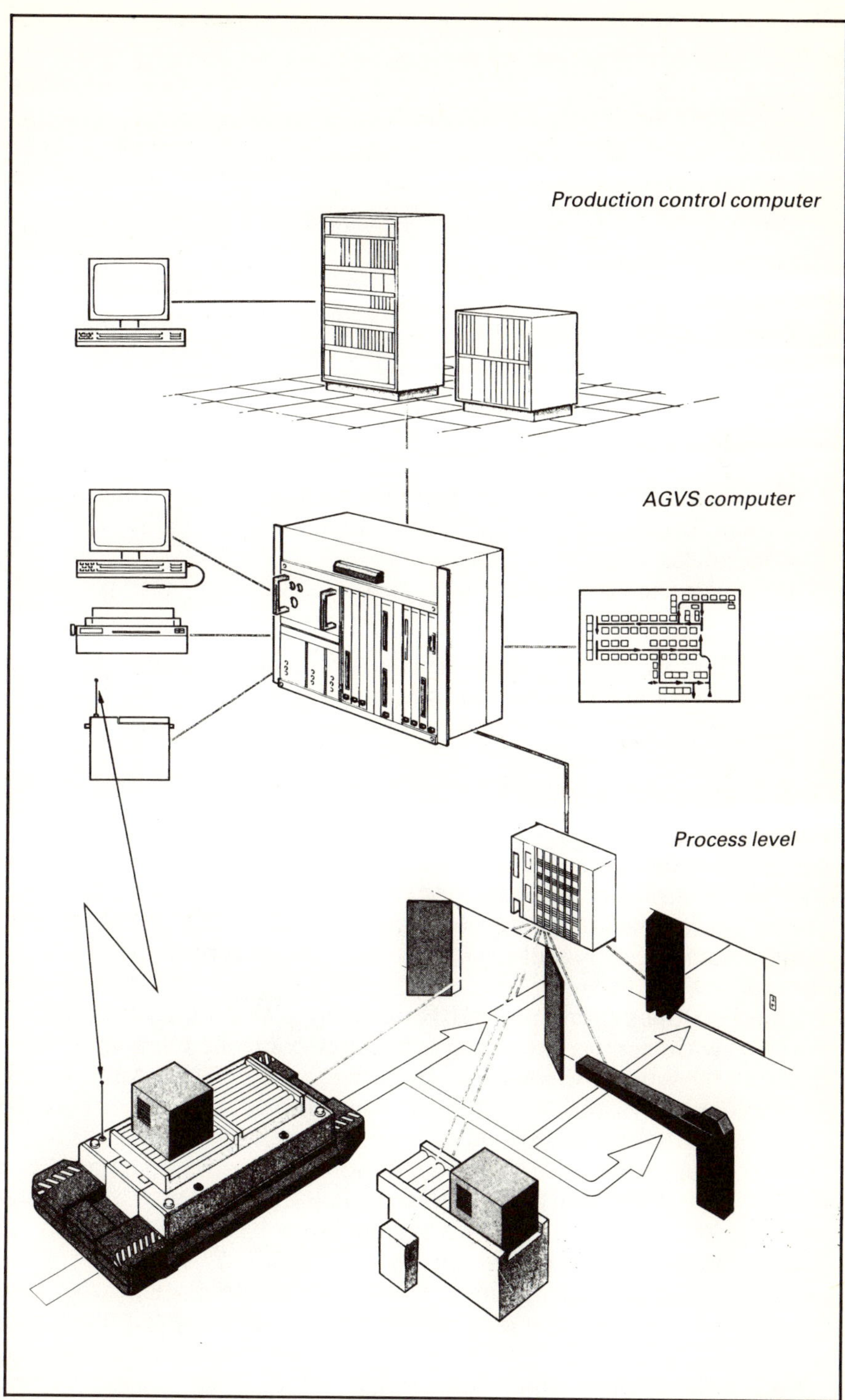

Fig. 10   MBB multiTrans computer hierarchy

iterative establishment of meaningful inspection intervals.

- The modular structure of control systems, wherever possible, in standardised form must enhance the possibility of the electronic specialists to familiarise themselves with new control systems. Today, even small facilities are equipped with control systems, which sometimes incur considerable problems for operator and maintenance with respect to software support.

Measures to reduce costs

The essential measures with respect to cost gain in maintenance (for unchanged equipment load) may be broken down into three areas:

- Optimisation of new facilities to be procured during the definition and design phases, i.e. realistic requirements on the part of the operator, maintenance-orientated design and equipment on the part of the supplier.
- Improvement of the facility 'facilities management' during installation and production to improve maintainability taking into account economic aspects.
- Improvement of inspection reports to increase the plannable portion of repair work.

The effect of any of these measures will be noticeable today or tomorrow but only after months or years. However, we are every day aware of the urgency of such measures.

Concluding remarks

Today it is mandatory for users of highly automated material flow systems to provide the maintenance areas with adequate knowledge in order to ensure that maintenance costs are substantially influenced. The organisational structure of a functional circuit 'facilities management' provides a model to effect this as it represents a closed information circuit.

Based on the experiences MBB made by planning, realisation and working with highly integrated computer-controlled manufacturing systems, a new division was founded, the MBB Automation-Technology Division, to provide the market outside the MBB Company with those experiences.

One of the first steps of this new division was to develop a new AGV system, the MBB multiTrans. Based on the experiences in maintenance of AGV systems, the MBB multiTrans has today no electrical installation in the floor; that means that all the active parts of the inductive guidance are mounted on the vehicle itself – the data transmission between the different vehicles and the system computer is made by 'radio system'.

Fig. 10 shows the computer hierarchy, Fig. 11 shows the vehicle and Fig. 12 shows the system itself – made for the first application in the

Fig. 11   MBB multiTrans

Fig. 12   Overall system

MBB Stade plant as an overall material flow system for the composite manufacturing centre.

The reduction of maintenance costs, in the author's opinion, can only be achieved in a joint effort by all areas engaged in planning, construction, utilisation and support.

References

[1] Handke, E. and Handke, G. 1980. Flexible Fertigungssysteme ändern Materialfluß-Konzepte. *Zeitschrift Materialfluß*, MF-Report, Spezialausgabe.
[2] Dronsel, M. 1979. Technische und wirtschaftliche Probleme der Fertigung in Flugzeugbau. *Prod. techn. Kolloquium '79*, TU Berlin.
[3] Handke, G. 1979. Funkgesteuertes Lager- und Transportsystem versorgt spanabhebende Fertigung. *Zeitschrift Fördern und Heben*.

MOBILE ROBOTS – KEY ELEMENTS FOR THE INTEGRATION OF TRANSPORT AND HANDLING FUNCTIONS

R. D. Schraft, M. Schweizer and J. Schuler
Fraunhofer-Institut für Produktionstechnik und Automatisierung (IPA), West Germany

To illustrate the integration of transport and handling functions a mobile robot is described. It is characterised by a combination of an automated guided vehicle incorporating a platform for carrying pallets and an industrial robot which is mounted on the vehicle. Applications for mobile robots are shown for workpiece loading and unloading of machine tools and transport and interchange of machining tools in the tool magazine of a machining centre. Finally, the installation tested in the laboratory of IPA in Stuttgart is described. There, the task of the mobile robot is to reload the tool magazine of a machining centre with presented tools.

For production companies who wish to use economical production systems in order to remain competitive, machining, information flow and material flow must be flexibly automated and computer-assisted[1]. As a further development in large batch production machining, devices and processes have recently started to use, for example, numerically controlled automation so as not to lose the flexibility necessary in these areas. Information and material flow, which were initially the 'poor relatives', have now also begun to be included in a similar fashion. Computer-guided or assisted production control systems on the one hand and, for example, inductively guided vehicles for transport and industrial robots for handling on the other are characteristic milestones in this development.

Automation of material flow in flexible production systems

Workpiece handling and positioning facilities, either integrated or next to the machine, have thus already been installed in a number of instances. The machining and handling functions are either additive, i.e. two independent appliances combined (Fig. 1), or integrated, i.e. linked to a single appliance.

The idea of employing a mobile handling appliance arose as a result of the perceived lack of exploitation in many cases of stationary, machine-orientated handling appliances due to the relatively short handling times and the inability to service several machines from a single stationary appliance. The combination of handling appliances with automatic conveyors to enable the former to operate as a mobile unit is only one possible solution. Material and workpieces are still positioned by another conveyor system.

The integration of the primary system functions of conveyance and handling in a single appliance (Fig. 2), on the other hand, makes use of the technical capacity available in the conveyors, not only for moving the handling appliance itself but also for the second essential flow function in the system, namely the transport of materials (e.g. workpieces, tools and parts) from and to the point of application (e.g. the machine tool). The conventional close linking of conveyance and handling tasks:

- transport of batch to machine,
- input of single parts into machine,
- discharge of single parts from machine, and
- transport of batch from machine

is thus implemented here with third- and fourth-degree material flow tasks being carried out by a single appliance.

Mobility of industrial robots

Industrial robots as programmable handling facilities can also be classified in accordance with their degree of mobility, irrespective of the possibility of combining them with workpiece positioning devices. They may be differentiated between floor-bound, standing or mobile, floor-independent and suspended types. Stationary industrial robots are usually floor-bound upright appliances which are used for a variety of applications in production engineering. Floor-independent industrial robots are used particularly for tool handling (e.g. spot welding of vehicle bodies).

Mobile industrial robots are made up of industrial robots and automatic conveyors. Apart from their own movement in the axes of the industrial robot there should also be powered automatic mobility in at least one axis other than those in the appliance. The degree of freedom of mobility may be divided up into one-, two-, and three-dimensional mobility (Fig. 3).

Industrial robots with one-dimensional mobility are characterised by the translating, usually horizontal, drive axis. This principle has already been put into practice in a number of instances with the aid of a straight track (Fig. 4)[2-6]. Disadvantages of this travel principle are

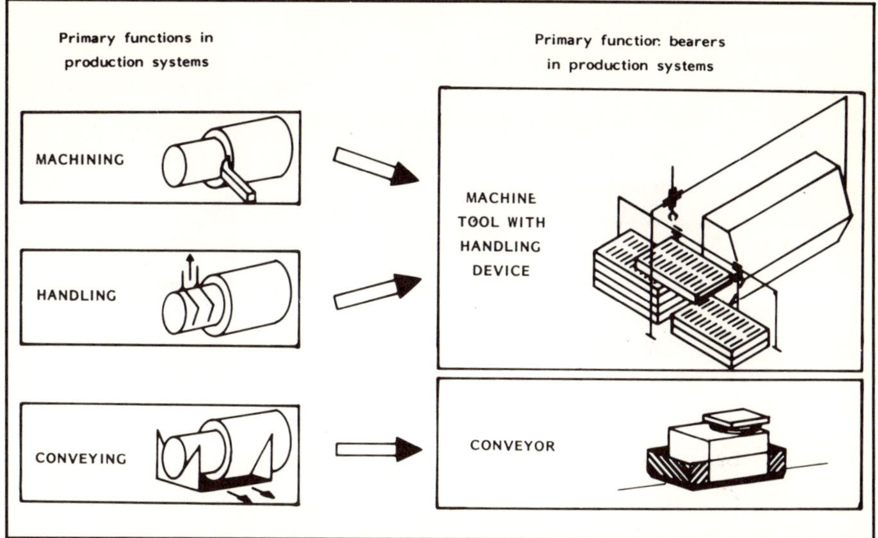

Fig. 1   Linking of machining and handling functions

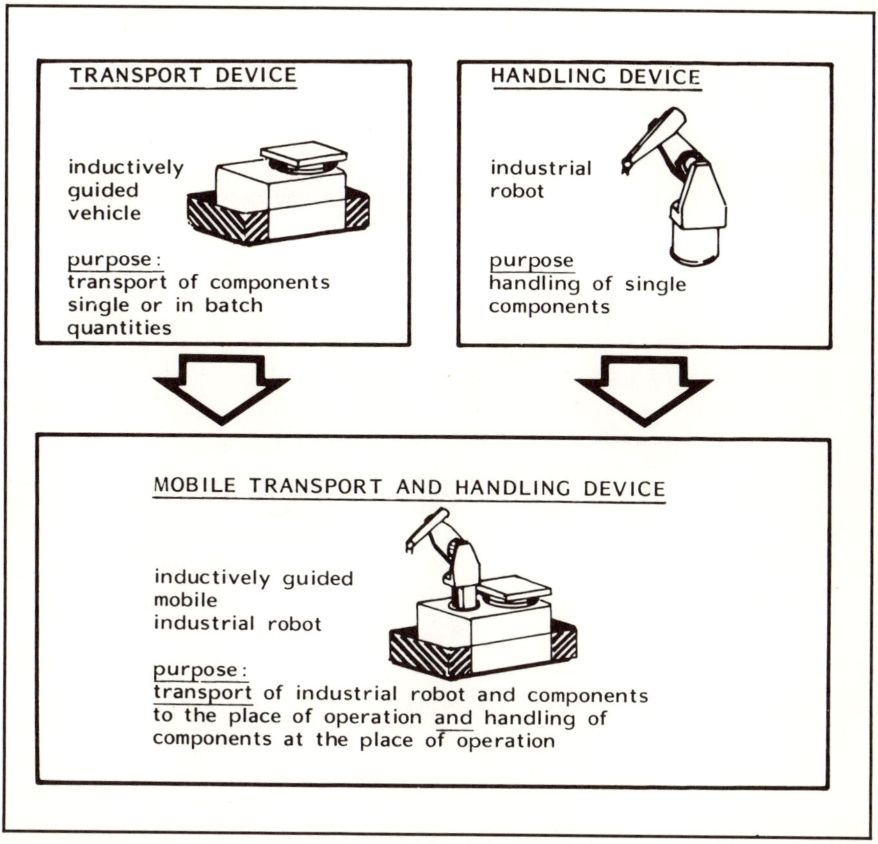

Fig. 2   Integration of transport and handling functions

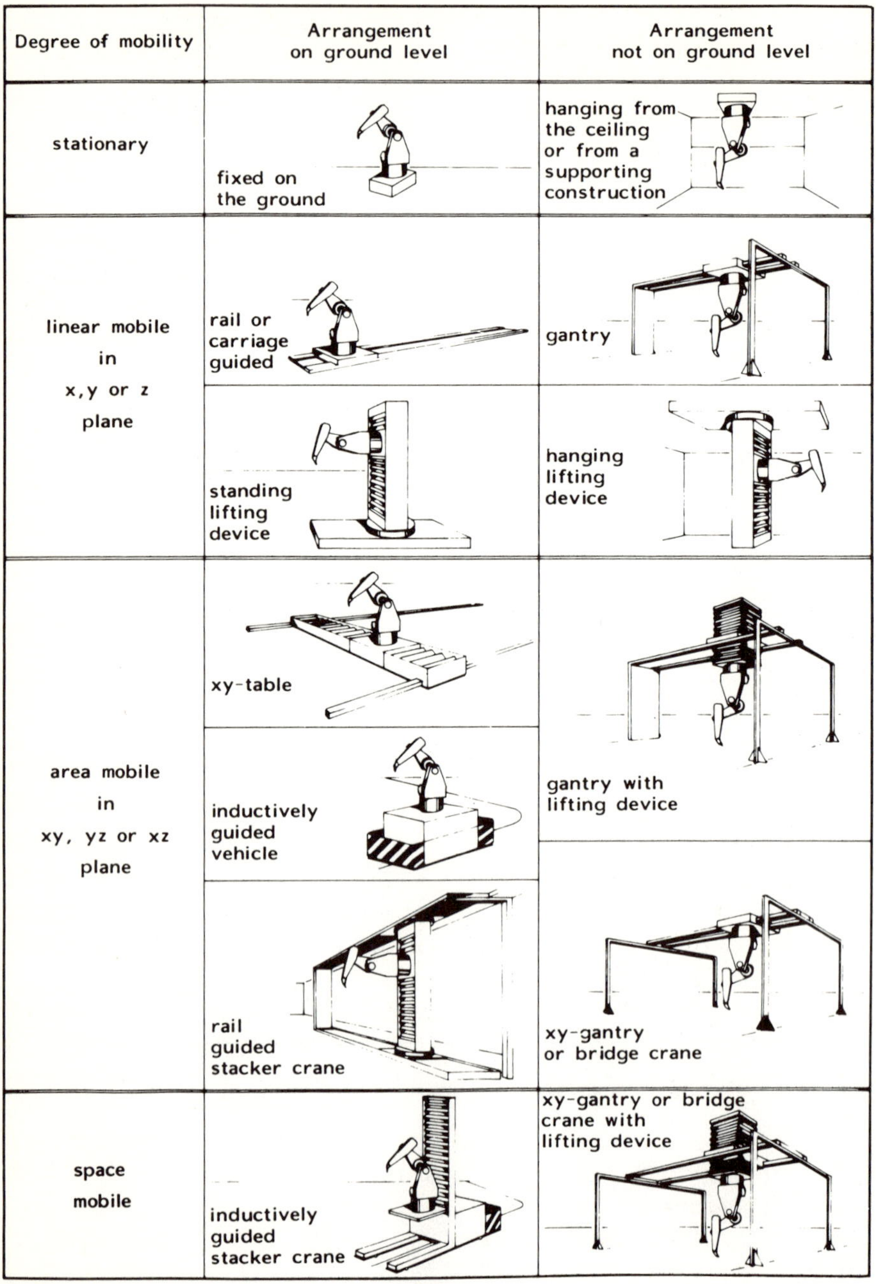

Degree of mobility	Arrangement on ground level	Arrangement not on ground level
stationary	fixed on the ground	hanging from the ceiling or from a supporting construction
linear mobile in x,y or z plane	rail or carriage guided	gantry
	standing lifting device	hanging lifting device
area mobile in xy, yz or xz plane	xy-table	
	inductively guided vehicle	gantry with lifting device
	rail guided stacker crane	xy-gantry or bridge crane
space mobile	inductively guided stacker crane	xy-gantry or bridge crane with lifting device

Fig. 3 Classification of industrial robots according to their degree of mobility

Fig. 4 Rail-guided industrial robot for commissioning tasks (ITW/Reis)

inflexibility with regard to changes in direction and disruption to other internal material flow operations. Solutions using floor-independent one-dimensional systems and gantries with suspended robots are also known[7,8].

Two-dimensional mobility on the floor is achieved by placing the industrial robot on a compound table, although here, too, the travel path is restricted to a certain degree. Battery-powered vehicles following a wire laid in the ground offer considerably greater freedom. Floor-independent solutions for total coverage with industrial robots can be implemented with the aid of a cross gantry or – for operation in the vertical plane – by linking the robot(s) to a shelf operation appliance. In both cases, however, the large percentage costs for achieving mobility with the industrial robots make utilisation prohibitive except in a few instances.

Three-dimensional mobility can be attained without the robot(s) being bound to the floor either with a modified gantry crane or with a cross gantry with crab and lifting mechanism. Floor-bound solutions offering sufficient positioning accuracy and stability (e.g. with inductively guided shelf operation systems) do not appear feasible at the moment.

Taking account of the possibilities available at present, industrial robots with two-dimensional mobility appear to be most suitable for extending industrial applications.

Requirements for industrial robots with two-dimensional mobility

The requirement profile for industrial robots with two-dimensional mobility results basically from the integration of the following part tasks:

- Transport of industrial robot and workpieces, tools, jigs, etc.
- Handling of workpieces or tools.

From the description of a mobile industrial robot for use in flexibly automated production areas it may be seen that such robots are ideally suited for versatile, autonomous operation, if possible without manual intervention. The independence of mobile robots *vis à vis* energy supply and control is desirable, and connection to a fixed supply system with the aid of automatic coupling must be ensured.

In order to keep down investment costs, modular construction using conventional components should be utilised. If it is assumed that many industrial robot applications will involve the servicing of a series of immobile stations, then the costly functions and 'intelligence' of the system should preferably be located in the mobile integrated appliance(s), of which there is usually only one or a few at most per system, rather than at every station.

The conveyor should not hinder any of the other internal material flow operations, and for this reason the travel path should be flush with the floor, with fixed elements at the points of application being either under or off the floor. Apart from the transport operation itself, the interchanging of palleted material with other conveyance and storing facilities must also be carried out automatically.

The execution of handling tasks by mobile industrial robots requires a good degree of stability both with regard to the change in centre of gravity as a result of an extended arm and to the dynamic forces arising during the handing operation. In addition, the positioning accuracy attainable at present, both in respect of fixed stations and accompanying handling and conveyed material, must also be maintained.

Design of an inductively guided mobile industrial robot

A conventional freely programmable industrial robot fixed to the transport device is used as a handling appliance. From a kinematic point of view, articulated arm appliances of compact construction with large working area are most suitable.

The conveyor is designed as a floor-bound concept which, particularly for large-area coverage, has the advantage over floor-independent principles of being less costly. Automatic floor-bound conveyors which are suitable for 'chassis' for industrial robots or material transport can have either sensor or inductive path guidance. Vehicles with sensor guidance are still currently in the testing stage but will offer great potential for large-area mobile coverage when they have been developed sufficiently[9,10].

As far as the choice, course and change of path are concerned, inductively guided floor vehicles are ideally suited for automated non-continuous transport of material, even in interlinked material flow structures. The drive, load transfer and take-up operations, controlled by computer program, make such vehicles suitable for warehouse work, commissioning and flexible production systems. When combined with the above mentioned advantages, the laying of the conduction cable where it will not disrupt other conveyor systems and the possibility of transmitting control commands to the vehicle without contact are the optimal pre-conditions for the use of an industrial robot as a 'conveyor'.

Implementation and construction design

Figs. 5 and 6 show the construction of a mobile industrial robot currently being tested at the Fraunhofer Institute for Manufacturing Engineering and Automation (IPA), Stuttgart. It consists of the following components:

- Inductively guided vehicle as 'chassis' for the permanently mounted industrial robot.
- Freely programmable handling appliance in the form of a conventional five-axis industrial robot.
- Load take-up facility for loading and carrying workpieces and tools to be handled by the industrial robot.
- Control and power supply for vehicle and industrial robot and positioning aids for independent, fully automated 'serving' of several work stations.

For the execution and constructional design of the system as a whole, standard components are not available for all parts. Apart from the required adaptation of the individual components to each other

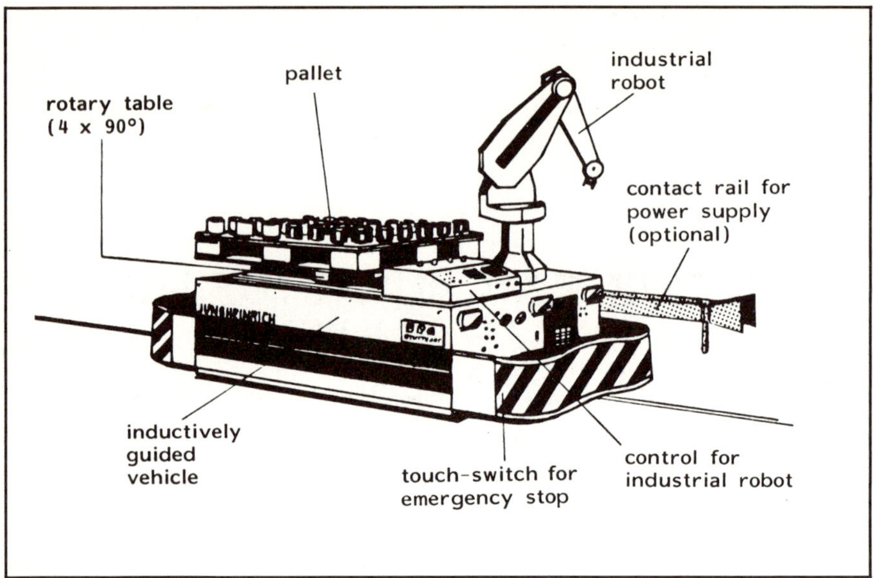

Fig. 5   Inductively guided mobile robot

Fig. 6   Inductively guided mobile robot (IPA prototype)

some special designs are also necessary to ensure full functioning, although these are kept to a minimum.

The vehicle serving as an inductively guided floor vehicle for the industrial robot and the platform has forward and reverse travel at the same speeds. The positioning accuracy of approximately ±10mm attainable with this vehicle is not adequate for the desired handling access to the industrial robot. To increase the positioning accuracy of the industrial robot with reference to a fixed point to ±0.5mm and ensure sufficient stability, the vehicle is provided with a platform to carry the robot which has four rigid feet that are lowered into the corresponding floor positioning elements at the point of application. The relative plane movement necessary for fine positioning is achieved by means of a 'floating' platform support which is locked into position when the vehicle is moving. The platform carries the industrial robot with control and material pick-up facility. In the main travel direction at the front of the vehicle is a pallet pick-up facility, behind which is the industrial robot and its controls. The robot is positioned off-centre on one of the long sides of the vehicle, preferably on the side nearest the machining or stopping point. From this position it can reach the pallet carried along with the vehicle which is used for transporting and positioning the workpieces or auxiliary devices needed for handling. The pallet can turn in 90° increments on a turntable so that the robot has access to the entire pallet area, quadrant by quadrant. The pallet is automatically loaded onto and off the vehicle by raising and lowering the platform at fixed pallet positioning stations.

The robot and vehicle control systems are separate and connected via adapted interfaces. Information and commands from the fixed central control point are inductively transmitted to the mobile system at defined transmission points. Work programs for the industrial robot can also be read in this way. The industrial robot and the floor vehicle with its auxiliary devices are each powered by a battery. The mobile industrial robot can thus work independently for a limited period without an external power supply source. With the aid of a wiper attached to the vehicle and a fixed conductor bar, the robot can be connected to the electrical power network at those stations where independent operation is not required and all batteries can be recharged there at the same time.

Application areas for mobile industrial robots

Mobile industrial robots with inductive guidance are particularly suitable for those areas with fixed handling devices where automation could not previously be carried out to an adequate extent. This applies in particular to cases where the capacity of stationary industrial robots was not sufficiently exploited or where robots could not be used at all for lack of space. The linking of the two major material flow functions of conveyance and handling by means of an automated transport and

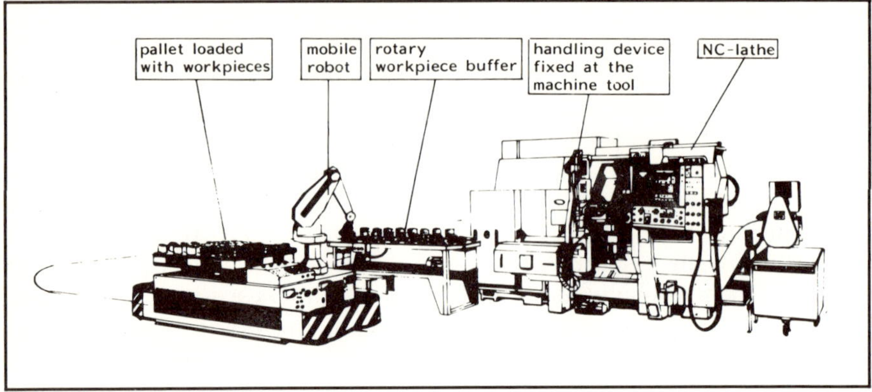

Fig. 7   Workpiece supply of a lathe with mobile robot

handling device designed to carry out both these functions means that such mobile robots could also be used effectively for workpiece and tool handling tasks particularly in flexibly automated production systems for shifts with low manning and no supervision.

Workpiece handling applications

When loading a workpiece store (Fig. 7) the vehicle approaches the production cell in question after automatically taking up a pallet laden with unfinished parts. After fine positioning of the mobile transport and handling device, the industrial robot swaps the unfinished workpiece carried with the vehicle for a finished one held in a store next to the machine in a defined sequence of handling operations. The workpiece magazine is restocked while the machine is working on the new workpiece. After the handling operation the vehicle travels with

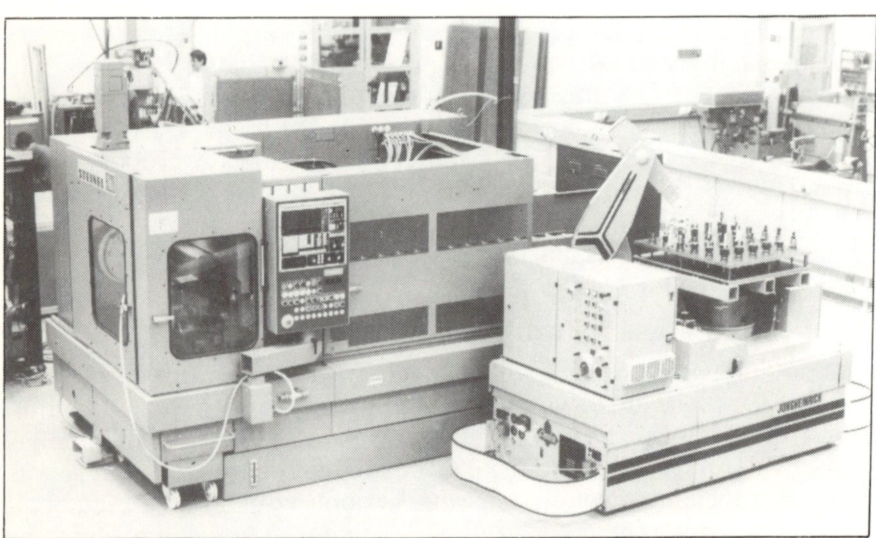

Fig. 8   Tool supply of a machining centre with mobile robot (IPA prototype)

the machined workpieces either to the next work station or back to the store, as the case may be. Especially in the case of short machining times, this workpiece supply and removal process, which is extensively independent of the machine cycle, demonstrates the powered automatic mobility advantage of integrated handling and transport devices using industrial robots.

At the Fraunhofer Institute for Manufacturing Engineering and Automation (IPA) in Stuttgart, a further example of parts handling, namely the feeding of a machine centre with machining tools, has been tested (Fig. 8). Small batch sizes call for frequent adjustments, particularly when operating with machining centres, along with a large amount of machine resetting. From a control point of view the necessary flexibility is obtained by calling up the appropriate workpiece-specific machining program. But even large machine-integrated tool magazines do not always have sufficient capacity. The tool reloading operations can be carried out by a mobile industrial robot with the vehicle carrying the tool pallet. In the in-line commissioning operation (Fig. 9) it is also possible to compile the necessary tool sets. This means that in addition to the feeding of workpieces to the machining centres by an industrial robot, an entire production system can be serviced in this way (Fig. 10).

Tool handling applications

A further application with the potential likely to be realised in the future is that of tool handling (i.e. actual guidance of the tool by a self-propelled robot). A condition for this is that the reactive forces resulting from the machining operation are not too great, a require-

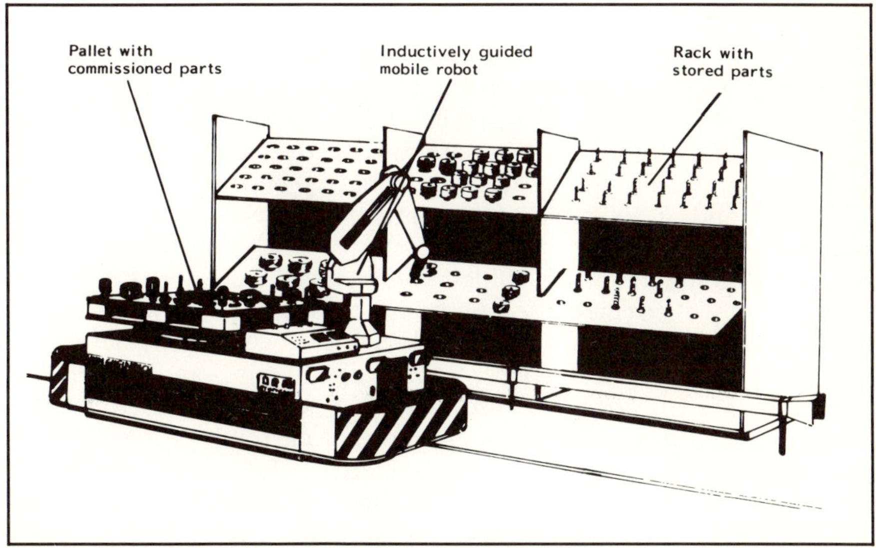

Fig. 9   Commissioning of parts with mobile robot

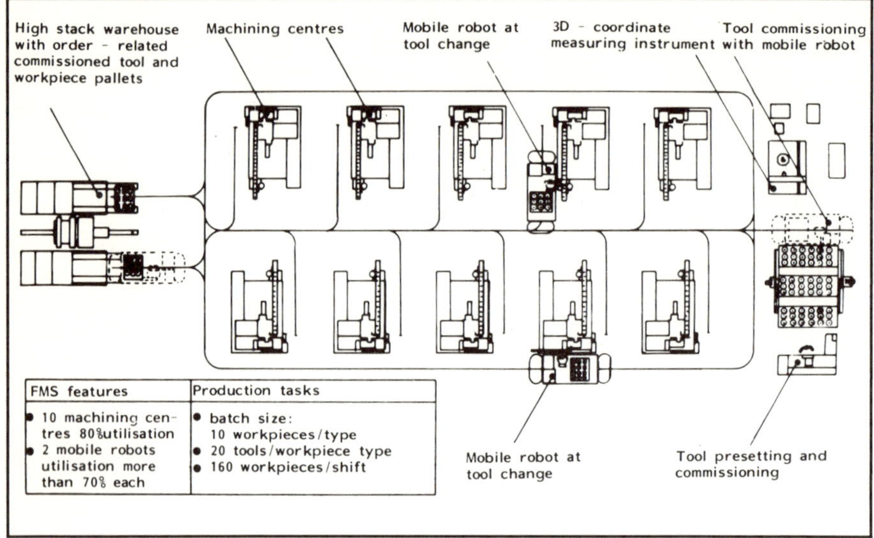

Fig. 10   Flexible manufacturing system (FMS) with workpiece and tool supply via mobile robots

ment which also applies to stationary robots. The actual tool (e.g. a welding gun) is taken up by the robot arm, whereas auxiliary items (e.g. welding transformer and inert gas flask) are carried on the mounted pallet (Fig. 11). In this way large fixed parts can be welded and coated with multi-station machining and, provided that the controls are properly arranged, it is also possible to work with several appliances operating simultaneously.

Concluding remarks

The various restrictions and marginal considerations make it necessary to employ mobile industrial robots alongside stationary ones. These mobile robots may be subdivided into those able to move in one, two or three dimensions. The integration of the main material flow functions of 'conveyance' and 'handling' is becoming increasingly significant and an example of such integration can be demonstrated with a two-dimensionally mobile robot. A newly developed system consisting of an inductively guided floor vehicle, freely programmable handling appliance (industrial robot) and transport pallet which can be automatically picked up by the vehicle has been considered here, with a discussion of its overall conception, design features and conceivable applications. The possible uses of this self-propelled robot for workpiece and tool handling operations are currently being tested in practice. Together with the anticipated further developments in handling appliances, this new concept shows considerable and decisive promise for success, given reliable operating conditions, particularly in the field of material flow in flexibly automated production systems.

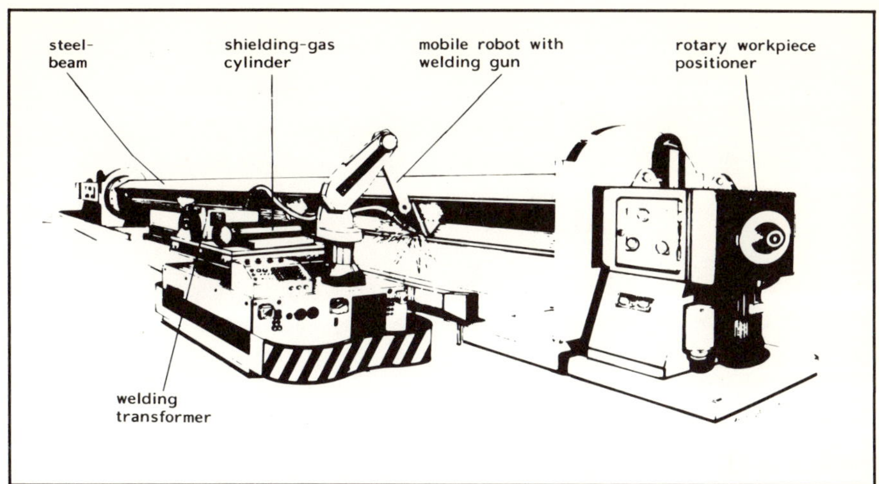

Fig. 11   Welding of large workpieces with mobile robot

Acknowledgement

Some of the work on which this paper is based was subsidised with funds from the Federal Ministry for Research and Development.

References

[1] Warnecke, H.-J., Roth, H. P. and Schuler, J. 1984. Perspektiven des Einsatzes flexibler Fertigungssysteme in Deutschland. In, *Flexible Fertigungssysteme 17*, IPA Working Conference 1984, pp. 9–25.

[2] Der Roboter muß nicht auf der Stelle treten. *Roboter*, 1:22–23, 1983.

[3] Lichtbogen-Schweißroboter. *Technische Rundschau*, 40:33–34, 1981.

[4] Norlin, B. 1982. Roboter erweitern Maschinennutzungszeit. *VDI-Nachrichten*, 26(16):10.

[5] Kwang-chuan, T., Hsing-dao, Ch. and Hsiao-tsu, Ch. 1981. Loading a lathe with the aid of a robot. *The Industrial Robot*, 8(2):98–99.

[6] Engelberger, J. F. 1982. The use of industrial robots for loading machine tools. In, *Int. Metalworking Congr.*, Leipzig.

[7] Anon. 1980. BMW buys KUKA portal robots. *The Industrial Robot*, 7(4):259.

[8] Weisel, K. and Katoh, H. 1975. Beachheads for robotics. In, *5th Int. Symp. on Industrial Robots*, pp. 1–10.

[9] Marce, L. et al. 1980. A semi-autonomous remote controlled mobile robot. *The Industrial Robot*, 7(4):232–235.

[10] Imai, H. et al. 1984. Advanced automated transportation systems. In, *Proc. 3rd Int. Conf. on Flexible Manufacturing Systems*, pp. 175–184.

FLEXIBLE ASSEMBLY THROUGH AGVs IN THE AUTOMOTIVE INDUSTRY

S. G. Sanden
AB Volvo, Sweden

This paper highlights some reasons for using AGVs in assembly lines.
What kind of advantages or disadvantages are involved and how they
can be justified are discussed. In some cases, it is more expensive
than a conventional line system. However, people still decide in
favour of an AGV system. Some installed systems are analysed and
the reasons why they were installed discussed.

Today, AGVs are used in almost all possible assembly configurations,
such as line systems, parallel systems and taxi systems. It is not the
intention of this paper to discuss the advantages and disadvantages of
these different systems, as the factors involved warrant separate
treatment. What can be concluded however is the fact that Volvo's
AGV system can be used in all three systems, and when a suitable
layout is chosen the configurations can be altered when the installation
is completed.

AGVs are now being used in several assembly applications, such as:

- Subassemblies
  - instrument panels
  - seat assembly
  - bumper assembly
  - engine assembly
  - gearbox assembly and marriage with engine

- Bodyshop
  - welding, automatic/manual operation
  - door/hood/bonnet fitting
  - inspection area
  - rework area

- Paint shop
  - inspection area
  - rework area

- Final assembly
  - trim lines
  - automatic robot station, e.g. window glueing
  - extra assembly stations for deluxe versions, etc.
  - inspection area
  - rework area
  - test area

Why use an AGV system?

All these applications have some very important common factors which justify the extra expense incurred with the purchase of an AGV system. One of the common factors is that it is easier to produce (due to simpler balancing) the models that the market demands at the time. Another benefit is the elimination of excess stock of unwanted models.

Another common factor is that with model changes there is a requirement to change production flow. With an AGV system, this can be done quickly and at much less cost to implement the new production flow. The system only requires the following adjustments:

- The addition of more vehicles.
- The addition of extra loops.
- Modification of computer programs instead of hardware changes.

Finally there is the third and perhaps the most important common factor; namely, the fact that the level of complexity is lowered and the availability of the production system increased with an AGV system compared to a conventional system. Today, this is a very important advantage when completely new models are introduced in an existing plant, as the number of models and versions is rapidly increasing and giving enormous problems to the production staff in the conventionally built automotive assembly plants around the world.

Advantages

This section looks at some examples from the different applications within the assembly process with the specific advantages for these systems.

Instrument panel assembly

- Logical modules depending on work intensity.
- Buffer between modules.
- Change of sequence possible.
- Minimising loss of time at work balancing.
- Possible to implement and train people with a step-by-step approach.

Paint shop – wet grinding

- Possible to vary cycle time without loss of production.
- Correct working height.
- Open/free floor-space.

Stripe/sunroof – assembly

- Stationary workpiece for fitting of stripes.
- Two systems incorporated into one.
- Possible to vary cycle time at taxi stations.

Final assembly

Since Henry Ford introduced the production line philosophy, no major changes were introduced within the car industry until the early 1970s. In 1974, the new Volvo Kalmar factory was opened and met with many sceptical comments, such as: "Another luxurious Swedish experiment." Today, approximately 12 years later, the results are as follows:

- 25% reduction in assembly completion time.
- 39% reduction in product quality problems.
- 57% reduction in inventory turnover time and improved management of work-in-progress (WIP).
- 5% reduction of employee turnover.
- 4% reduction of absenteeism, below the Swedish national average.

What this type of factory does is place the people in the centre and let them decide the speed of the line – not the other way around. This also emphasises the teamwork (with production cells), job enrichment, and job rotation, which all give higher quality.

Volvo car factory, Ghent – A step-by-step implementation

The Volvo Ghent production facility is the latest installation using the new generation of AGVs and their computer control. To date, 134 AGVs have been installed.

When the factory was built in the early 1960s it was based on conventional techniques. In 1976 it was modernised using AGVs (54) in final assembly. Several subsystems have been added which again confirms Volvo's belief that an AGV system has a great advantage compared to a conventional production line system. The factory now has a production capacity of $2 \times 35,000$ cars a year. The following types of AGVs are used:

- Transport carrier, welding shop with roller table.
- Tilt carrier with 90° turning fixture.
- High carrier with a lifting bridge for the marriage operation.
- Engine dress carrier with a fixture on lift table.

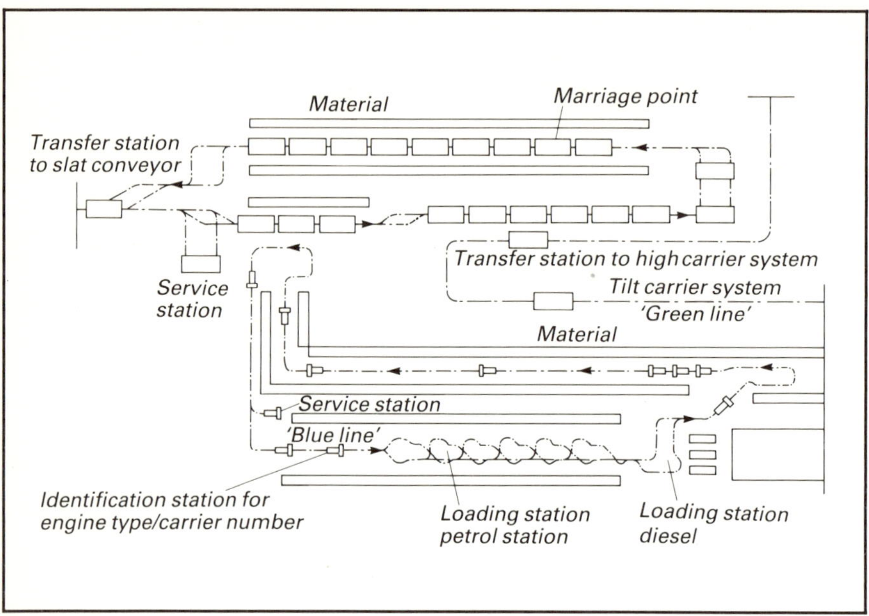

Fig. 1   *Layout for low carrier system – (carrier with a roller table for skids)*

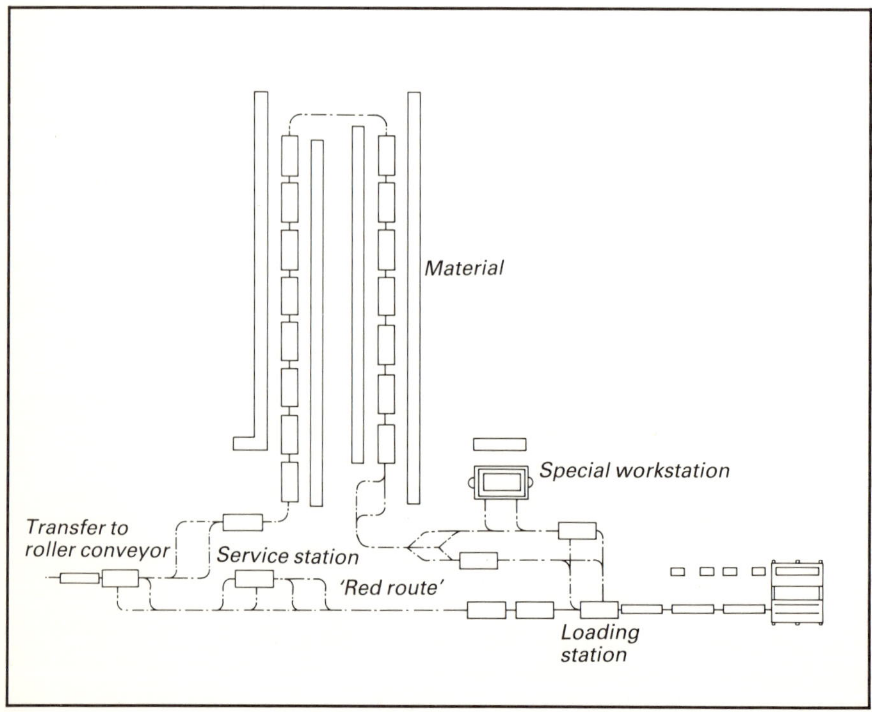

Fig. 2   *Layout of high-carrier system and motor/gearbox assembly system*

The parts of the AGV system to be discussed are shown in Figs. 1 and 2. The painted body arrives from a body buffer area on skids and is loaded on the RoRo carrier. It then follows the 'red route' where such things as cable tree and rubber list are attached. At this line there is also the opportunity to prepare the body for ABS brakes. The body is then transferred to a conventional line system, where the car body is more or less completed indoors and outdoors. Thereafter the body is transferred to the tilt carrier system ('green line') where the 54 AGVs are situated. Here the body is tilted to a 90° angle to provide the operators with a correct working position when working underneath the car body.

The tilting mechanism in each carrier gives the system a far higher redundancy towards a conventional line with fixed tilting positions. At this moment an engine is loaded onto the engine carrier and dressed to suit the type of car, which is at the same time produced in the tilt carrier system. (This 'blue line' is the loop of the engine carrier and it is lifted over to a lifting bridge on the high carrier.)

In the high carrier system the marriage of the engine and the body is carried out. Here again there is higher redundancy towards a conventional marriage. Also other operations such as wheel mounting, etc. are carried out. Finally, the car is lifted automatically onto a slat conveyor for inspection and test.

Included in the control system is a colour graphic monitoring system to see if a carrier has exceeded the allowed cycle time, whether it is loaded or unloaded, and if it has 'time out' between two workstations. Each system is controlled by its own computer to increase the availability of the system. The control of type of body/car is achieved through a separate barcode system.

Concluding remarks

Over the last few years major resources, capital and personnel have been allocated to improve the efficiency of assembly and manufacturing processes in the automotive industry. The general conclusion is to have more flexible assembly and manufacturing systems, and here the AGV system has been the key to many companies' solutions and success.

ACS – Autocarrier System AB, a member of the Volvo Group, manufactures and markets AGV systems under the product name of Autocarriers. The company has many years of extensive experience in supplying a complete turnkey installation, including all software developments and electronics to carry out installation and service of the system.

4.2

Warehousing

AUTOMATED WAREHOUSING USING AGVs AND NARROW-AISLE TRUCKS

I.A. Noble
Jungheinrich (GB) Ltd, UK

Some of the design considerations in planning a warehouse and the factors which influence the choice between various types of equipment are discussed. An automated warehouse system using AGVs and ANATs is described.

There are many factors and constraints which influence the design of an overall warehouse layout. The three most important factors are the nature of the operation, the building available and the budget. It would be ideal, of course, to have the time and budget to start on a green field site. Very often, however, the building is an existing one, and this immediately imposes constraints on the design.

The Concise Oxford Dictionary defines a warehouse as 'a building in which goods are *stored*'. This presupposes that goods need to be stored for a period of time. In order to store goods as economically as possible, then, the greater the volume of goods stored in a given building volume, the better. Storage density depends on a number of factors and the need for access will have a major influence on the storage system. These factors have generated the recent developments in high-bay, narrow-aisle technology. Should narrow-aisle storage offer the optimum solution, the methods of moving goods within the warehouse have to be decided. Some of the factors affecting that choice are discussed later in this article.

Bulk storage is often associated with order picking, which therefore has to be incorporated in the warehouse design. The nature and location of the picking operation are dependent on the particular application. The ratio of live picking store to bulk store and the rate of replenishment necessary will usually determine whether the picking area is located within the bulk stores or in a separate area.

The economics of the degree of automation to be used are very complex. They depend upon the nature of the 'ware', its method of containment (i.e. palletised, tote bins or otherwise), product carried on a pallet (i.e. single or mixed), volume of throughput, shift patterns, etc. The budget also has to be taken into account.

Choice between ANATs and fully automatic stacker cranes

When planning the design of an automatic warehouse, one has a choice of handling methods. Within the racking in particular, that choice is between stacking cranes and automatic narrow-aisle stacking trucks (ANATs). Stacking cranes can operate up to 30m high and can work extremely fast – up to 3m/s – within an aisle. Although cranes have been designed to work in more than one aisle, the transfer from aisle to aisle by conventional bogie is cumbersome and slow. However, a new generation of cranes which can negotiate curved rails and switch aisles more rapidly is now available.

Automatic narrow-aisle trucks (Fig. 1) on the other hand, are easily capable of transferring from aisle to aisle and, indeed, the pick-up/deposit station may be some way from the storage aisles. Trucks are practical when operating up to 12m high. Fully automatic narow-aisle trucks are a comparatively new development. There are at least three installations on the Continent and one is currently being installed in the UK, which is described later in the article.

The ideal warehouse for a crane is, therefore, one which is long, narrow and high with a high throughput volume. The higher the throughput, the more cranes and aisles are required, and the warehouse becomes wider due to the additional aisles.

The narrow-aisle truck can range over many aisles and a lower, more spread out building is preferable. Therefore the type of building and throughput requirements influence the decision as to which system to use. Tall buildings are not as expensive as low ones and cranes are slightly more expensive than narrow-aisle trucks. If the building is already in existence, the choice is made. If a new building is to be erected, the amount of land available may dictate a high-bay building, although very high buildings can cause planning consent difficulties. If money is to be loaned for the project, the financiers will often dictate that a new building will be of a resaleable/leasable shape, so that their risks are partially covered. Very narrow tall buildings have far fewer alternative uses and financiers often dictate a more usable spread out building.

Inter-store transport

Having determined the shape of the warehouse building and the method of storage and retrieval, the transport mechanism to and from the bulk store, if necessary, must be considered. In many applications roller conveyors will be the most suitable method. However, if the distances are great and the throughput not so great, then an automated guided vehicle (AGV) system might be a more economical and flexible

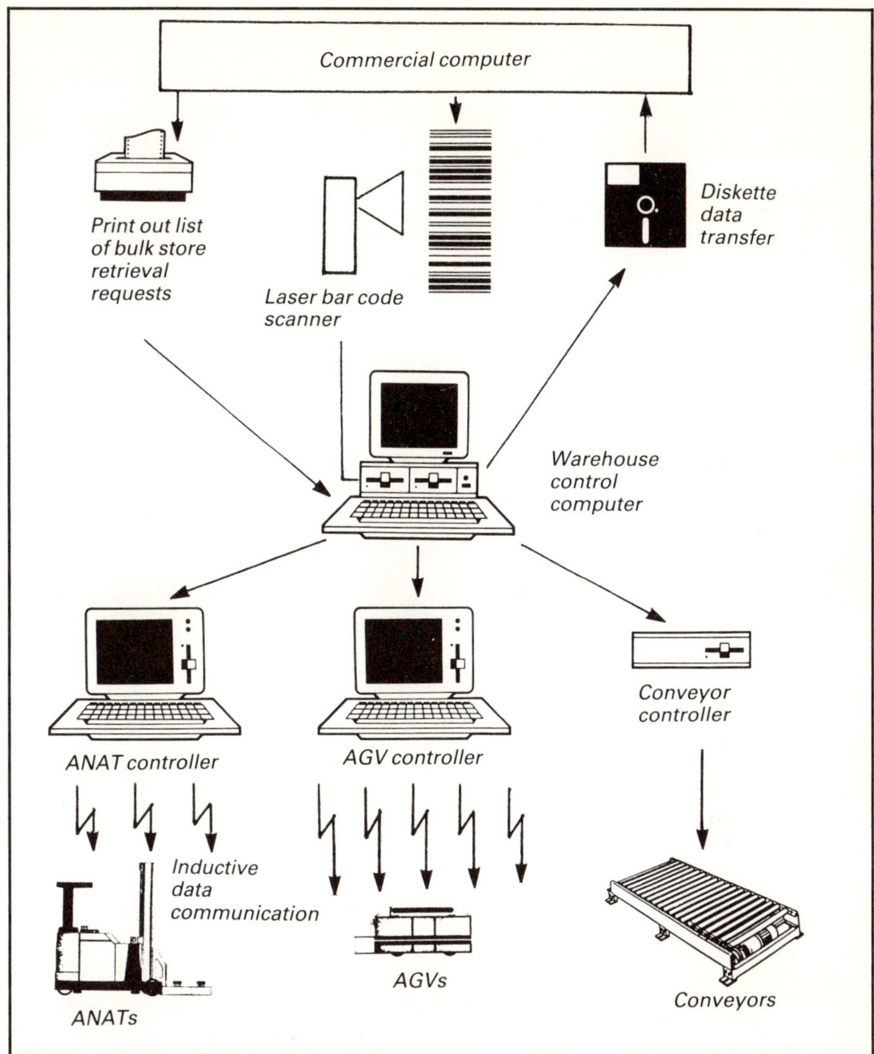

Fig. 1 Warehouse control system hierarchy

alternative. AGVs can operate down aisles and on routes used by other trucks and even pedestrians. They usually offer a space saving and are relatively easy to fit into an existing layout.

Load identification

A fully automatic handling system requires a load identification system, of which there are many in existence. It is often possible for the load to be identified at the entry to the system by the input booking operator. The load can then be monitored through the system and its location and identity memorised electronically at any one time. Such a system, however, does have a risk involved, because a malfunction either by accident or by human interference can lead to many loads being lost to the control computer. For this reason, individual load identification is desirable. Two systems in common use are 'identity by bar code label' and 'identity by a magnetic coded device'. The bar code is cheap and each label may be treated as disposable, whereas the magnetic code label, although more robust, has to be treated as reusable by reason of its cost.

Bar code labels can be read by an operator using a bar code wand or a laser scanning gun. Laser bar code label scanners can be set up to read a bar code label as it is passed in front of the scanner. Such scanners are capable of reading labels with a very high degree of positional tolerance. The labels can be read at any distance up to 1m away from the scanner, with the label lying flat on the pallet or on the vertical side of a load. It can be positioned at up to 30° from the horizontal, depending on the length of the code. Provided the label is placed so that the bars are horizontal and the scanner scans vertically, then the label can be scanned several hundred times as the pallet is passed in front of it.

There are many different bar code formats with different characteristics. It is not necessary to restrict the size of the code; in fact the larger the actual code, the easier it is to read. Some codes are numeric only, while others are alpha-numeric.

Description of an ANAT/AGV warehouse

The particular warehousing operation now described was designed with the following conditions and constraints in mind:

The company is international with worldwide manufacturing plants. All goods entering the warehouse are palletised on a standard 800 × 1200 Europallet and stretch-wrapped. The purpose of the warehouse is to store in bulk several months' supply of finished product. The product is then issued as required to the order picking areas for making up into individual orders for distribution to retail outlets in the South of England. The product requires three separate storage conditions, namely controlled temperature and humidity, ambient and chemical storage. The operation works on an extended single day shift and during this time approximately 300 pallet loads are put into bulk storage and 300 are retrieved. A further 50 pallets go directly from goods-in to the order picking areas. The building used is an existing building with a headroom restriction of 6m. The area required for order picking and the height restriction dictated that the picking area should be separate from the bulk stores. The volume of bulk storage necessary is 7000 pallet locations.

Choice of equipment

Transport between the goods input conveyor, the various bulk stores, order picking areas, finishing and marshalling is effected by five AGVs with roller conveyors transporting one pallet at a time. Approximately 1,200 pallet movements between stores take place each shift.

Storage and retrieval within the bulk stores is effected by three fully automated narrow-aisle trucks equipped with telescopic forks. Each truck normally operates within a distinct territory, but these territories can be merged so that all storage areas can be covered by one truck in the event of breakdowns or servicing of other trucks.

The trucks are guided by inductive wires laid in the floor. Each aisle is identified by a recognition device at the start of the aisles. Location down the aisles is effected by a combination of reflective rack markers and wheel encoders on the trucks. The height selection is by means of an encoder which operates on the mast. There is an automatic door which isolates the cool and ambient stores, and this door is activated by the trucks. The whole of the area in which the ANATs work is designated as a personnel-free zone and the trucks are disabled automatically whenever any access door is open. The trucks are provided with a rider position so that they can be manually operated for maintenance, driving to the battery charging area, or out-of-shift emergency operation.

Transfer between AGV, ANAT and order picking areas is by roller conveyors which are controlled by a central programmable logic controller.

Control systems

Fig. 2 shows the warehouse control system hierarchy. Product details on each pallet, together with order requirements, are stored within a separate commercial computer system. As each load enters the system, its identity is keyed into the commercial system which then prints a bar code identity label. The commercial system generates on a daily basis a print-out of all the withdrawal orders from the bulk stores. A daily summary of store occupation and pallet movements is transferred by floppy disk from the process computer to the commercial computer.

Operations within the warehouse are controlled by a DEC Micro PDP 11/73 computer. This computer maintains a database in which the occupation of all the bulk store locations is maintained. Details of pallets entering the warehouse are given to the process control computer from a bar code label laser scanner at the input station. This scanner reads the information as each pallet passes into the system. Thus the input data is transferred from the commercial system to the warehouse control system. The bar code label contains a unique pallet identity and a destination code. On receipt of this information, the warehouse control computer gives instructions to the AGV controller and the ANAT controller for the transport of pallets within the system.

Retrieval requests from the bulk stores are manually entered via a

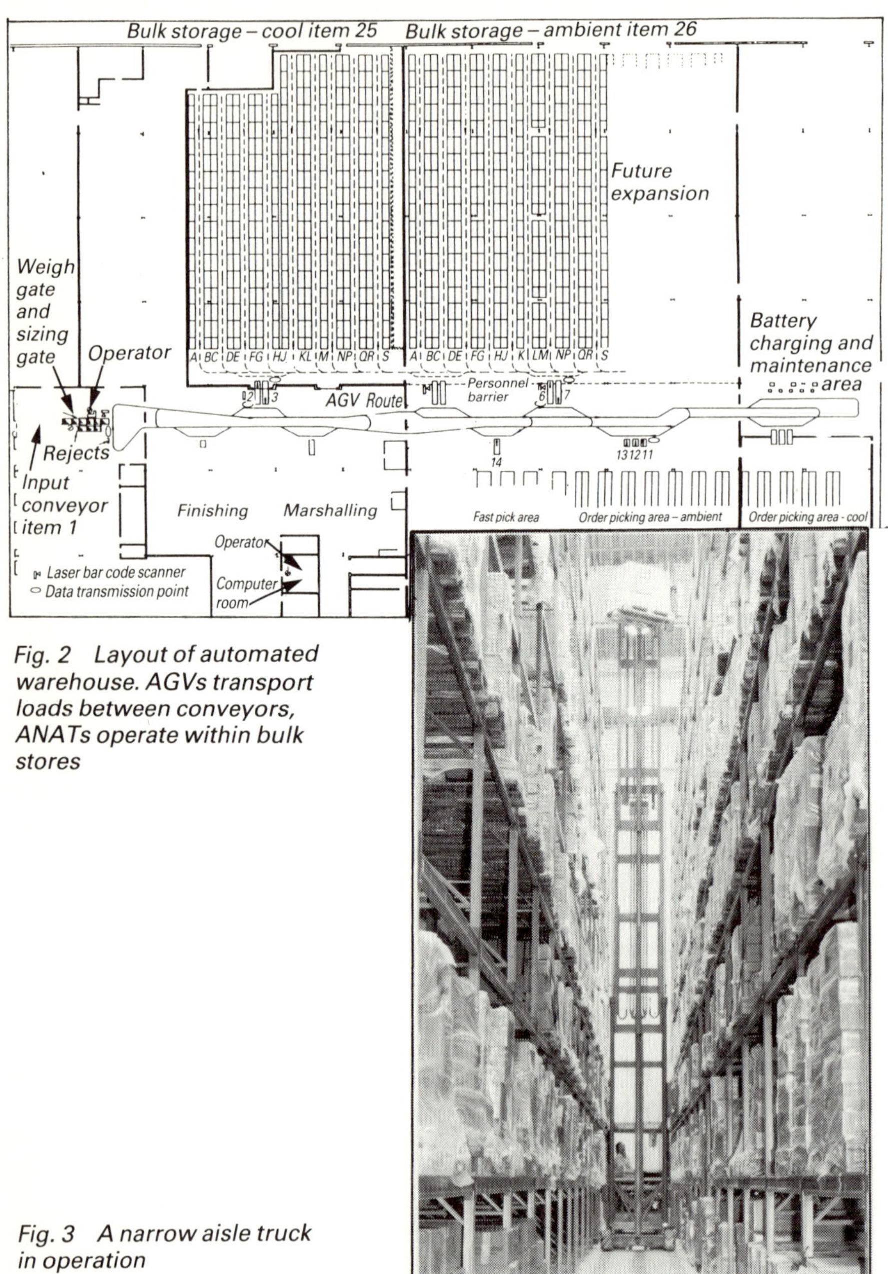

Fig. 2   Layout of automated warehouse. AGVs transport loads between conveyors, ANATs operate within bulk stores

Fig. 3   A narrow aisle truck in operation

keyboard into the warehouse control computer. These requests are passed down to the vehicle control computers for execution.

Interrogation facilities

It is possible to identify the location of any individual pallet within the bulk stores or the occupation of any location. The percentage occupation of each store is constantly logged. The number of movements into and out of the stores is also logged. It is possible to redesignate areas of the stores at any time. Non-chemical items can be stored in the chemical zone and ambient items in the temperature-controlled areas, although not vice versa.

Method of operation

Fig. 3 shows the layout of the automated warehouse. Lorry loads of pallets of goods identified by man-readable labels arrive at the goods receiving dock. These pallets are unloaded by conventional fork-lift truck and loaded onto a powered roller conveyor. Loads pass along this conveyor and their size and weight are checked at a sizing gate. They then stop at an identity point. If the pallet is acceptable, an operator keys the pallet identity into the console linked to the commercial system computer. If a load is not acceptable, it is rejected down the reject conveyor for rebuilding or repalletising. The commercial system prints out a bar code label which includes a unique identity and a destination. This label is attached to the side of the load and serves as a means of data transfer to the warehouse control computer. The information is read as the pallet passes a bar code label scanner which is connected directly to the warehouse control computer.

Destinations from the goods in conveyor are:

- Bulk store area – cool.
- Bulk store area – ambient.
- Order picking area – cool.
- Order picking area – ambient.
- Fast order pick.
- Finishing.
- Marshalling.

The AGV system receives a request from the central control system to service the input station. The AGV controller signals a vehicle via an inductive loop communications system at the queue point. The truck travels to the input station, receives the load and takes it to the requested destination.

The AGV travels around the route following the guide wire. Any changes in direction are decided upon by comparing the information received from the floor installation with a map programmed into the onboard controller.

On arrival at the required position, the AGV is in communication with the controller. Should there be another load requiring transport, the AGV deposits its load, and instead of returning to the start point, transports that load to its required destination. This is a double cycle and wherever possible the system operates in this manner.

The ANAT input conveyor indexes the load to the ANAT pick-up point. At this point, a bar code scanner collects the identity of the load and the control system informs the ANAT controller of:

- The rack in which to store the pallet.
- The position along the rack.
- The position within that bay (there are three pallets per bay).
- The height (level) at which to store it.

Wherever possible, the control system instructs the ANAT to retrieve a load in the same aisle to optimise the ANAT operation. The ANAT receives its instructions via a data communications loop at the pick-up/deposit station.

If for any reason the scanner fails to read the label or the control system does not accept the identity, the ANAT picks up the load and places it directly onto the output conveyor for the AGV to transport it to the reject point.

The control system keeps a file on the instructions issued and informs the AGV when the next pallet is output from the ANAT system.

Other than the initial input to the system, loads may be moved within the system between the following positions:

- Bulk store to order picking.
- Bulk store to marshalling.
- Bulk store to fast picking
- Order picking to marshalling.
- Order picking to finishing.

Individual loads can be moved within the bulk store zones by manual input to the ANAT controller or between zones via the reject point and the input conveyor. At the queue point, an AGV can be issued with a transport command that bypasses the input section.

AGV emergency stop

If the AGV bumper is activated, the truck stops within the bumper length. The truck remains stationary for 7 seconds and sounds an audible alarm. After 7 seconds the truck restarts for the original destination. If the obstruction is still in the way, the truck tries to restart twice and then remains stationary until an operator comes to remove the obstruction.

In the event of an AGV being stopped by emergency stop or by the key-switch, that AGV will stop and await a manual restart. It retains its destination memory, but the operator may decide to have the load rechecked, in which case he/she reprograms the truck to travel to a position prior to the input conveyor (reject point), where the load is removed, checked for security and put onto the input conveyor by manually operated fork-lift truck. The control system issues another location for that bar code and overwrites the current stock record.

Priority retrieval system

It is possible to call up pallet numbers on a priority basis. Such requests go to the front of the queue of retrieval requests. The ANATs retrieve these requests, performing double cycles – or single cycles should there be no incoming pallets. It is, therefore, possible to speed up the retrieval cycle by closing the input station and forcing single cycles.

Out-of-shift manually controlled retrieval

Should a small number of pallets be required in an emergency out of normal shift times, it is possible to interrogate the process control computer to establish the storage locations of the required pallets. These pallets can then be retrieved by manually driving an ANAT truck from the battery chargers into the bulk store.

AUTOMATED MATERIAL HANDLING IN A DISTRIBUTION CENTRE

U. Wältken
VVA – Vereinigte Verlagsauslieferung GmbH, West Germany

System design and configuration as well as practical experience of a highly automated distribution centre for books, records, toys and cassettes is presented. Major elements are a large, fully automated, high-bay warehouse for 60,000 pallets; a distribution system for pallets comprising 37 driverless trucks; a unique automatic buffer-store and depalletising area; three picking bays for 6,000 articles, equipped with gravity roller shelves and sensors for automatic replenishment; and a complex overhead monorail system as a picking aid.

Vereinigte Verlagsauslieferung GmbH (VVA) in Gütersloh is part of the West German Bertelsmann Group. It serves as a distribution centre for the programmes of the Bertelsmann book clubs and of 180 publishing houses. Its functions include reception of goods, storage and dispatch of consignments to bookshops and club members, as well as processing of orders, invoicing, accounting and customer service.

In 1961 the first European high-bay warehouse was built by VVA with a storage capacity of 4,500 pallets and manually-operated high-rack stackers. As business and the need for storage capacity grew, the company rented and acquired more conventional warehouses in its area, finally achieving a complex of seven different distribution centres – each with its own internal organisation.

In 1976 VVA decided to build a completely new warehouse in order to integrate all the separate organisations. It was supposed to handle both the club and the publishers' programme, which meant more than 45,000 articles ranging from books, records and cassettes to toys, video tapes and leisure articles. Based on the experience with the existing organisations, VVA defined the following main objectives:

- Creation of capacity for long-term expansion as well as possibilities for growth into new market segments, especially leisure and/or audio-visual items.

- Improvement of productivity by automation where possible and economically feasible; higher utilisation of storage capacity; higher exploitation of central functions, and reduction of internal handling and transport costs.
- Better service for the company's clients through improved working quality, reduction of order-processing time and the offer of new or additional services.

Consequently a VVA team of engineers and experts was formed and studies as to general layout and location were undertaken. As a result, VVA opted for:

- A location in Gütersloh close to the suppliers (i.e. book printers and record manufacturers) and head office. (This meant that VVA could retain its staff, regarded as its main asset.)
- A high-bay warehouse as a central store for all products.
- A common goods-reception and expedition area.
- Separate picking zones for club and publishers' consignments, because of their different structures.
- An integrated transport system to supply the picking areas with pallets from the central store.

Target figures for the system were 60,000 consignments and 200 tons of freight per day with approx. 500,000 articles. At the same time it was planned to keep the full workforce of 1000 employees.

Fig. 1   Overall view of the distribution centre

The building

The building (Fig. 1), with a ground area of 55,000m^2 and a total area of 65,000m^2, was designed according to the requirements of the operating facilities. In this way the building matched the internal operational structure perfectly. Furthermore, it can be extended at three sides to give room for future expansion. It was built in two phases: first the high-bay warehouse, goods-reception, expedition and club picking area, and then the publishers' picking area and the offices.

The general concept is a one-storey building, 4m high, divided into three fire sections. This results in a spacious working area which can be adapted easily to any changes in working process when required later. Out of this low building two higher parts protrude: the high-bay warehouse, with a height of 35m, for automatic pallet storage, and a three-storey building for the picking of the club orders (13m high, two floors at a height of 5 and 9m).

The back part of the flat building is used for the publishers' order picking. In the middle of that zone is placed an office section for the fulfilment and accounts departments, so that they are in direct contact with both the articles and the dispatch organisation they deal with.

Due to a relatively low picking rate, the publishers' order picking is carried out conventionally and will not be described in detail. This article focuses on the mainly automated materials handling in the club order section.

Material flow and administration

The material flow was designed in principle to be U-shaped. Thus the loading court, which stretches all along the front of the building, serves as a connection between internal and external transport (Fig. 2). Incoming goods are unloaded, checked, counted and prepared (if necessary) for automatic handling. As the entrance to the central store is located directly behind the goods-in area, all pallets can be stored with a minimum of transport. Finished consignments are sorted by order and destination and set out for truck loading on the other side.

Both order picking sections are replenished from either end of the central warehouse. Therefore each side is fully equipped with an ID-point and a pallet storage and retrieval system. The front end serves the club section and the rear end the publishers' section. The main part of the internal transport runs along the length of the high-bay warehouse (HBW).

In order to separate personnel movement from the material transport a separate track was created on the right-hand side of the building. Adjacent to the track are the social amenities to the right and the entrance to the respective working areas on the left.

Some of the material transport facilities are linked together and process computer controlled. These are the HBW, the pallet transport system and the integrated club replenishment system. Other areas and

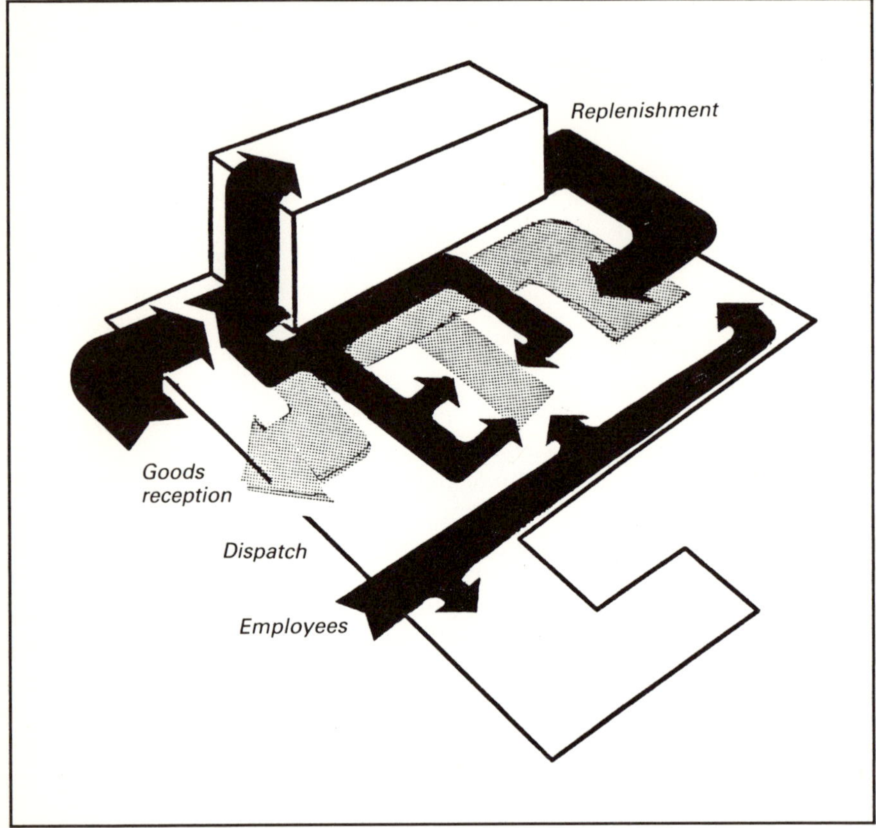

Fig. 2   Material flow scheme

systems are controlled manually and not connected to the process computer.

In summary, goods reception, storage, pallet transport and dispatch have been centralised and serve both parts of the business, whereas replenishment and picking areas are tailored to the respective requirements. Three independent process computers are in charge of controlling these systems. The main units, a Siemens double R30, with a main memory of 512kbyte, controls on-line the following functions:

- Data input at four ID-points.
- Selection of pallet locations in the central store.
- Administration of all storage bins in the HBW (60,000) and the buffer store (6750).
- Communication of orders to the automatic guided vehicle (AGV) system (37 vehicles).
- Allocation of orders to all high-rack stackers in the warehouse (14) and in the buffer store (8).
- Data protection and automatic restart.

A process computer, type Siemens R10, controls the trolleys of an overhead monorail system for the order picking for club members and a double PDP 11/34 administrates and controls the AGV system.

High-bay warehouse

The HBW as the central store proved to be the most economical method for pallet storage. Cost comparisons between a conventional warehouse and the HBW over a period of ten years indicate clearly the cost advantage of the automatic system (Fig. 3).

The warehouse consists of 14 aisles, with three fire sections, and has a capacity of 60,330 pallets in the standard format of 800 x 1200mm. It measures 130m long, 65m wide and 35m high. One process computer-controlled crane per aisle attends to the pallet transport.

One of the basic rules in the new store is: 'One article per pallet'. As this rule leads to a variation in pallet height, two different bin sizes were created in the new store, one measuring 1.75m which covers 70% of the total capacity, and the other 0.80m for the rest. In order to achieve correct bin selection, the height of pallets is checked optically before storage.

The handling of pallets is fully computer controlled and the appropriate locations are selected according to such considerations as:

- System stability.
- FIFO (first in, first out) principle.
- Temperature zones.
- Fire sections.
- Club or publishers' programme.

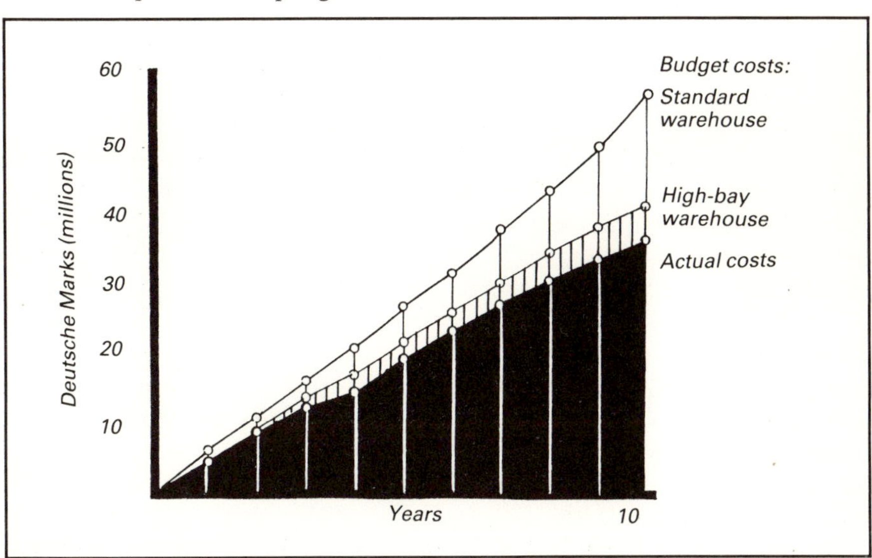

Fig. 3   Operating costs comparison for standard warehouse and high-bay warehouse

Fig. 4   *High-bay warehouse: storage and retrieval system*

However, turnover rates are not considered, as this would mean a continuous relocation all the time.

The central warehouse is equipped with pallet conveyors for storage and retrieval at both ends, so the cranes carry out two jobs at either end. At the front side (Fig. 4) pallet lifts raise the entering pallets to the storage level of 3m. On the floor level each aisle is equipped with a chain conveyor and turntable for pallet retrieval. When the pallet has reached its final position on the turntable, it can be picked up automatically by carriers of the AGV system. It can then be transported to any place in the building without human interference.

At the rear end of the warehouse there is no direct link to the AGV system. Instead a circular pallet conveying system with 16 special unloading points allows any quantity of books to be taken off, while the rest remain on the pallet. These are then transported back into the warehouse.

AGV system

The pallet transport from a central store to various places in a spacious low building can be very expensive. When carried out with conventional systems many pallet trucks and drivers are necessary. Furthermore such systems are not flexible, often involve passageways being blocked, and require a lot of capital.

An AGV system offered VVA a solution. It is flexible and more economical as far as labour and capital are concerned. The results of a cost comparison over a period of ten years are shown in Fig. 5.

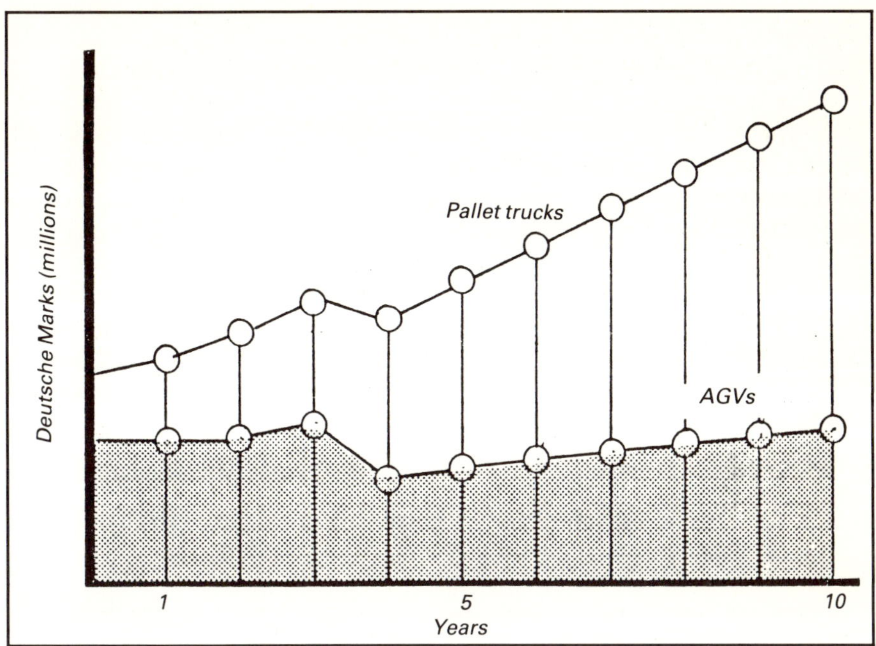

Fig. 5   Cost comparison between AGV systems and normally operated pallet trucks

Fig. 6   Automated guided vehicles

For the transport of the heavy standard pallets VVA chose the Swedish autocarrier system offered by ACS together with a double PDP 11/34 as a control unit. The system has been designed for 45 carriers; at present 37 are used. The track is 2,100m long and leads from the front end of the warehouse in several loops to 61 loading/unloading points in the various picking areas. At any of these loading/unloading points, pallets can be ordered from or sent back to the central store by means of a VDU connected to the process computer. The total system capacity is 360 pallet transports per hour.

The autocarriers (Fig. 6) are battery powered and can operate during two shifts. When automatically recharged at night in a recharging station, the battery remains on board the vehicle. All carriers can also be operated manually and loaded or unloaded by means of a standard fork-lift truck.

Club distribution system

Within the club distribution system three types of orders are distinguished:

- Those for direct mailing to club members – mostly small parcels with up to three different articles.
- Replenishment for the bookclub shops (at present, 280 shops in major towns), which receive consignments twice a week – normally up to several hundred kilograms.
- Those which concern fast-movers, new books or bulky prepacked articles – normally medium-size consignments selected from a limited range of articles.

Each type of order has its own picking system in the three-storey building. On the ground floor fast-movers, etc., are picked directly from pallets. This is still done conventionally, as the picking rate for these articles is extremely high.

On the second floor the bookshop consignments are picked. As these orders comprise many articles, and in order to save packaging material, foldable plastic containers are used for shipment. Total capacity here is 3000 containers of 30kg per shift. Picking is done by means of a trolley and a picking list organised according to rising shelf number. Empty containers are presented on a circular conveyor throughout the central part of the picking area, with a separate conveyor for finished consignments underneath. In that way all employees can walk along the shelves continuously in the same direction and push off finished containers anywhere, take an empty one at the same place, and continue with the next order.

On the third floor orders are picked which are relatively small – at a rate of up to 25,000 parcels per shift. Conventional picking would lead to low productivity due to long waits between each pick. Thus, in order to achieve a higher rate of productivity, VVA's central EDP

Fig. 7   Overhead monorail system

system selects orders which are as similar to each other as possible, into groups of 12. For each group of orders, invoices and picking lists are then printed accordingly. In the middle of the picking area there is an overhead monorail system with 140 self-powered trolleys (Fig. 7). At the starting point each trolley is loaded by the system with 12 pre-erected cartons and the attached orders. Each then drives through the picking area and stops wherever a pick has to be made. The employee in charge of that picking zone fills the appropriate cartons and signals to the process computer when the job is finished. The trolley is then automatically transported to the unloading area. Finally, the cartons are closed and sealed and conveyed directly into a van for dispatch.

Depending on the order structure and the total volume, the average number of stops per trolley is between two and three per trolley run.

Replenishment

Floors two and three are built in the form of a long narrow platform with picking shelves on both sides. All articles of the programme are identically presented on both floors, with books on one side and records on the other. The shelves consist of a series of 4m deep steel frames equipped with a small gravity roller conveyor for each bin. This arrangement offers the following advantages:

- Maximum density of presentation with sufficient capacity per bin; here 24 articles on a length of 1.2m.

- Excellent picking facilities, due to the fact that all articles are always located on the front side and bins legibly marked. Furthermore the front of the shelves is constructed in an ergonomic curve to make picking easier.
- Refilling can be carried out automatically by high rack stackers with special equipment.

In the replenishment process a number of books stacked up to 10 and grouped in a long row are fed onto the conveyor of the appropriate bin. Every one of these units carries near the end a reflector, and every bin is equipped with a photocell detector supervised by the process computer. When a bin becomes empty, contact is made and the process computer initiates a replenishment action.

In order to ensure quick refilling, there is a buffer store for replenishment units on each side of the picking platforms. When a picking shelf is empty an appropriate unit is transferred automatically from the buffer store (Fig. 8). The contents of the buffer store are in turn supervised by the process computer. When the preplanned level of stock attains a preset minimum, a pallet of that article is called for out of the central store and an autocarrier ordered to bring it to a special unloading point (Fig. 9). There an operator is instructed via a VDU to build a number of replenishment units, i.e. to take off a certain quantity of articles from the pallet and arrange them in the prescribed way. When the operator acknowledges the job, a high rack stacker with a sort of telescopic device picks up the unit and transfers it to the buffer store. Automatically the pallet with the remainder is then brought back to the warehouse by autocarrier and the articles are returned to storage.

Experience with automated material handling

During the start-up phase, VVA experienced a lot of problems with the system. Although the company had prepared quite well for the day when it would transfer its business from the old premises to the new installation, the scale of the problems encountered came as a surprise. The fundamental problem was that, once the stock had been transferred into the new warehouse, the company were totally dependent on that 'machine', including the process computer, to bring the required number of pallets out again. When all systems failed to operate as planned in the beginning, a workforce of up to 250 people was employed to substitute temporarily for the automatic operation.

The time needed for problem diagnosis and correction had been greatly underestimated. It took 1000h of operating time to analyse all the major problems and another 4000h to solve them. Major problem areas were:

Fig. 8 Buffer store, transfer stacker and picking shelves

Fig. 9 Unloading point for replenishment

- *High-bay warehouse*: Although the performance of the individual stackers measured according to a DIN specification was achieved at an early stage, total system performance including that of the storage and retrieval system and process computer control was substantially below the design value.

- *Process computer system*: During the start-up phase the operating system collapsed when loading heavily. Further to that, the average response time at the ID-points and the loading/unloading points was greatly exceeded. Siemens, the company responsible for the computer package, had to extend main memory and implement a new operating system to achieve design performance.

- *AGV system*: Although the carriers were built very solidly, the on-board electronics caused difficulties due to bumping on floor cracks and also due to temperature problems. Furthermore, it proved necessary to reduce the carrier speed at all communication points in order to ensure correct message transfer from the central computer. That meant that two extra carriers had to be put into operation to compensate for the lack of speed.

- *Buffer store and replenishment system*: As many components of the buffer store and the high rack stackers were not standard, but designed especially for this job, it took a lot of detail work and tuning to get the system operational. Positioning and wear in particular caused breakdowns, and although design speed has now been achieved, VVA is still not satisfied with the amount of breakdown time.

Concluding remarks

Having implemented the system and overcome the start-up problems, VVA can say that both it and its partners (i.e. suppliers, contractors, etc.) have learnt a lot during this project. One thing VVA would recommend very strongly to other companies considering automation is to be careful, when choosing partners, that they are able and willing to get the system working. They should be aware that the implementation of automated material handling on a big scale needs a lot of preparation and precaution, especially, regarding the start-up phase.

VVA has also learnt the value of a good contract. In addition to the standard clauses, the company improved on:

- Description of all components. It cannot be too detailed or precise.
- Agreement about performances. It is not enough to specify standard test procedures: it is vital to agree on a test program under normal operating conditions, with a specified operator crew.
- Agreement about efficiency. It is best to establish a contract wherein the system has to perform with a guaranteed efficiency when maintained in a specified way.

Finally, it should be emphasised that, although material handling in the club section is highly automated, one element, the actual picking, is still carried out manually. VVA is studying the possibilities of automating this process using industrial robots, but the calculations to date show clearly that the human worker as a complex handling instrument is still sometimes cheaper and more flexible in this area. This may, of course, change in the future.

LOAD TRANSFER TECHNIQUES FOR AGV SYSTEMS

T. Müller
Fachhochschule Hamburg, West Germany

The possibilities of improved links between high-level storage shelving (HSS) systems and a remote-controlled transport system (RTS) are discussed. Decision criteria are handling rates, the load transfer technology, the type of vehicle introduced and, last but not least, the control technology, which is fully independent of the previous criteria.

However future-orientated a company may be, it can only achieve automated materials flow if it succeeds in integrating preproduction processes that are capable of automation in the total plan. A vital part of this integration is, for example, the direct linking of high-level storage technology (and automatic stores) with a remote-controlled transport system (RTS).

Conventional high-level storage shelving (HSS) systems, faced with a higher handling rate, usually make use of a continuous conveying system in the area in front of the store, as shown in the example in Fig. 1. Where necessary the incoming and outgoing stock points must be separated physically in a second plane, if of course the levels of materials flow cannot be contained within one plane. Connection with the manufacturing area is generally achieved by the use of floor conveying vehicles, in many cases by RTS (e.g. at points A and E in Fig. 1).

The question which arises inevitably from this situation is: to what extent can the continuous conveyor be dispensed with in the area in front of the store (e.g. see Fig. 2) by the use of the RTS vehicle to bring up the individual vehicle transfer position either directly or via a slave buffer?

Dispensing fully with a continuous conveyor in front of the store achieves four significant advantages:

- A considerable cost saving.
- Space can usually be saved.

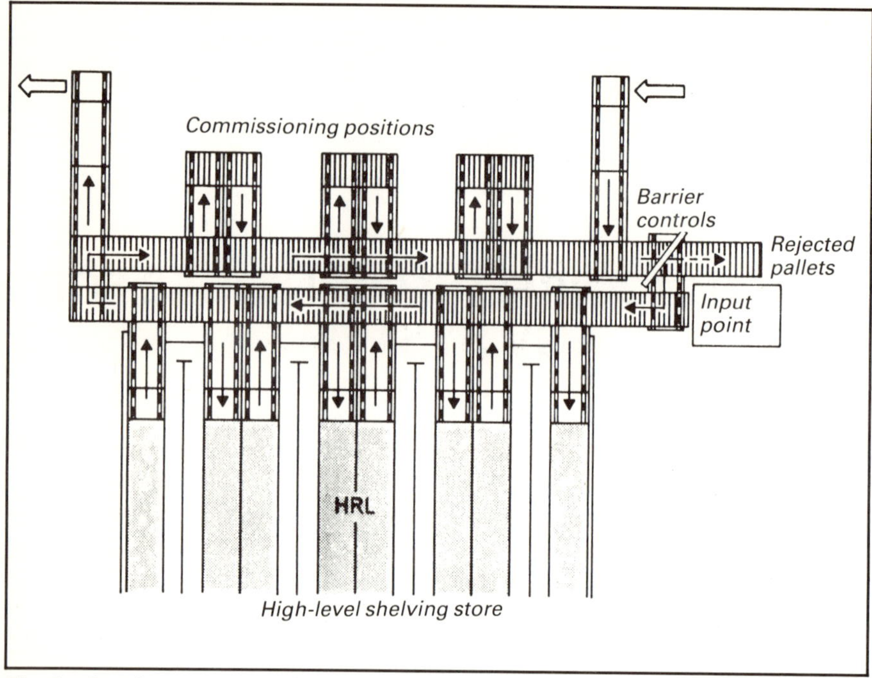

Fig. 1   Continuous conveying system with roller conveyor and chain conveyors

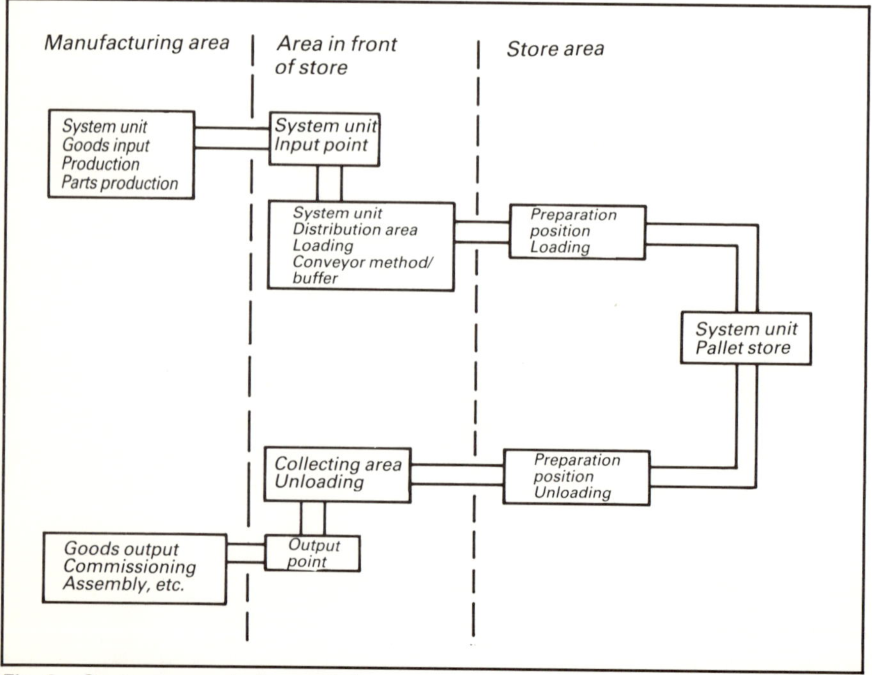

Fig. 2   System areas in front of the store

- Higher availability when there are no shutdowns or reduced operating levels on the associated conveying apparatus.
- Better access to the high-level storage shelving for maintenance and servicing.

Possibilities of linking HSS with RTS

The load transfer between HSS and RTS can be subdivided into two basic alternative layouts: parallel tracks and intersecting tracks (Fig. 3).

The concept of an intersecting track requires a vehicle which can reverse at a speed that matches as closely as possible its forward speed. Usually differentially driven vehicles are used. If, as in Fig. 3, the intersecting track has inward and outward branches, then a loaded pallet entering in the main direction of flow will turn through 180° for unloading. This can be significant for automatic record reading from the pallet.

The principal method of off-loading for intersecting tracks is the lifting platen (in some cases, lifting beam), which is compatible with all loading devices. Where gravity roller conveyors and chain conveyors are used, account must be taken of the appropriate transport position for the loading devices. German Standard pallets are conveyed most

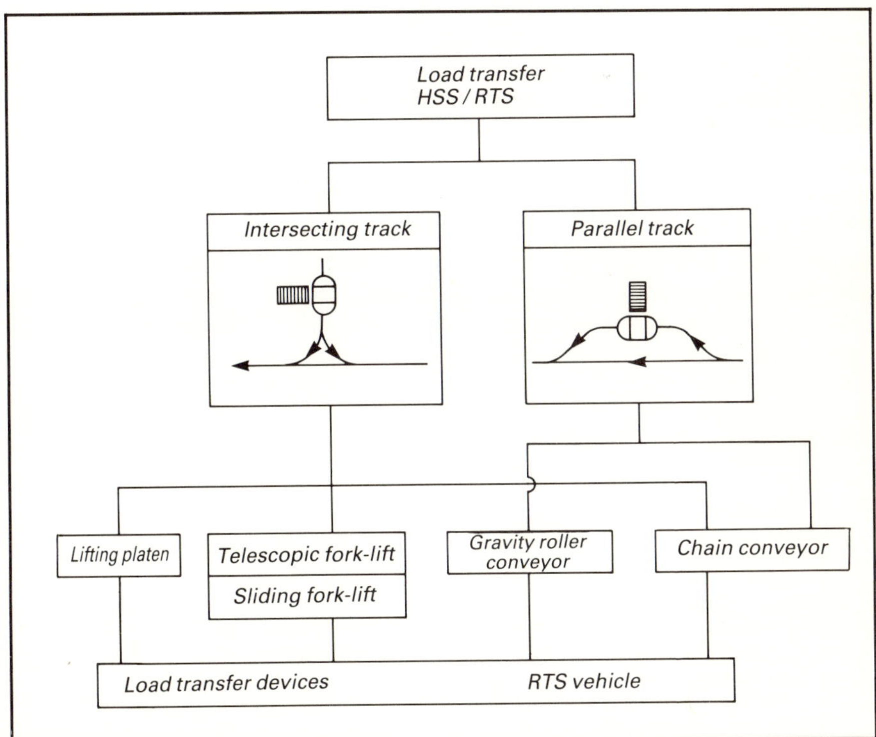

Fig. 3   Principal layout alternatives

appropriately in a longitudinal position on gravity roller conveyors and in a transverse one on chain conveyors. Mesh box pallets can be conveyed on chain conveyors in both longitudinal and transverse positions. Pallets with small (circular) feet can in most cases be transported only with lifting devices. These limitations on the part of the loading devices are particularly important for the automation of material flow, since the use of reduced pallets is only economically justifiable in exceptional cases.

Load transfer in a parallel track layout requires only forward-moving vehicles. Usually gravity roller conveyors and chain conveyors are used by the side of the vehicle and station. In the case of chain conveyors, the load (e.g. German Standard pallet) stands longitudinally on the vehicle. The vehicle width is therefore determined by the pallet width (e.g. 800mm). This arrangement is more preferable for the construction of the route and its width than the use of gravity roller conveyors on the vehicle, which would involve, in the case of German Standard pallets, a vehicle width of 1200mm. This greater width would mean an increase in the transfer time of one pallet movement of about two seconds. In addition to these criteria, the interfaces for load transfer at the HSS and at the manufacturing end are also important. The parallel track alternative allows high handling rates. In addition to the transport layout, the loading device and the control system are determining factors in the cycle time (i.e. the handling time per load transfer station). Table 1 details these factors.

Each load transfer station is linked to one or two wheeled vehicles for the loading and unloading processes. The cycle time is thus the limiting factor of an individual interface. The total handling rate for a store zone is calculated from the number of load transfer stations, the layout and the stochastic material flow movements. Accordingly, efficiency losses at the load transfer stations are unavoidable, but these can be held inside certain limits by the use of an appropriate layout and an efficient control system. Concrete data relevant to the specific case can be seen through a detailed simulation. In general, however, traffic bottlenecks can be avoided by careful analysis.

Table 1   Factors influencing the cycle time for load transfer

---

Transport layout

Means of loading
- *chain conveyor*
- *roller conveyor*
- *lifting platen*
- *telescopic fork-lift/sliding fork-lift*

Control system
- *lock-and-block system*
- *double-function strategies*
- *switching times and communication times*
- *efficiency of controlling computer*
- *positioning time for vehicle*

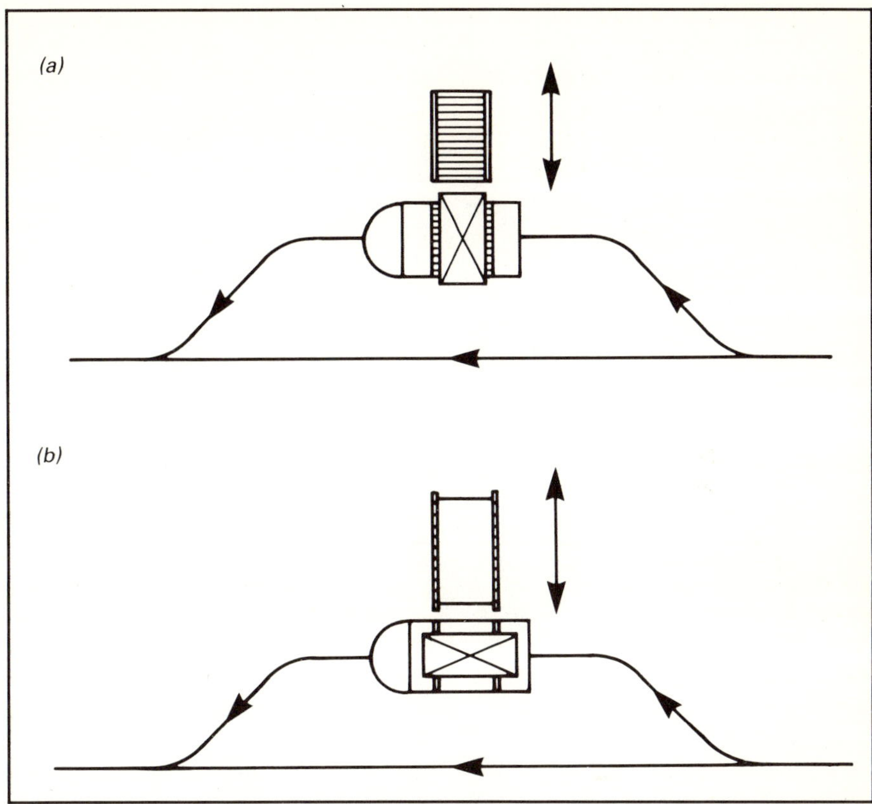

Fig. 4   Load transfer alternative for parallel track: (a) roller conveyor and (b) chain conveyor

Fig. 4 shows parallel track alternatives. Here cycle times in the region of 30 seconds for the load transfer of a pallet can be achieved. This kind of installation would be used mainly for unloading levels (manufacturing supply) with high handling rates. An example for Fig. 4 would be the distribution centre for a publisher. Fig. 5 shows the alternative using a chain conveyor for production supply in a car factory. To maximise utilisation of the conveying capacity on the transport circuit, the spacing safety device between vehicles is designed to be visually flexible. The stopping precision of the vehicles is maximally ±15mm. To bridge the spacing between the chain conveyors of the vehicle and the station a side shift is built into the side of the vehicle.

Figs. 6 and 7 show the simplest alternative for load transfer in the intersecting track layout. The capital required for this type of load transfer would be about DM 5000. This alternative for load transfer can be used equally for loading and unloading. To carry out both, the vehicle must run to an additional intersecting track. The limiting

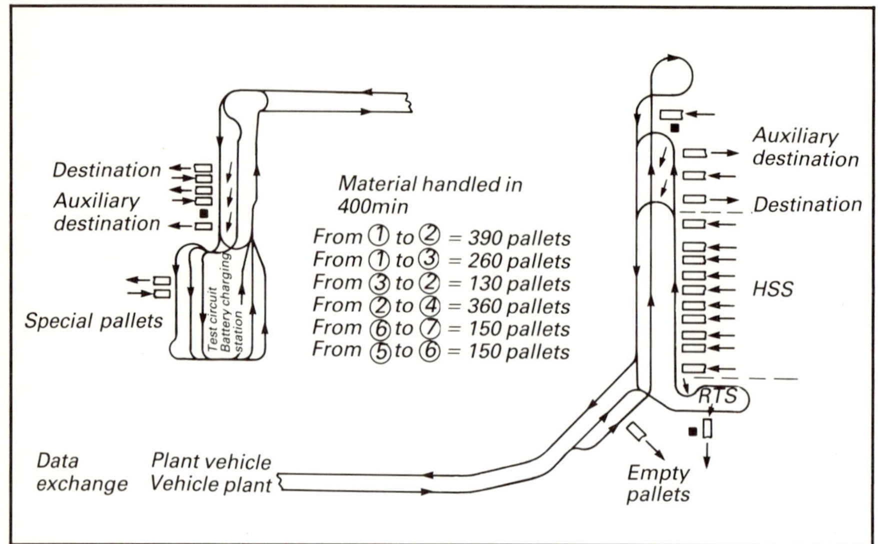

Fig. 5   RTS-HSS interface for production supply

output is maximally 15 double functions (30 loading activities) per vehicle per hour. To ensure that the wheeled vehicle and the RTS do not lock against one another, a certain safety spacing should be maintained. As a result of this, the vehicle's fork-lift must extend approximately 2×50mm (left and right) further than in the store area. Otherwise, all corridors for the vehicle must be laid out about 100mm wider. If the vehicle has a gravity roller conveyor and lifting platen fitted (see Figs. 8 and 9), then instead of the above-mentioned width

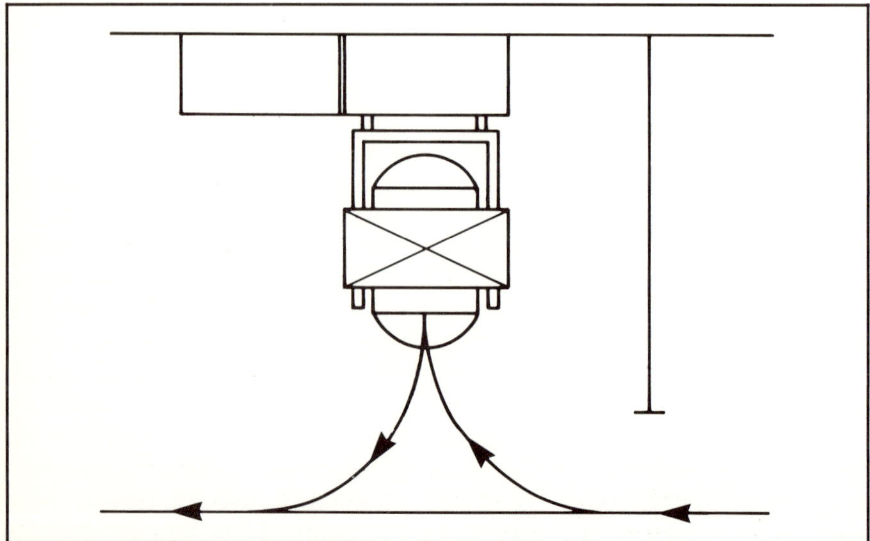

Fig. 6   Intersecting track lead transfer: vehicle with lifting platen

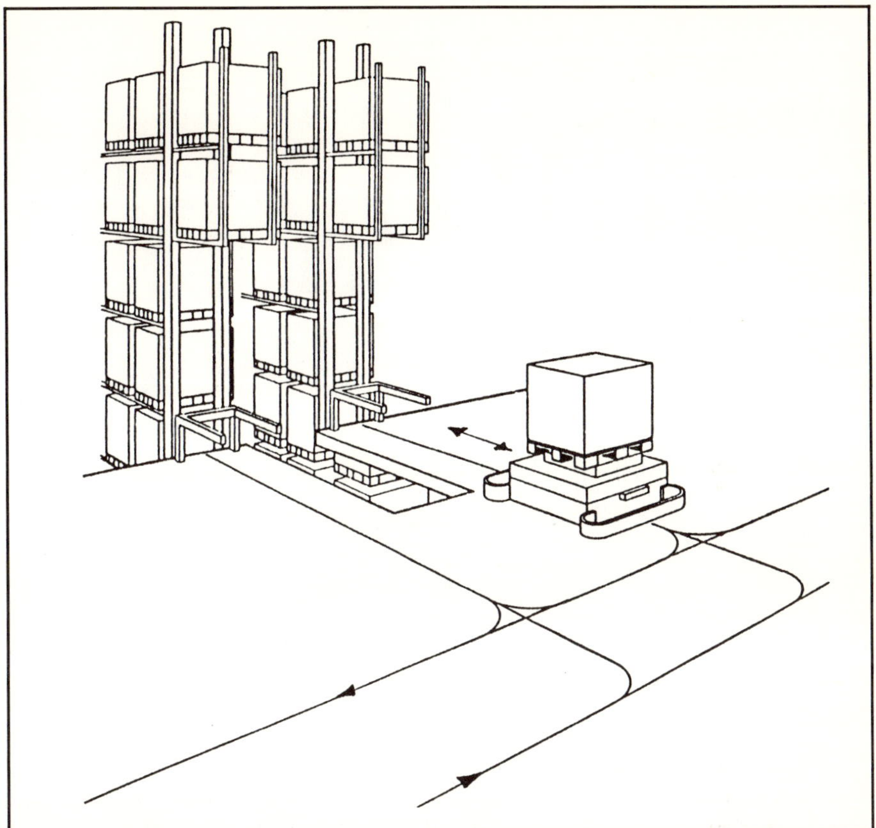

Fig. 7 Example of load transfer system of Fig. 6

measures, directly before the lowering of the load the gravity roller conveyor can be activated. A fourth possibility is to fit the fork support of the load transfer station with a side shift; this variant is shown in Fig. 10.

A shorter cycle time is possible if, as in Figs. 8 and 9, the vehicle has a roller conveyor and lifting platen fitted, because then a double function is possible within the intersecting track. Organisationally this generates priority driving orders, which allow a loading and unloading function between two neighbouring vehicles. Between loading and unloading only a short distance is to be covered, as a result of which the limiting output increases to a maximal 22 double functions per vehicle per hour. The capital required for this load transfer is approximately DM 30,000.

Fig.10 shows a load transfer with a vehicle with a roller conveyor and lifting platen and a stationary platform at the HSS. The intersecting tracks have only one entry branch (in contrast to the system in Figs. 8 and 9). Because of this, entry to the intersecting

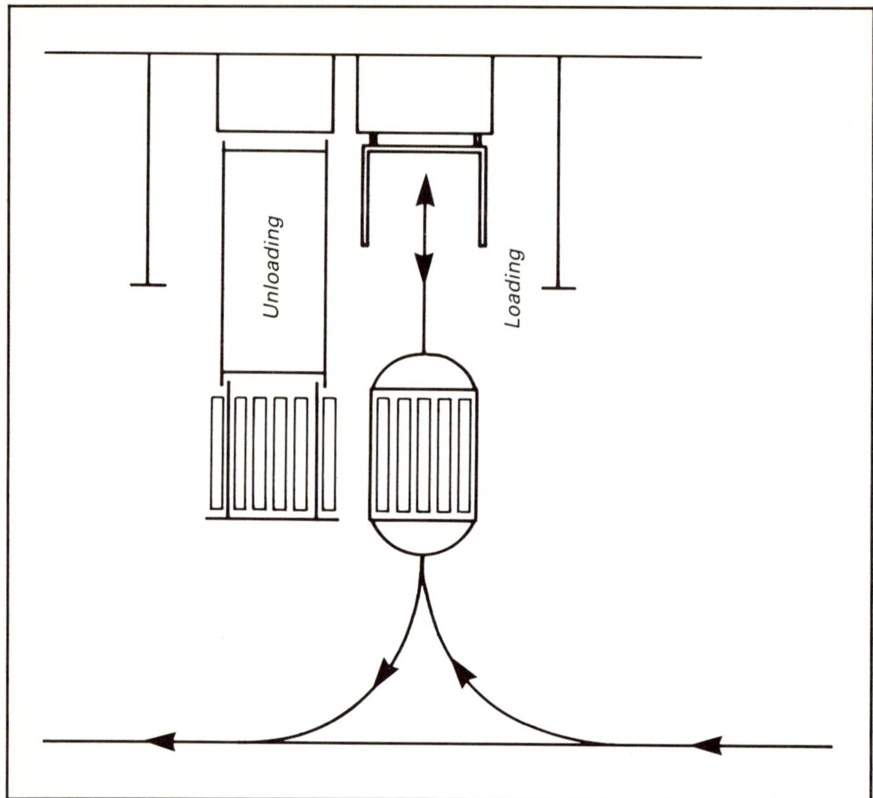

Fig. 8   Intersecting track load transfer: vehicle with lifting platen and roller conveyor; station with roller and chain conveyors

tracks is 'backwards' in relation to the main direction of flow. This has the advantage that pallets are not turned for unloading, which can be important for the subsequent materials flow when using automatic record reading from the pallets and mesh boxes with opening flaps. There is also the advantage that floor information is always read from the same side of the vehicle. A disadvantage is the somewhat longer cycle time necessary because of the directional change at entry to the intersecting track. The roller conveyors on the vehicles are used in conjunction with stationary roller conveyors in the manufacturing area.

Removing the need for movement between loading and unloading completely, as in Figs. 11 and 12, allows a maximal limiting output of 26 double functions per vehicle per hour. The capital needed for a complete load transfer station such as this is in the region of DM 50,000.

In the link between HSS and product output, load delivery at floor level is often required at the product output end. The load can be removed from the roller conveyor by using RTS stacker trucks with

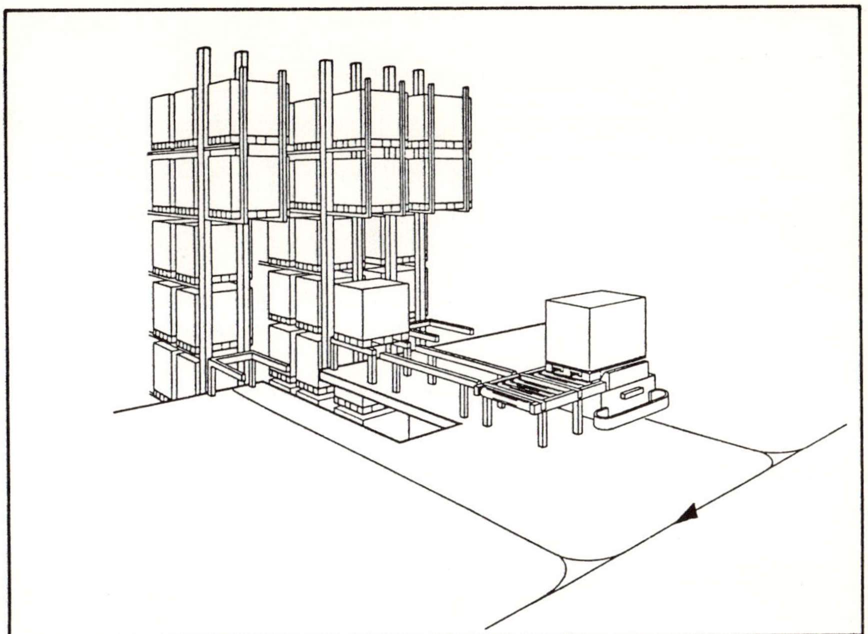

Fig. 9   Example of the system of Fig. 8

Fig. 10   Another example of the system of Fig. 8

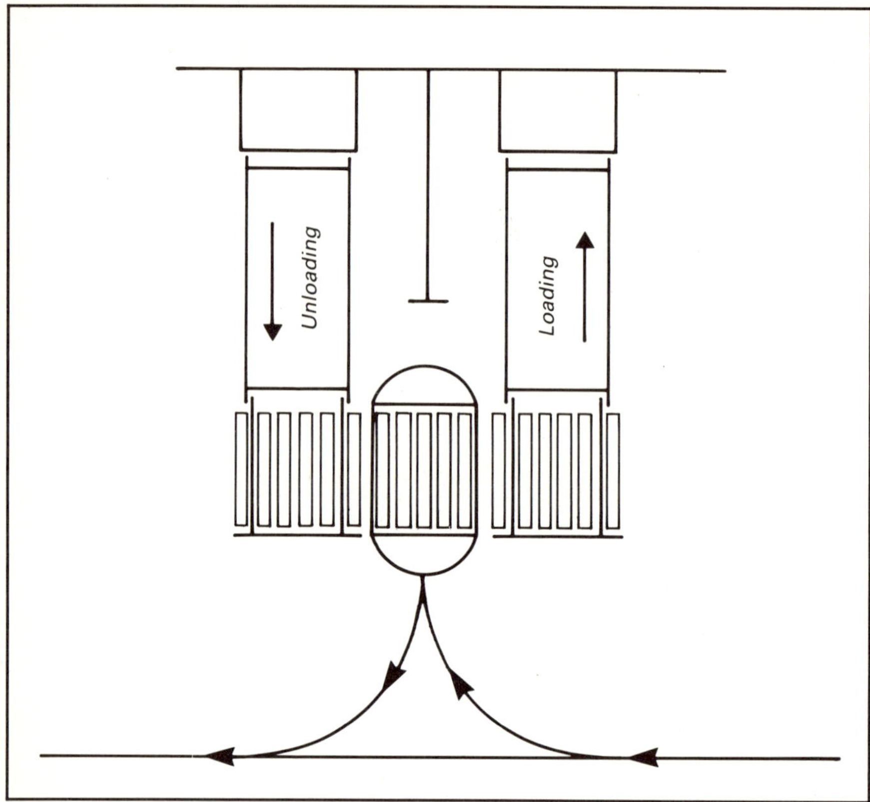

Fig. 11   Intersecting track load transfer: vehicle with roller conveyor; station with roller and chain conveyors

raised fork-lift (see Figs. 13 and 14). Loading in the high-level shelving generally involves a continuous conveyor on a second level. If the unloading level has a roller conveyor segment available at each RTS transfer station, then from time to time reloading can be carried out by the RTS. In this case, the segment of roller conveyor takes over the alignment of the pallet before the transfer to the chain conveyor. At load transfer the vehicle immediately leaves the main truck and turns into branch 1 as shown in Fig. 14. Then the bogie rotates as shown by curve 2 and the vehicle is turned through 90°. The vehicle then reverses as indicated by 3. The vehicle finally leaves by branch 4. In this as in other similar store applications with overlapping paths in the area to the front of the store, 'virtual' courses are increasingly used (i.e. the curved path is programmed into the microprocessor of the vehicle by geometric tables). Stopping points can be specified as floor installations in the conventional way or can be filed incrementally as pulses in the computer. The limiting output of these transfer stations is

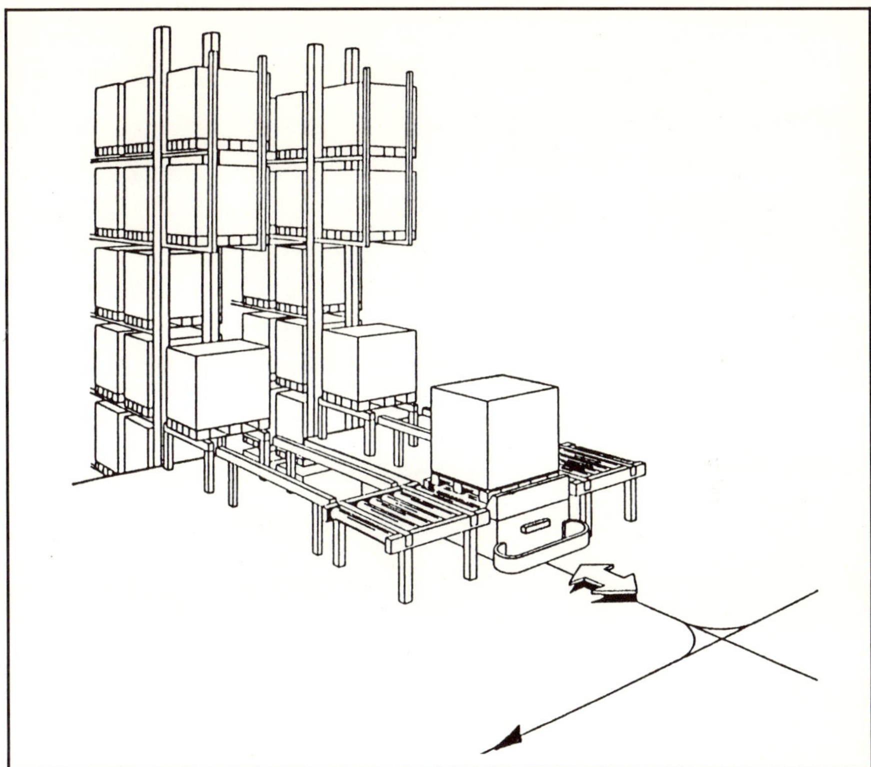

Fig. 12   Example of the system of Fig. 11

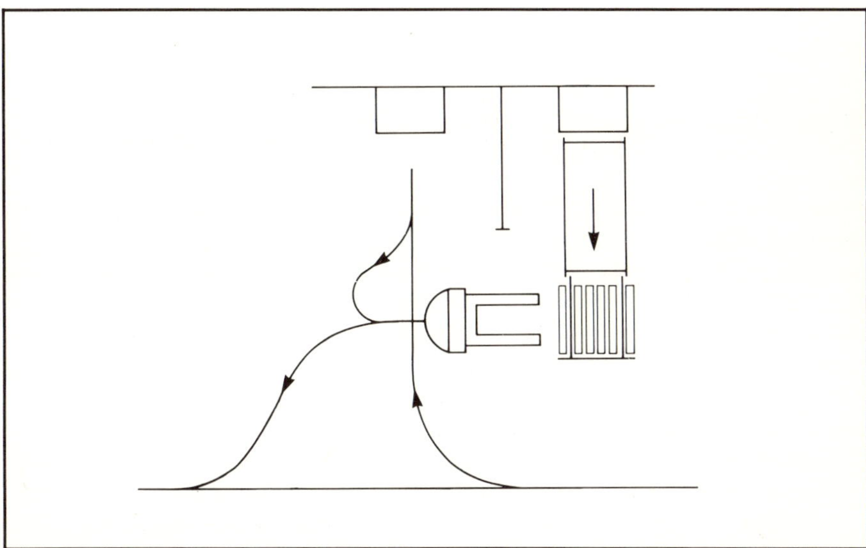

Fig. 13   Intersecting track load transfer: stacker truck; station with roller and chain conveyors

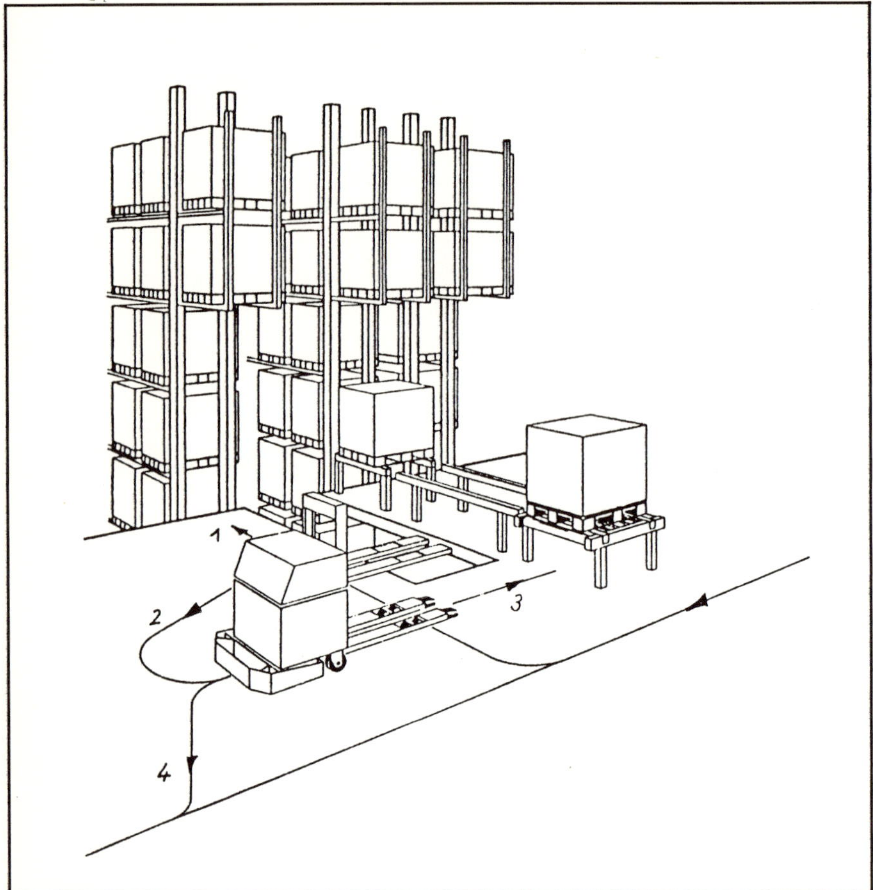

Fig. 14   Example of the system of Fig. 13

a maximum of 35 single functions per vehicle per hour. The deciding factor in the space requirements for the travel layout, particularly for reversing RTS stacker trucks, is the steering system. For larger vehicles, the optimal steering behaviour is given in Fig. 15. Fig. 16 shows in comparison the steering system and required turning circle for bogie steering.

An economic alternative for carrying out double functions is possible if the vehicle has lifting beams available and roller conveyor segments are installed in the area in front of the store (see Fig. 17). Here it is necessary for the vehicle to unload the pallet on the loading side, before a pallet is taken onto the vehicle on the unloading side. There is a consequent slight lowering of the handling rate compared with that of the load transfer in Fig. 11.

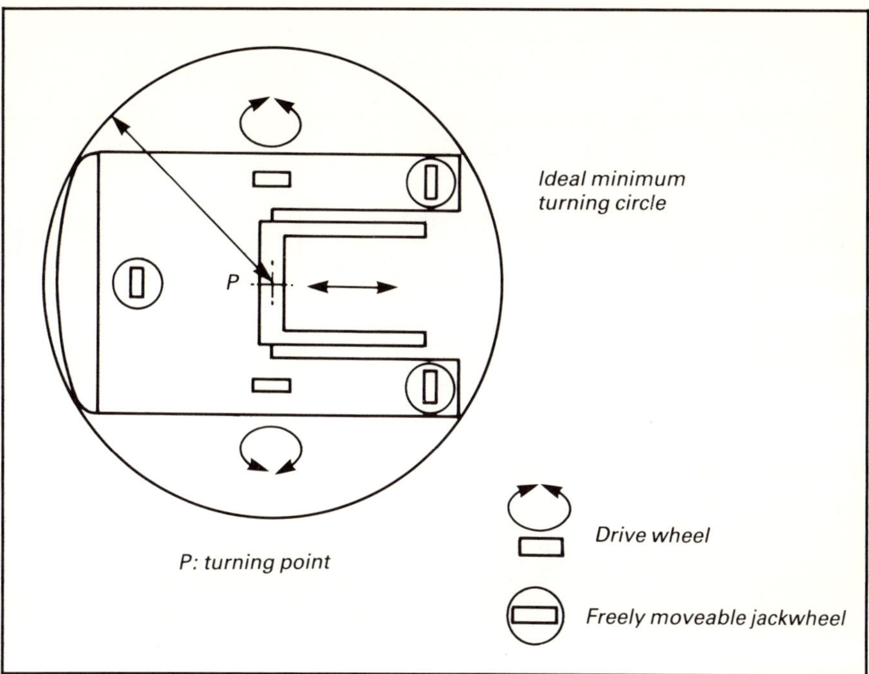

Fig. 15 Steering system of a differentially steered remote-controlled sliding tower stacker

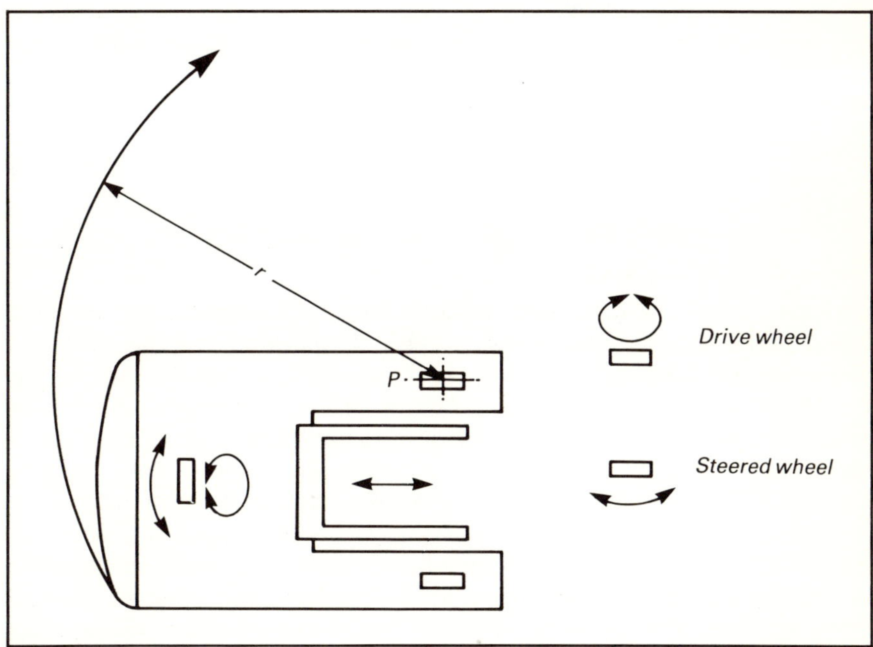

Fig. 16 Steering system of a remote-controlled sliding tower stacker steered by pivoted logic

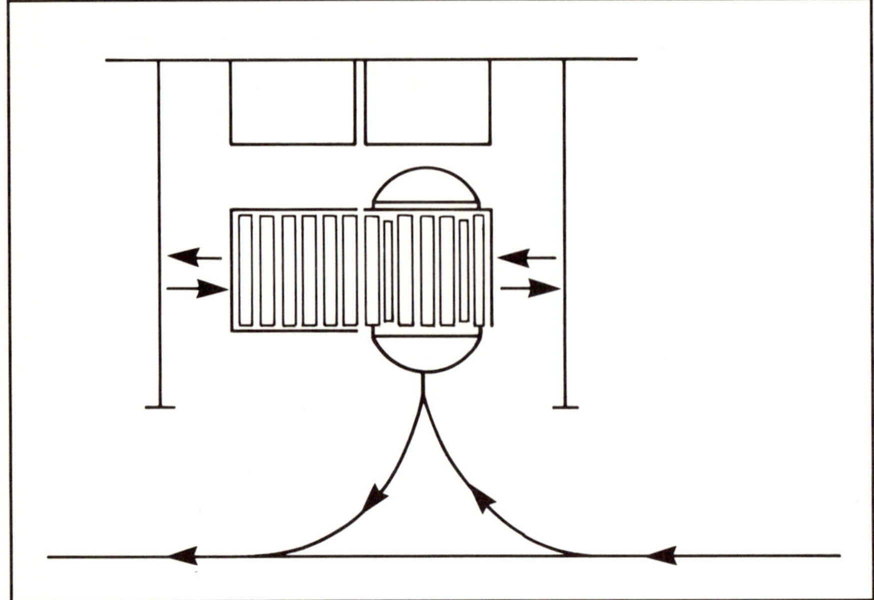

Fig. 17 Intersecting track load transfer: vehicle with lifting beam; station with roller conveyor

When transporting large load components, curved paths in front of the store should be avoided if possible on space-saving grounds. Economic paths for transportation would then best be achieved by four-wheel steering of the vehicle (see Fig. 18). After turning the steering wheels and the driving wheels respectively, the vehicle leaves the guide wire (i.e. the main flow direction) for a short while for the load transfer. These load transfer stations are suitable for both loading and unloading.

Basic control system

The control of the transport sequences of RTS vehicles in front of a store linked to a manufacturing area is a demanding exercise involving the minimisation of empty journeys and the best possible use of the transport capacity of the equipment. The exercise for the central transport computer can be described as handling a large number of different transport activities with different fetching and carrying objectives by a stochastic quantity of instructions. For this reason an efficient transport computer is vital, one which includes similarly efficient data transfer. This means, for example, that drive instructions as double functions should be sent to the vehicles as fast as possible through a number of communication points in the total layout. This assumes a dialogue between vehicle and transport computer at each load transfer position. The question as to whether the transport

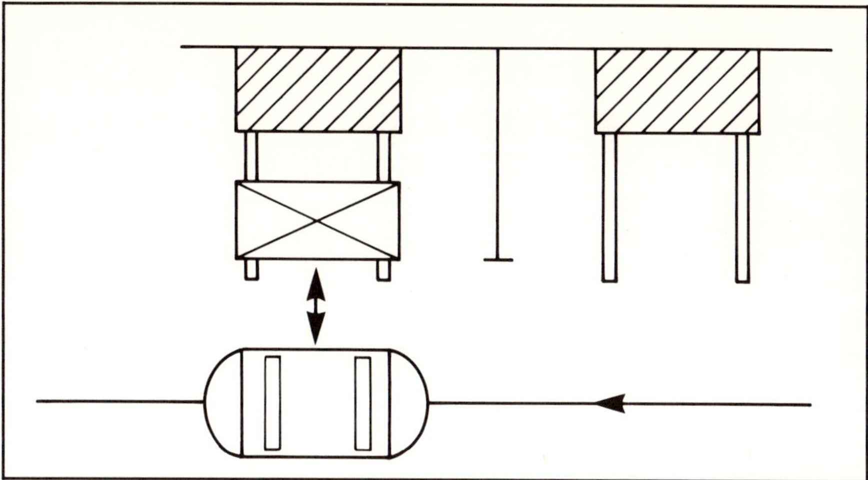

Fig. 18   Load transfer with four-wheeled steering vehicle

controls (section block controls) should be centralised or decentralised functions is largely irrelevant here, as essentially for the present the strategically correct handling of the whole transport exercise is accomplished by track optimisation and strategies.

In contrast with this, the transport sequence in RTS assembly systems is quite different. In this case all transport exercises are similar and the load on the equipment is close to constant, so that on the basis of the constant material flow the transmission of instructions can be centralised at a waiting station.

5

Developments

This final chapter comprises six papers on various aspects of future development. Automating an existing design of fork-lift truck is another approach to achieve an AGV system which is presented in the first paper. Next comes a description of an experimental free-ranging AGV which inherently offers greater flexibility of path change, but requires somewhat more intelligent scheduling and control. Then follow three papers on more advanced AGV systems, two for transportation and one also incorporating a robot arm, all using novel means of guidance. The chapter ends rather surprisingly with a paper about a revolutionary type of wheel which would allow an AGV to move in any direction, not just bidirectionally.

AUTOMATED GUIDED FORK-LIFT TRUCK SYSTEM

H. Takanami
Komatsu Forklift Co. Ltd, Japan

Komatsu has developed an automated guided fork-lift truck system
specifically for automatic conveyance and handling of materials in
refrigerating warehouses. The system and its merits are discussed.

At present, Komatsu's automated guided fork-lift truck is widely
employed in various applications. Komatsu has now adopted this
fork-lift truck in an automatic material handling and conveyance
system for exclusive use in refrigerating warehouses. This system
enables handling and conveyance operations to be rationalised with
higher efficiency under the strict conditions required by such environments.

Outline of the system

An automated guided fork-lift truck running along guidance cables
laid in the floor in the refrigerating warehouse is used to load material
from the external goods reception area and to deliver the load to the
far end of the specified warehouse starting lane. When a load is
required for dispatch it is taken from the nearest occupied site along
the specified lane and taken to the external dispatch department. On
completion of the operation the fork-lift truck returns to a waiting
origin outside the warehouse.

To illustrate the above, Fig. 1 shows a simple single-lane layout. In
practice, much more complicated layouts are arranged to meet
customer requirements.

In the storage lane loads are placed in stacks. The number of stages
of a stack is previously determined for each unloading position
(typically three or four stages).

A lane is specified by the on-board control panel. This can be
mounted on the fork-lift truck itself, or in some place outside the

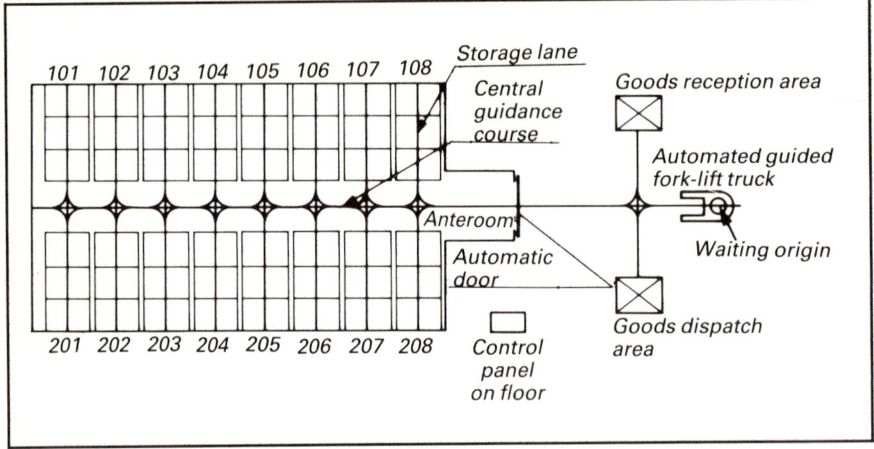

Fig. 1 *Example warehouse layout*

refrigerating area. When a lane is specified, either of the following two operation modes can be selected:

- Continuous mode – when some lane has been filled up (or emptied), the automated guided fork-lift truck goes to the lane of the next number automatically in order to continue with the warehousing (delivery) operation.
- Single specified mode – the warehousing (delivery) operation is performed only once in the specified lane.

The door of the entrance to the refrigerating warehouse automatically opens and closes according to the movement of the automated guided fork-lift truck. The truck stops temporarily in front of the door, and passes through it after confirming that it is open.

Features of the system

The system can be realised only by laying guidewires in the warehouse floor. Compared with the construction of the refrigerating warehouse itself, the AGV construction period is short and construction cost is moderate.

The interior of an automated refrigerating warehouse is basically the same as that of a normal warehouse, and requires no special fixed equipment such as racks, etc. Therefore it is possible to perform material handling by the use of attended as well as automated fork-lift trucks. Loads can be piled up directly on the floor, and therefore dead space can be reduced and high storage efficiency obtained.

As Komatsu's battery-operated fork-lift trucks are used as a base, the system has high durability and excellent reliability. Also, since a fork-lift truck has many functions, versatility is obtained at low cost, with resulting high operation efficiency and labour saving.

Finally, it should be noted that the system allows the automated

Table 1   Comparison between fork-lift truck system and other automated systems for refrigerating warehouses

	Cost	Storage efficiency	Reliability (durability)	Versatility
Computer-controlled fork-lift truck system	No special fixed equipment is required Construction is simple The existing tried and tested fork-lift truck is used as a base	Since loads are piled up directly on the ground, dead space is reduced Attended fork-lift trucks can be used together	Since the fork-lift truck is tried and tested, less trouble occurs No equipment is required to be kept under the condition of low temperature If required, attended fork-lift trucks can be used	Extension and transfer are simple It is possible to perform usual warehouse operations Fork-lift trucks can be switched from computer to manual control
	(Cheap)	(High) About 60%	(High)	(Excellent)
Other automated equipment	Fixed equipment is required Construction is complicated It is necessary to construct according to the shape of the warehouse	Since fixed equipment is used, dead space is large Attended fork-lift trucks cannot be used	Since equipment is fixed under the conditions of low temperature, frequent trouble occurs When any trouble occurs, all functions are suspended	It is impossible to perform any function except the functions of the equipment It is difficult to extend the existing installation
	(Expensive)	(Low) About 30%	(Low)	(Inferior)

guided fork-lift truck to be easily inspected and maintained outside the refrigerating warehouse.

A comparison of the fork-lift truck system and other automated systems for refrigerating warehouses is given in Table 1.

Extensibility of the system

- Two or more automated guided fork-lift trucks can be operated according to the scale of operation and the scale of a warehouse.

- This system can be extended to a total system which is capable of automatic conveyance and automatic inventory control by introducing an inventory control computer.

- It is easy to transfer and rearrange the stored articles within the warehouse.

- Storage lanes can easily be extended and/or resituated according to any extension or other modifications of the warehouse.

- Change-over to attended operation from automatic operation can be made by single-switch control.

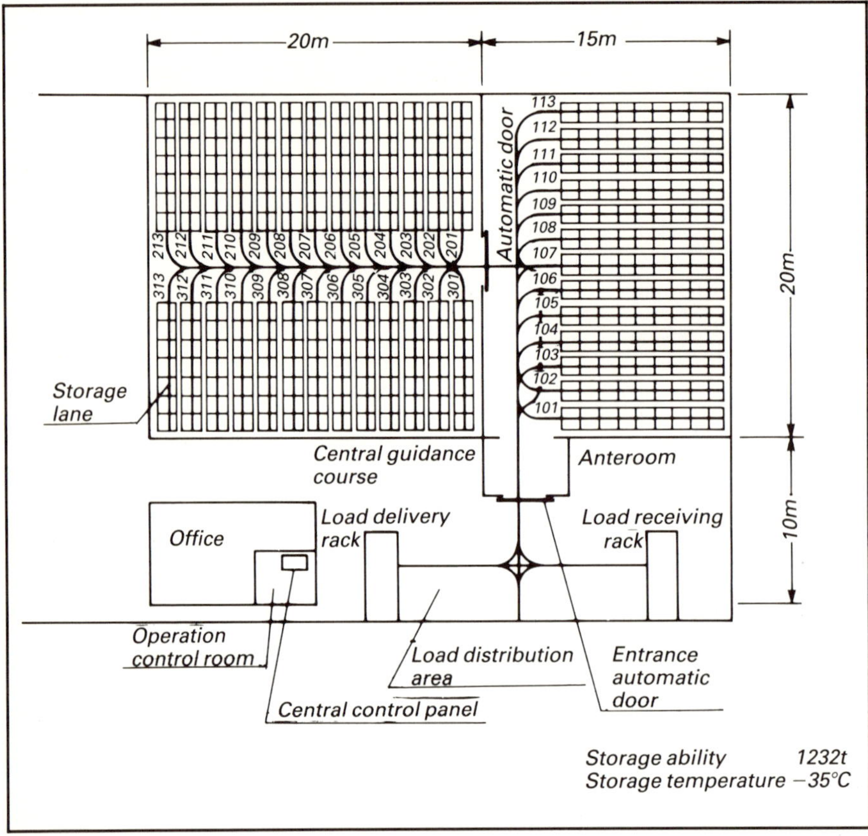

Fig. 2   Layout of a warehouse system actually in operation

An example of the operation

The example taken here is the refrigerating warehouse of Zen Choshi-shi Suisan Kako Kogyo Kyodokumiai. This involves simultaneous operation of two Komatsu fork-lift trucks, type FBISR. The layout of the warehouse is given in Fig. 2.

The conveyance capacity of this system is large, because of the simultaneous operation of two trucks. The central control panel guides the simultaneous operation of the trucks and assures their smooth operation in a common area.

The specifications of the system are as follows:

- Applicable model: Komatsu's reach fork-lift truck FB15R; two units. (For −35°C refrigerating warehouse with 4.3m high mast.)
- Travelling system: guidance control by underground guidance cable.
- Operation mode: continuous mode or single specified mode.
- Operation method: control by the central control panel (change-over of warehousing and delivery operation, setting of operation

mode, specification of lane for unloading and loading of material, etc.).

- Number of storage lanes in warehouse: 39.
- Storage quantity: 1182 pallets.
- Package form: 1.2m × 1.0m pallets (fork insertion gap: 9cm); wooden or corrugated cardboard boxes piled up; height or package dimension: 1.1m – 1.3m.
- Lane spacing: 1.43m.
- Travelling speed: about 0.5m/s (variable).
- Storage efficiency: about 60%.

FREE-RANGING AGV AND SCHEDULING SYSTEM

S. P. Walker, S. K. Premi, C. B. Besant and A. J. Broadbent
Imperial College of Science and Technology, UK

Free-ranging automated guided vehicles (AGVs) follow software
path networks without physical guidance, in comparison with
conventional AGVs that follow buried wires or painted lines on the
floor. Thus when combined with intelligent scheduling and control,
free-ranging AGVs inherently offer greater flexibility with respect to
path network changes and efficient use of floorspace. The design,
path programming and control of the Imperial College free-ranging
AGV (ICAGV) are described. The scheduling philosophy employed
to control a fleet of such free-ranging AGVs is also discussed,
together with its implementation on an external supervisory
computer. Finally, principal areas of future work are outlined, some
of which have been initiated.

The use of automated guided vehicles (AGVs) is now widespread in
warehouse and assembly applications. Existing vehicles invariably
follow paths which are physically determined, typically by the burial of
a signal-carrying wire, or the painting of a visible line on the floor[1].
Installation of these paths can be costly and disruptive, and subsequent
modifications are troublesome. Further, the cost of the paths can
inhibit use of the high density of paths which would maximise the
efficiency of the installation.

Free-ranging AGVs do not require paths to be physically defined.
Routes along which vehicles are allowed to travel are stored in a
computer in the form of spatial coordinates. Vehicle travel along these
routes is controlled essentially by continuous odometric position
determination. The ability to specify routes of essentially no cost
means that dense, highly redundant networks can be used, with paths
limited only by availability of space. Further, networks can be
modified easily as needs change, without the expense and disruption of
cutting passages for more guidance wires. A fuller review of the
advantages of free-ranging AGVs has recently been presented[2].

Conceptual design

The development of the AGV transport capability was undertaken as a single, integrated activity. It involves both the ability of an individual vehicle to perform its required tasks, and the higher level coordination of many vehicles in executing a continuing sequence of transportation demands. The need is then apparent to determine the relative amounts of 'intelligence' to be allocated to the supervising controller and to the vehicles, and thus to decide what degree of capability and discretion is to reside in them.

The solution adopted was for the vehicles to receive journey instructions in the form of a list of node (path junction) spatial coordinates, with corresponding arrival times at each node. The vehicle is able to follow the route so defined. Using odometry it continuously calculates its current position and heading, compares these with those required to reach the current target node, and corrects for lateral and angular deviation. Throughout the journey it calculates its expected arrival time at the next node, and modifies its speed as necessary to arrive on time. In essence, once the route and timetable is specified, complete responsibility for and discretion regarding its execution is delegated to the vehicle.

In the supervisory computer resides all topological information concerning the path network, and information about routes and timetables of all journeys in progress. A transportation demand is received, requiring a certain vehicle to travel between two nodes. The supervisor selects a route and timetable, such that the new vehicle does not interfere with journeys already in progress, and transmits this to the vehicle.

The vehicle

Electrical and mechanical design

The vehicle specification adopted was that it should transport loads of up to 1000kg at a maximum speed of 2m/s. It is 'differentially steered' by two independently driven wheels, which are fixed at each end of the central axle, as shown in Fig. 1. Differential steering provides high manoeuvrability, permitting forwards and backwards motion, and rotation about any point on its central axis. Four heavy-duty castors are attached at each corner, which are spring-loaded to provide suspension for the vehicle.

Each drive wheel is driven by a 700W, 48V dc servo motor and associated switching power amplifier. Incorporated in each motor is an electromagnetic brake and tachogenerator, as shown in Fig. 1. The tachogenerators provide speed feedback to the amplifiers, which control the speed of each drive wheel.

A 1000 pulses per revolution incremental optical encoder is coupled to each drive wheel, as shown in Fig. 1. These encoders can detect a

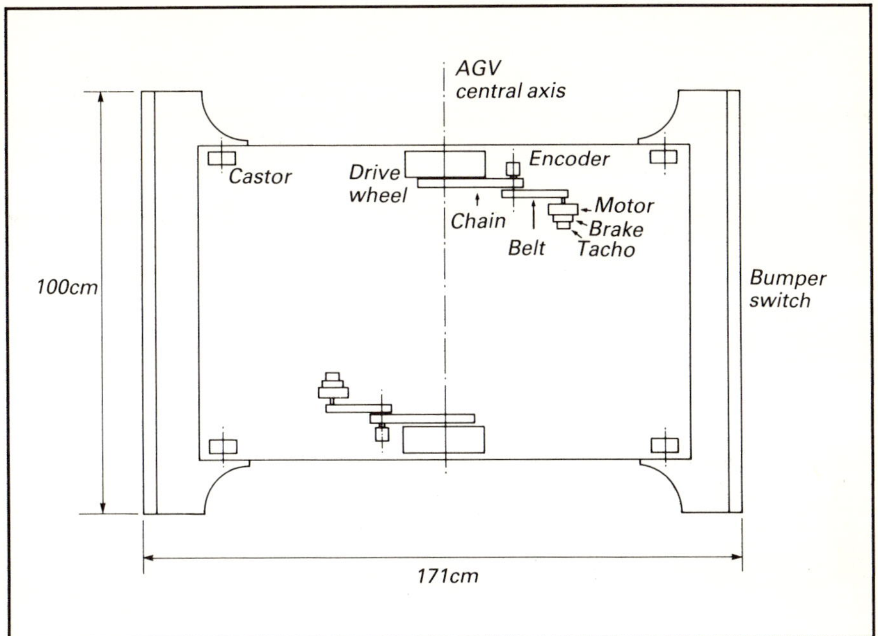

Fig. 1   The ICAGV mechanical structure

circumferential movement of each drive wheel of less than 0.1mm, and are used by the on-board odometry-based navigation system. Four 12V , 40Ah car batteries wired in series provide traction power for the vehicle, while a separate 12V, 35Ah battery via a specially developed voltage convertor provides power for the on-board control system. The voltage convertor is protected against power surges, and is capable of informing the on-board control system of a 'flat battery condition' on the traction or control system battery.

Rubber bumpers incorporating sensitive micro-switches are fixed at the front and rear of the ICAGV, which when touched bring the vehicle to rest. An ultrasonic 'proximity sensor' will shortly be incorporated which will allow obstacles to be detected prior to collision.

At present, communication with the vehicle is via a specially developed hand-held programming pendant, the structure of which is shown in Fig. 2. The pendant essentially acts as a hand-held VDU into which commands are entered and transmitted to the vehicle. It is envisaged that eventually communications will be via an FM radio or infrared-based data link.

Position and heading determination

The ICAGV uses odometry, via optical encoders coupled to two 'measuring wheels', which at present are the drive wheels, to determine continuously the following parameters, which are then used by the on-board navigational scheme:

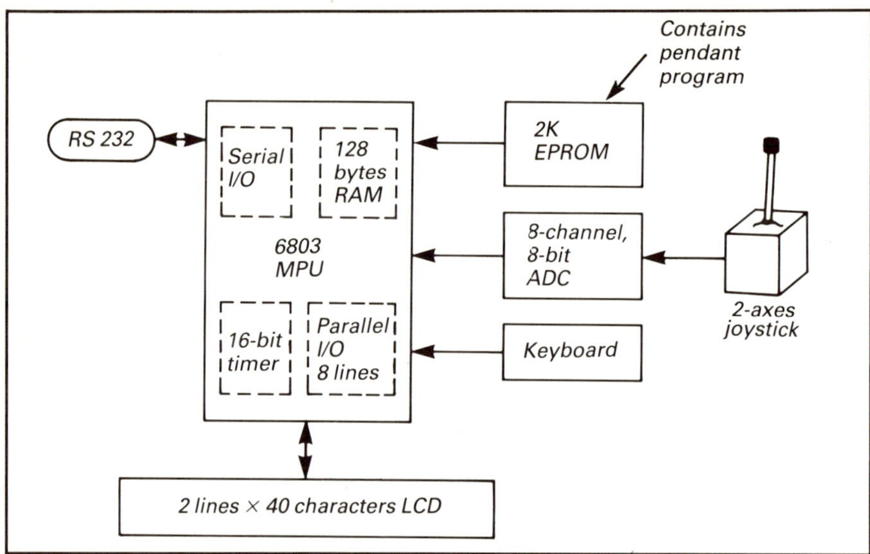

Fig. 2 *The ICAGV programming pendant structure*

- The current heading, relative to an arbitrary reference heading (τ_i).
- The current Cartesian coordinates, relative to an arbitrary datum (x_i, y_i).
- The lateral deviation from the currently commanded straight-line path (p_i).
- The distance travelled along the currently commanded straight-line path (q_i).

The variables τ_i, p_i and q_i are used for path and speed profile control, as will be discussed later. The variables τ_i, x_i, y_i, p_i and q_i are calculated using a floating-point arithmetic chip incorporated in the on-board control computer every 0.02s.

Odometry inherently suffers from cumulative errors. Experiment and theoretical analysis confirm that a means to determine absolute position and heading is required to correct odometry periodically. Various secondary systems are possible [1], such as optical or ultrasonic imaging of the surroundings, triangulation using three fixed passive or active known points in the surroundings, strategically fixed laser rangefinders and strategically fixed reference laser lines. Most secondary systems, apart from the reference laser lines system, yield only position information. Thus to obtain the more crucial heading information the positions of two points on the vehicle must be instantaneously determined, or the position of one point on the vehicle must be determined twice, the second measurement occurring shortly after the first. Work has already been carried out on a laser-based active triangulation system and work on reference laser lines is in progress. The system that emerges as the most suitable will be incorporated.

Paths programming and control

The vehicle follows a sequence of straight-line paths which take it from its current position to the desired destination. These paths are selected by the supervisory computer. A sequence of trapezoidal speed profiles is used to follow the commanded sequence of paths, as shown in Fig. 3, and motion along a constant-radius circular arc, at a fixed cornering speed, is used to achieve smooth path transitions. During motion the ICAGV uses odometry to monitor the distance travelled along each commanded path, and to calculate continuously the vehicle position and heading. From these, lateral deviation from the path and heading deviation are determined and minimised through proportional control.

Hardware and software configuration

The control system is essentially a parallel computer with specially designed interfaces. This computer structure is chosen because it offers expandability and concurrent processing. It consists of two MC68B09, 8-bit microprocessor-based processing cards, termed the 'Delegator' and 'Multipurpose' cards, together with a specially developed interface card, as shown in Fig. 4. The processing cards, plus any future cards, communicate with each other via the IEEE-488 bus. The Delegator card contains the bulk of the on-board software, including communications with the programming pendant, paths storage, the path-following scheme, vehicle control and the implementation of odometry. The Multipurpose card is a 'slave' processing card to the Delegator card. It essentially obeys the IEEE-488 commands sent by the Delegator card such as 'transmit both the encoder counts', or 'Set the drive-wheel speeds to the following values'. If either bumper is hit, the Multipurpose card informs the Delegator card. As the name implies, the Interface card links the controller processing card to the vehicle mechanics, as shown in Fig. 4.

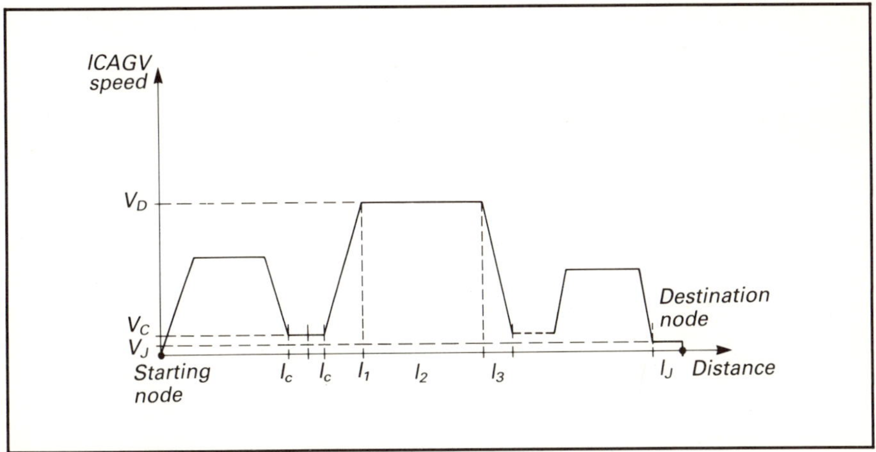

Fig. 3   The ICAGV speed profile for a sequence of commanded paths

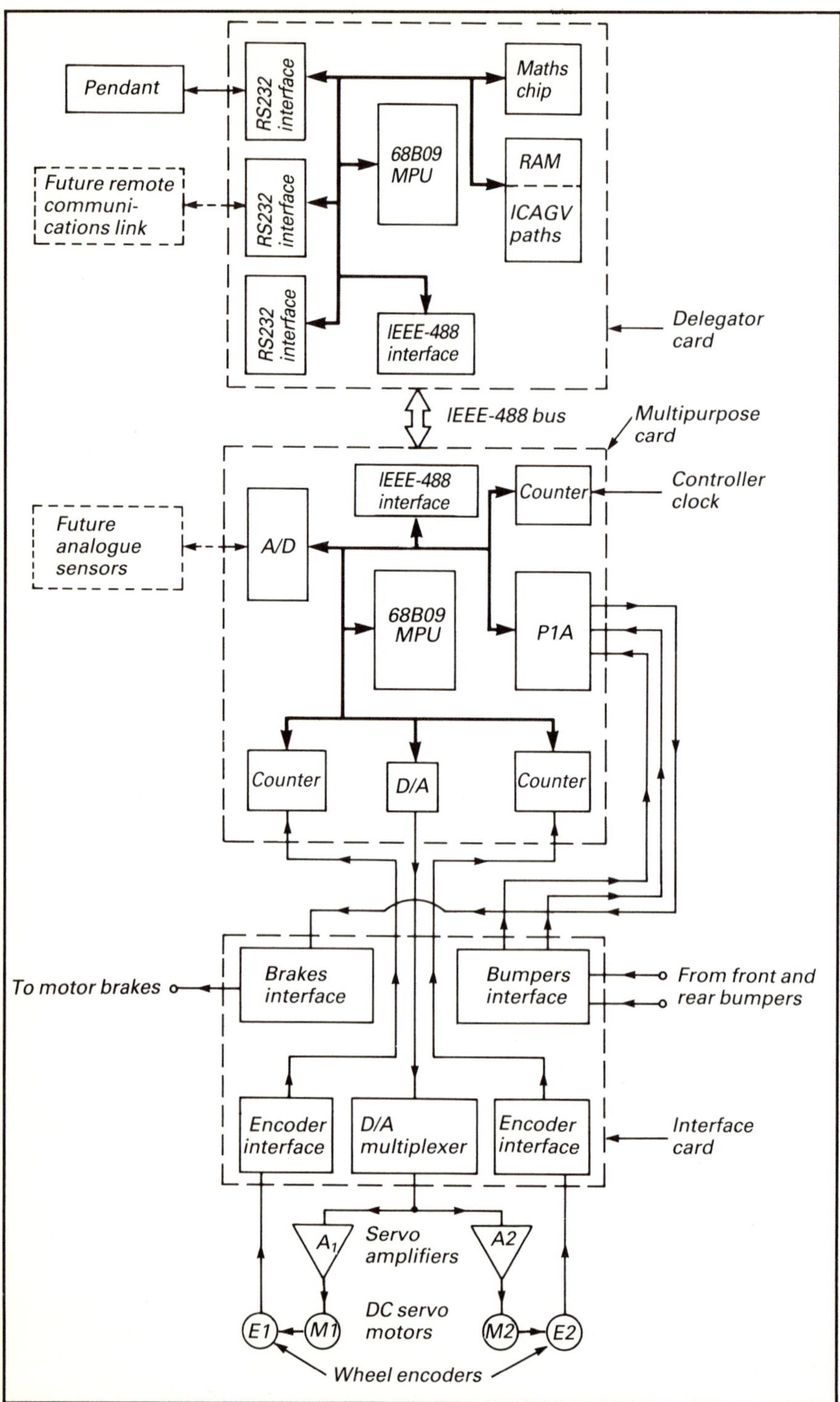

Fig. 4   A simplified diagram of the ICAGV on-board controller

Fig. 5 *Structure of the alogorithm used to schedule a fleet of ICAGVs*

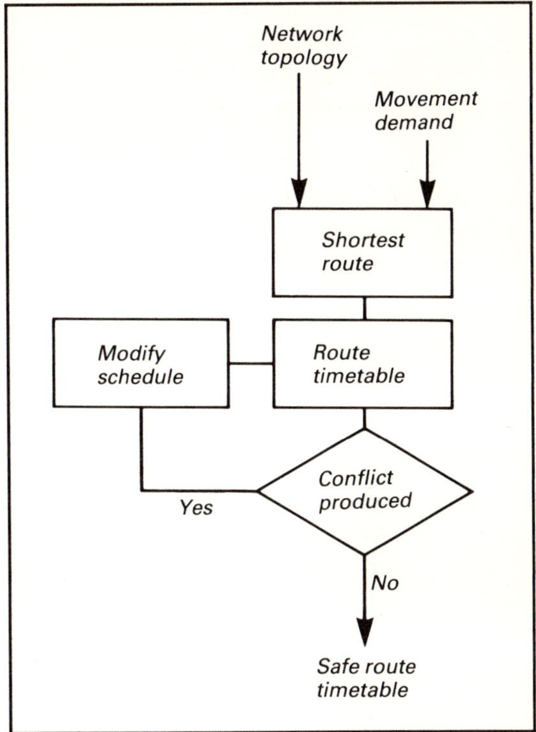

Supervisory control of an AGV fleet

The supervisory controller developed here receives a transportation demand and generates a timetabled route for a selected vehicle. For present purposes the demand is best thought of as arising from an external source such as a human operator or an FMS controlling computer.

Problem reduction

Scheduling a fleet of AGVs in a non-conflicting manner is a complicated real-time problem. The problem can be reduced conceptually as follows. A number of vehicles are currently performing journeys in a non-conflicting manner. A new demand is received, and a vehicle must be selected and then scheduled to satisfy the demand so that it does not conflict with the existing journeys. Hence the scheduling problem devolves to that of inserting an extra vehicle into an existing population of vehicles performing non-conflicting journeys.

Algorithm

Fig. 5 shows the structure of the algorithm used to schedule a fleet of AGVs. There are four main parts in the schedule, and these will be described separately.

Shortest route selection. The initial assumption is made that the best route for a vehicle is the one which involves travelling the minimum possible distance between the two end points (while of course staying on defined paths). This geometrically shortest route is determined by using the Dijkestra optimal minimum spanning tree algorithm [3].

Timetable generation. The heart of the scheduler is a matrix which describes the node occupation times for such vehicle. When a new route has been calculated as above, the node occupation times for each node in that journey are calculated from the vehicle acceleration and maximum speed values.

Conflict analysis and route modification. By comparing the node occupation times for the new journey with the node occupation matrix for all existing active vehicles, one can determine whether the new journey conflicts with any journeys already scheduled. Three types of conflict are possible:

- A collision between two vehicles travelling along the same path in opposite directions.
- A collision between two vehicles travelling along the same path in the same direction but at different speeds.
- A collision between two vehicles attempting to occupy the same node simultaneously.

Head-on conflicts are overcome by finding a different route for the new journey, avoiding the path on which the conflict was predicted. The shortest route in the absence of this path is found. Junction and 'catching-up' conflicts are overcome by delaying the new vehicle. This has the effect of preventing the new vehicle from catching up with the existing vehicle, or allowing the existing vehicle to vacate the node before the new vehicle arrives.

With the conflicting route appropriately modified, a new timetable is calculated for the new journey, and a conflict check is again performed. This process is repeated until a successful route is found. On occasions it is impossible to schedule a new vehicle when all the possible routes are blocked by stationary vehicles.

Issue of instructions

Once a non-conflicting route and timetable for the new vehicle has been determined, this is then transmitted to it. The information is in the form of a string of node coordinates and corresponding required arrival times. Provided the vehicle reaches each node at the correct time, and no unforeseen events such as breakdowns occur, it will satisfy the transportation demand without coming into conflict with any other vehicles.

Principal future developments

Fleet induction

A simulator which mimics the control and operational performance of a fleet of vehicles is being developed.

The simulator has two distinct uses: Initially it will serve as a development tool to test and improve the scheduling algorithms. Ultimately it will be used for analysing the performance of a fleet of ICAGVs in real applications before installation. As a development test it will be used to ensure the robustness of the system. The scheduler must be able to cope with vehicle breakdowns, blockages, and the simulator will enable the best path networks, the optimal number of vehicles, the most efficient position for load/unload stations, and other such factors to be determined.

Dynamic rescheduling

The scheduler as described above has no facility to reschedule once it has commenced a journey. Such a facility is necessary to enable the vehicles to avoid obstructions which may appear on their path, and to allow for high-priority journeys which must take the shortest route, forcing the journeys of the existing vehicles to be modified to avoid them. The ability to reschedule the vehicles dynamically is currently being implemented.

Concluding remarks

It is felt that the free-ranging AGV and scheduler described above is a very promising concept which should meet the increasing needs for flexible and efficient materials handling in many different environments.

Installations of this type are not likely to be significantly, if indeed at all, cheaper than conventional AGVs. Savings resulting from the removal of the need to bury guidance wires will be compensated for by the costs of the greater intelligence required of both vehicles and the supervisory computer. However, much denser, more redundant networks will make more effective use of space, allowing greater material throughputs and higher traffic densities. Network modifications to meet changing needs will simply require the entry of different topological data.

An intelligent, flexible AGV system as discussed here will provide materials handling capability worthy of a sophisticated modern warehouse or manufacturing plant.

References

[1] Premi, S. and Besant, C. 1983. A review of various vehicle guidance techniques that can be used by mobile robots or AGVs. In, *Proc. 2nd Int. Conf. on Automated Guided Vehicle Systems*, Stuttgart, West Germany, 1983.

[2] Broadbent, A. et al. 1985. Free ranging AGV system: Promise, problems and pathways. In, *Proc. 2nd Int. Conf. on Automated Materials Handling,* Birmingham, UK, 1985.

[3] Tanenbaum, A. S. 1981. *Computer Networks*. Prentice Hall, Englewood Cliffs, NJ, USA, pp. 38–40.

ADVANCED AUTOMATED TRANSPORTATION SYSTEM

H. Imai, K. Fujiwara
and
Y. Kawashima
Mitsubishi Electric Corp., Japan

One of the key subjects of manufacturing automation is how to transport materials in a factory. An advanced automated transportation system is described. This system consists of a robot vehicle with a handling mechanism, warehouses and communication terminals linked to higher computers by an optical fibre network. The robot vehicle can move in any direction without rotation of its chassis and can communicate with the terminals by the wireless method. C-MOS ICs are employed for the control circuits to save battery power of the robot vehicle.

Factory automation is one of the major issues of recent years. To realise this goal it is indispensable to develop a computer system for production control and automated or mechanised equipment, and to integrate them into a flexible system.

The individual technologies – machining, assembly, adjustment and inspection – have been developed for automation of the manufacturing process. It is to be noted, however, that the technology of work transportation is less advanced. This technology is necessary to establish a highly flexible manufacturing system by combining the above-mentioned automation components systematically.

There are many kinds of conveyor systems available for work transportation. Recently robot vehicles have been used between machine tools, but they are far from sufficient. Mitsubishi has investigated the work transportation system concept and developed a model system of higher function than previous designs.

Design philosophy

Mitsubishi first investigated the conditions necessary to realise a highly flexible system. It turned out that the system should be made up of robot vehicles having the following features.

Steering function

In highly automated production processes, the flow of workpieces is complicated in many cases, making it difficult to lay out loading and unloading positions regularly. It is therefore necessary that the robot vehicles have the ability to move quickly along often intricate routes.

Travelling route

For conventional types of robot vehicles, it requires quite a lot of work to set a travelling route or to change the route. Therefore it is difficult to practice rationalisation which includes alteration of travelling routes. Easy alteration of the travelling route is essential.

Positioning accuracy

Work must be transferred between robot vehicles and manufacturing equipment. If the positioning accuracy is poor, it is costly to improve the loading mechanism or to correct the position of the vehicle. To eliminate this labour and cost, the robot vehicles should be positioned accurately.

Energy saving

The manufacturing equipment becomes more expensive as automation advances and the up-time of the equipment must be raised. It is therefore necessary to make the down-time of the robot vehicles as short as possible – just to replace batteries. In other words, the vehicles should be of energy-saving design.

Communication function

In an advanced job-shop-type production, production information should be synchronised with the flow of workpieces. Thus the robot vehicles should be provided with an advanced communication function.

Prototype robot vehicle

Structure

The robot vehicle (shown in Figs. 1 and 2) is driven by dc batteries (24V, 100Ah). The vehicle has two driving wheels and two self-positioning wheels. Each driving wheel is driven by an individual dc servo motor with reduction gears. The vehicle motion is controlled by rotation of the driving wheels. A gyro unit is built into the centre of the vehicle to determine the direction of travel.

Fig. 3 shows a block diagram of the control unit. All of the vehicle operations are controlled by a microcomputer (M5L8085). Driving control is effected through a dc servo system and the number of pulses and pulse rate correspond to the amount of rotation and rotating speed of the driving wheel respectively. Wireless communication is performed via a VHF band.

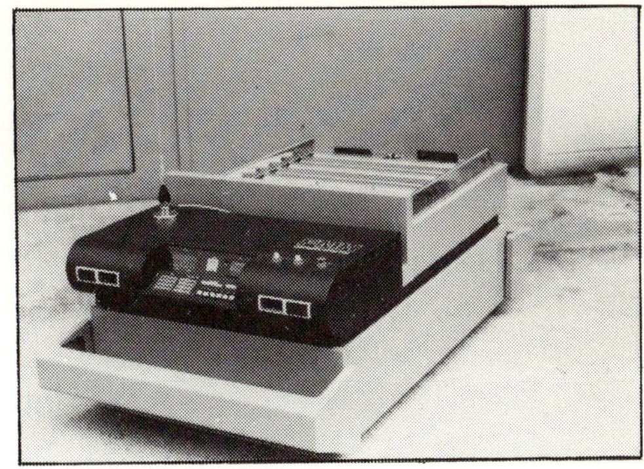

Fig. 1 View of the prototype robot vehicle

Modification of direction by gyro

The gyro has two degrees of freedom. The spin axis is parallel to the horizontal plane. A potentiometer detects angular rotation about the yaw axis (i.e. the direction of the vehicle). During teaching, an absolute direction with respect to the gyro is established and stored in the microcomputer memory. In automatic travelling the vehicle controller monitors the travelling direction and compares it with the previously stored data. Corrective pulses are then output to the driving wheels according to the difference. The vehicle measures its direction every one second in straight travelling and at the end of each turning procedure. In this way, the vehicle can eliminate directional errors caused by floor conditions, slip of wheels, backlash of driving mechanisms, etc.

Experimental results

The experiments using the robot vehicle were sufficiently successful to offer a prospect to realise a guideless robot vehicle for material handling use in workshops.

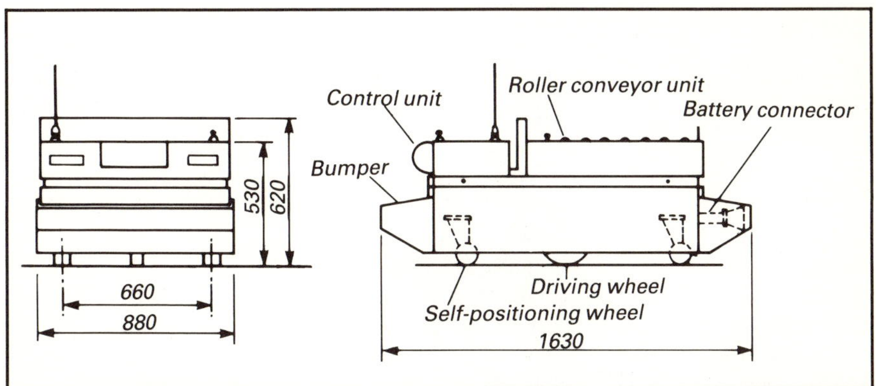

Fig. 2 Construction of the prototype robot vehicle

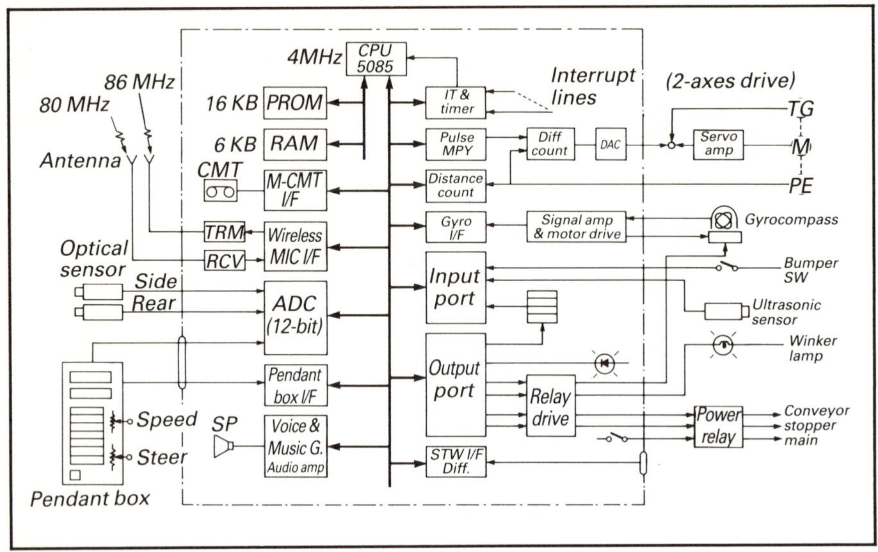

Fig. 3   *Robot vehicle control system*

The vehicle automatically travelled between the stations arranged at approximately 5m intervals, and at the stations, loading and unloading of workpieces with roller conveyors were well performed repeatedly.

The gyro kept sufficient accuracy in short-time travelling, but the drifting errors inherent in this gyro could not be disregarded in long-time travelling.

New robot vehicle

On the basis of the prototype, a new robot vehicle was developed. The main object of the development was to improve the travelling function and route tracking functions.

Fig. 4   *View of the new robot vehicle*

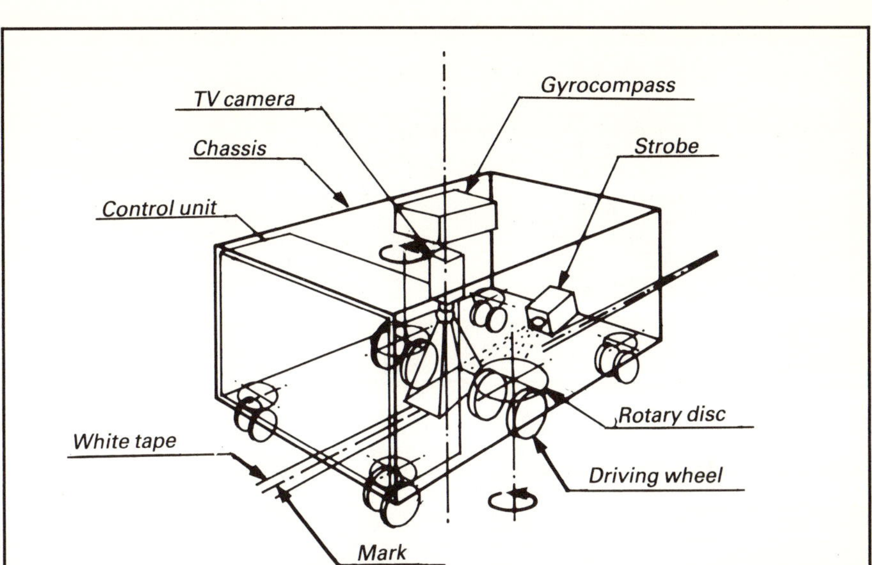

Fig. 5 Construction of the new robot vehicle

Structure

Figs. 4 and 5 illustrate the new robot vehicle. It has two rotary discs to each of which are attached two driving wheels – four driving wheels in all. The orientation of the rotary discs with respect to the chassis can be changed by controlling the revolving speed of these four wheels independently.

When the two wheels attached to a rotary disc are revolved in opposite directions to one another, the direction of the rotary disc can be changed without moving the chassis at all. Since the wheels are attached flexibly to the chassis so as to be in contact with the floor surface at all times, the robot vehicle can run stably on somewhat uneven floor surfaces. (Step difference up to 5mm on floor surface causes no problem to travelling performance.)

A TV camera continuously senses white tape laid on the floor along the travelling route. This camera is mounted at the centre of the chassis and no complicated coordinate transformation is necessary. To illuminate the white tape, a strobe is provided on the bottom of the chassis so that a stable image can be obtained even during travelling. A gyrocompass is also mounted onto the vehicle to sense the direction when the vehicle goes off the route temporarily.

Control system

Fig. 6 shows the hardware of the control system of this robot vehicle. The control system consists of the MCU (main control unit), RCU (recognition control unit) and VCU (velocity control unit). For each

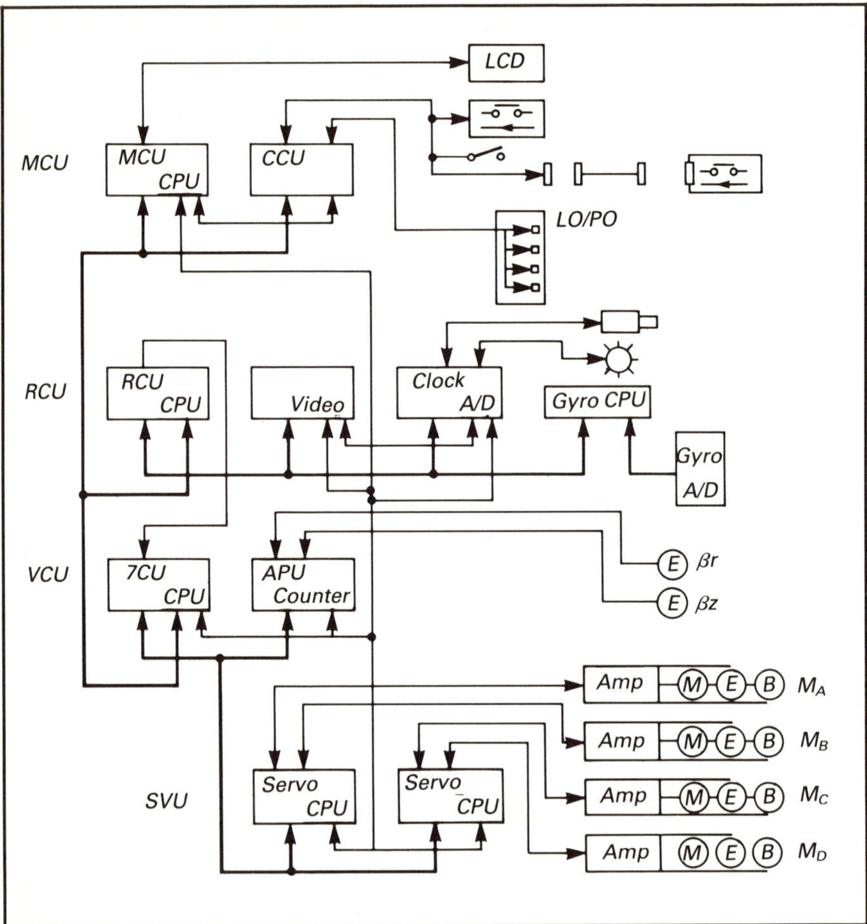

Fig. 6   Block diagram of control system

unit a microcomputer (C-MOS 8085) is employed, and almost all of the functions can be executed independently in both aspects of hardware and software. This enabled adjustment and test of each unit in parallel, which served not only for efficient development but also for easy maintenance. The units are connected with each other by a system bus, and various kinds of information are exchanged by parallel data transmission.

Most of the circuit elements used for these control units are low-power-consumption-type C-MOS ICs, and they are effective for energy saving. The description of each unit is as follows.

MCU. The function of this unit is to control the communication between the robot vehicle and the workstation. It receives the travelling instruction through a data transmission line, and transmits necessary instructions to the RCU or VCU. The unit also functions for input and output processing of the operation panel and for trouble judgement during travelling based on the information from the RCU and VCU.

RCU. This unit calculates the position of the vehicle by processing the image of white tape sensed by the TV camera, and reads various kinds of floor markings (Fig. 7). The strobe used for illumination at this time is synchronised with the CPU control cycle to obtain highly accurate image information quickly. At other places than the travelling route (i.e. where there is no white tape), this unit senses the direction of the chassis by receiving information from the gyrocompass.

VCU. This unit controls travelling and steering of the robot vehicle and receives the travelling instructions from the MCU and the positional information from the RCU. In the VCU the target revolving speed of each of the four wheels is calculated at a high speed for steering control and the results are transmitted to the servo control unit (SVU). In the SVU, dc servo motors attached to each driving wheel are controlled at the optimum revolving speed by two CPUs.

Travelling and steering

Because of its unique wheel layout and control, this robot vehicle accepts any motion within a plane. The typical operation modes are as described below.

Linear motion mode. The four driving wheels are turned to the same direction and are driven at the same revolving speed. The vehicle then moves straight. If the direction of the wheels is in line with the direction of the chassis, there is no difference from conventional vehicles. By this method, however, the vehicle can be moved in the preset direction of the wheels with no regard to the direction of the chassis. In other words, moving is possible in a diagonal, or lateral direction to the direction of the chassis (Fig. 8a).

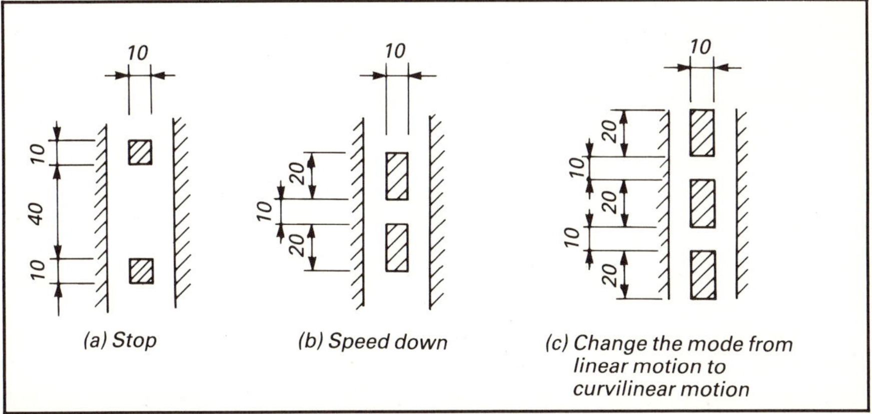

Fig. 7   *Various kinds of travelling instruction marks*

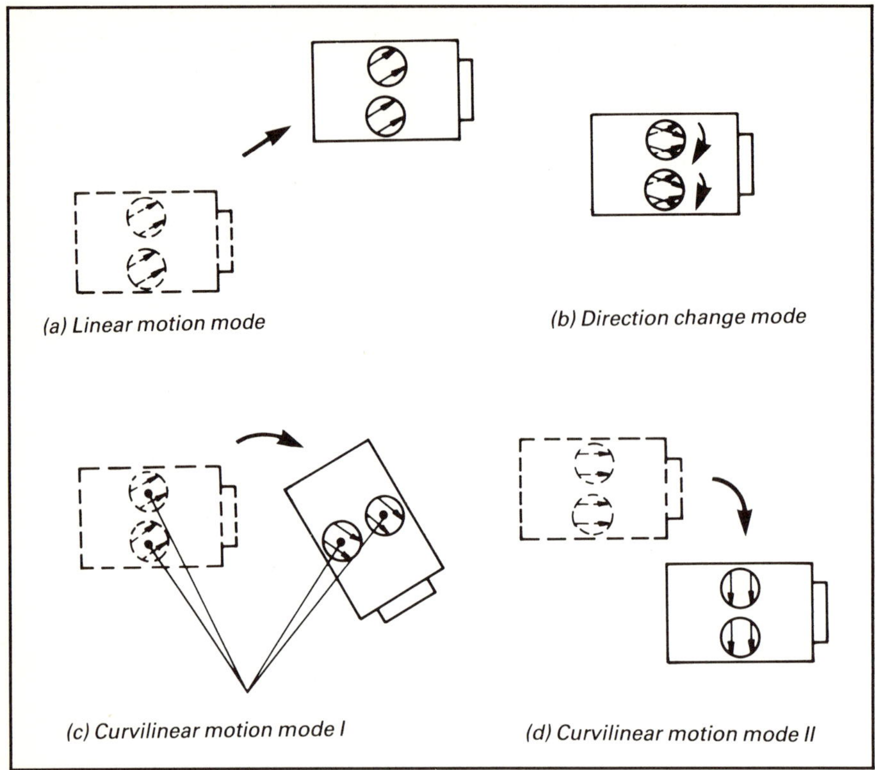

Fig. 8   Motion modes

Direction change mode. By revolving the two wheels attached to each rotary disc in opposite directions to each other and at the same speed, the direction of the rotary disc can be changed without changing the chassis direction. This mode is used to change the moving direction without changing the chassis direction after temporary stopping (Fig. 8b).

Curvilinear motion mode I. In this mode the vehicle can undergo curvilinear motion around a point selected freely by changing the chassis direction. The crossing point of the extension lines of the rotary shaft of the two sets of wheels is the centre of the curvilinear motion in this case (Fig.8c).

Curvilinear motion mode II. Here the robot vehicle undergoes curvilinear motion without changing the chassis direction. Such motion allows, for example, smooth change from straight travelling to transversal travelling (Fig. 8d).

Algorithm

The algorithm for the robot vehicle to follow exactly the reference line of the travelling route is as described below (see Fig. 9).

(1)   Control of speed in the X-direction. The revolving speed of the four wheels is controlled by taking the chassis speed in the X-direction as the reference. The command speed can be expressed by the following equations:

$$u_a = V \tag{1}$$

$$u_b = V \tag{2}$$

$$u_c = V \tag{3}$$

$$u_d = V \tag{4}$$

where u_a, u_b, u_c and u_d are the command speeds to each wheel and V is the travelling speed of the chassis.

(2)   Rotational correction. For rotational correction of the chassis, the feedback (u_ϕ) is calculated from the value of the deviation (ϕ) as follows:

$$u_\phi = a_1\phi + a_2\dot{\phi} + a_3\!\int\phi \tag{5}$$

where a_1, a_2 and a_3 are the feedback gains. The value of u_ϕ is deducted from the command value to the two wheels attached to the rotary disc at the left side of the chassis viewed in the moving direction, and is added to the command value of the two wheels attached to the rotary disc at the right side:

$$u_a = V - u_\phi \tag{6}$$

$$u_b = V - u_\phi \tag{7}$$

$$u_c = V + u_\phi \tag{8}$$

$$u_d = V + u_\phi \tag{9}$$

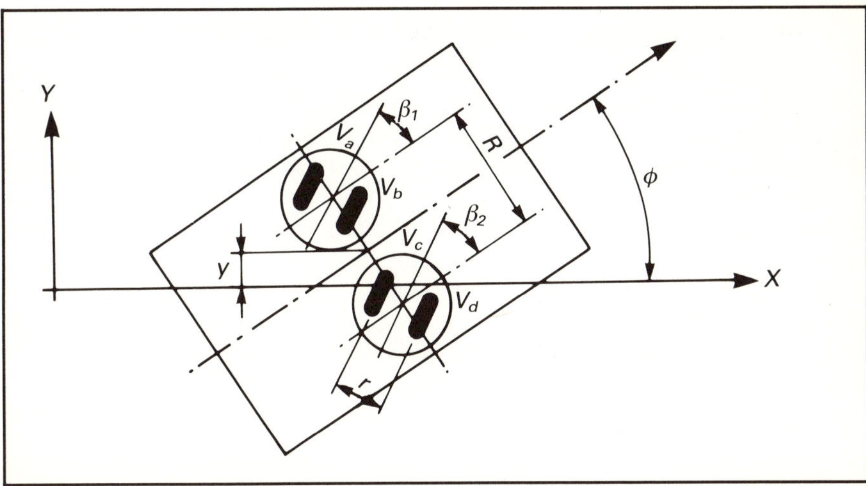

Fig. 9   Robot vehicle model

(3) Lateral correction (Y-direction). For lateral correction of the chassis, the feedback (u_y) is calculated from the value of the deviation (y):

$$u_y = a_4 y + a_5 \dot{y} + a_6 \int y \tag{10}$$

where a_4, a_5 and a_6 are the feedback gains. The value of u_y is added to the target values of the rotary discs (B_1, B_2) to give new target values (B_1', B_2'):

$$B_1' = B_1 + u_y = u_y \tag{11}$$

$$B_2' = B_2 + u_y = u_y \tag{12}$$

Then lateral correction of the chassis is made by directional control of the rotary discs.

(4) Control of direction of rotary discs. For directional control of the rotary discs, the feedbacks $(u_{\beta_1 + \phi}, u_{\beta_2 + \phi})$ are calculated from the difference between the directional target of the rotary discs $(B_1' = B_2' = u_y)$ and the actual value $(\beta_1 + \phi, \beta_2 + \phi)$ as follows:

$$u_{\beta_1+\phi} = a_7(\beta_1+\phi-u_y) + a_8(\dot{\beta}_1+\dot{\phi}-\dot{u}_y) + a_9 \int (\beta_1+\phi-u_y) \tag{13}$$

$$u_{\beta_2+\phi} = a_7(\beta_2+\phi-u_y) + a_8(\dot{\beta}_2+\dot{\phi}-\dot{u}_y) + a_9 \int (\beta_2+\phi-u_y) \tag{14}$$

where a_7, a_8 and a_9 are the feedback gains. The values of $u_{\beta_1 + \phi}$ and $u_{\beta_2 + \phi}$ are deducted from the command values to the left-side wheels in the moving direction, and are added to the right-side wheels:

$$u_a = V - u_\phi - u_{\beta_1+\phi} \tag{15}$$

$$u_b = V - u_\phi + u_{\beta_1+\phi} \tag{16}$$

$$u_c = V + u_\phi - u_{\beta_2+\phi} \tag{17}$$

$$u_d = V + u_\phi + u_{\beta_2+\phi} \tag{18}$$

The above-mentioned algorithm is suitable for the travelling function of the robot vehicle, and this steering method has the following advantage. Lateral correction of the chassis can be controlled only by the directional control of the rotary discs. In other words, lateral correction of the chassis can be corrected without rotating the chassis but simply by changing the direction of the rotary discs. Accordingly, the steering is more efficient than in the conventional method.

The research team proved the algorithm by computer simulation and also searched for the optimum feedback gain in the simulation. In the calculation the travelling system was expressed in the following manner by approximation:

$$\dot{x} = \{\frac{v_a+v_b}{2}\cos(\phi+\beta_1) + \frac{v_c+v_d}{2}\cos(\phi+\beta_2)\}/2 \tag{19}$$

$$\dot{y} = \{\frac{v_a+v_b}{2}\sin(\phi+\beta_1) + \frac{v_c+v_d}{2}\sin(\phi+\beta_2)\}/2 \qquad (20)$$

$$\dot{\phi} = \{-\frac{v_a+v_b}{2}\cos\beta_1 + \frac{v_c+v_d}{2}\cos\beta_2\}/2R \qquad (21)$$

$$\dot{\beta}_1+\dot{\phi} = \frac{-v_a+v_b}{2r} \qquad (22)$$

$$\dot{\beta}_2+\dot{\phi} = \frac{-v_c+v_d}{2r} \qquad (23)$$

where v_a, v_b, v_c and v_d are the wheel speeds, R is the distance between the two rotary discs, and r is the distance between the two wheels attached to a rotary disc.

Fig. 10 shows the results of the simulation and the data of the actual travelling.

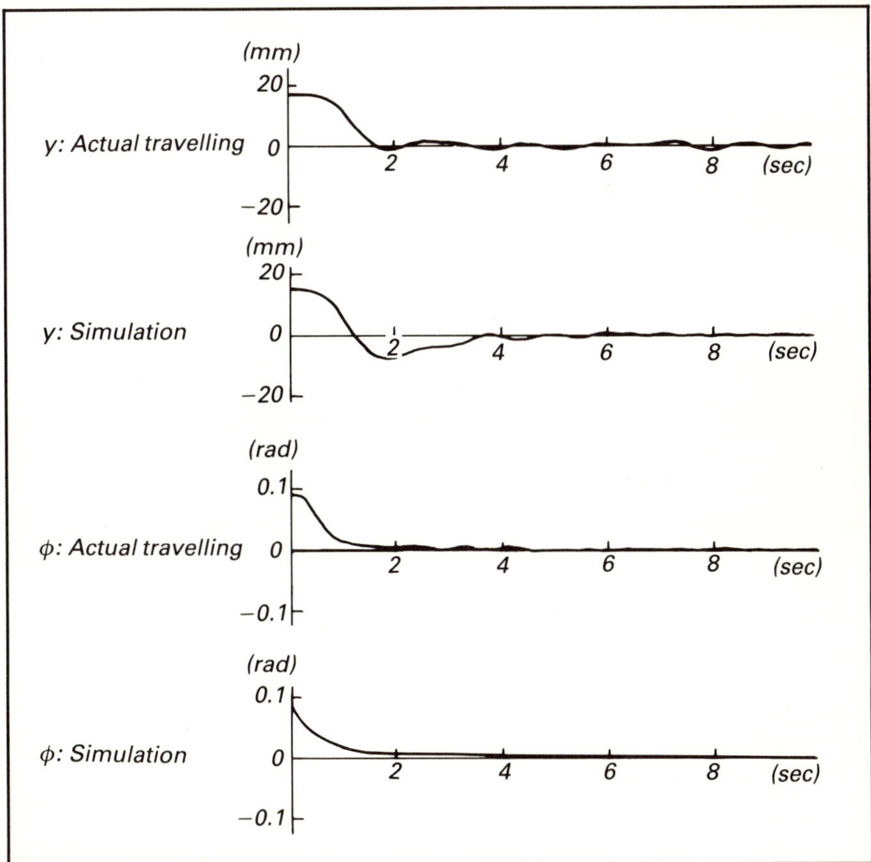

Fig. 10   Simulation results and actual travelling data

Transportation model system

Besides the robot vehicle, the essential elements of an unmanned transportation system are the work transfer function between the vehicle and the workstation, the communication function between the vehicle and the workstation and the communication network between the workstations. Using the latest technology for each of these elements, Mitsubishi has developed an advanced unmanned transportation model system (Fig. 11).

Transfer function

The vehicle is provided with a transfer device composed of a table to carry workpieces and a pusher at the upper part. After stopping of the robot vehicle at the workstation, the loaded workpieces are carried out to the workstation by the pusher. For loading the workpieces onto the vehicle, a pusher at the workstation side is employed. Fig. 4 shows the robot vehicle with the transfer device.

Communication function

A communication unit for projection and reception by infrared laser diodes and photodiodes is provided on both sides of the vehicle. The same communication unit is also provided at the workstation side, and non-contact communication of operation and other instructions is made by these units when the vehicle stops at the workstation. The system is serial, and the transmission protocol for the communication between control and host computers is MSEC, based on SECS, the communication standard of the USA. The transmission speed is 2400 byte/s.

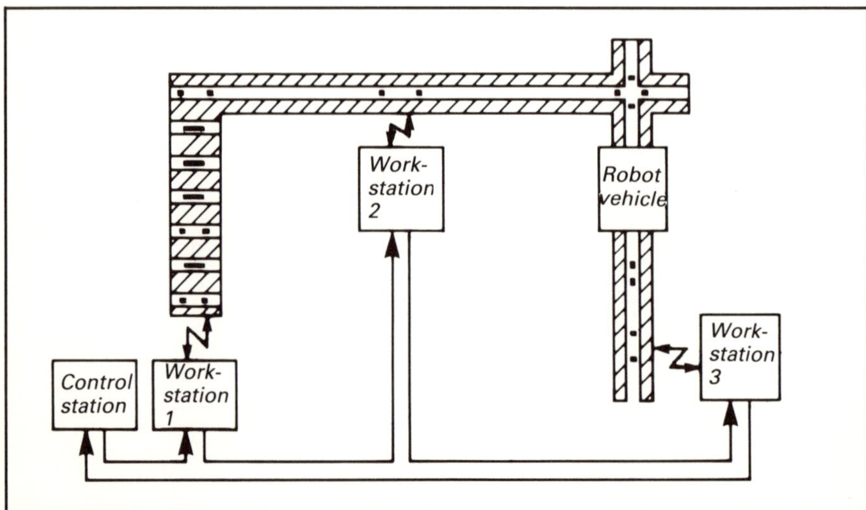

Fig. 11   Transportation model system

Communication network

For this model system, one control station and two or more workstations are connected through an optical fibre link. The control station monitors the condition of each workstation and issues workstation travelling instructions to the vehicle and work transfer instructions.

Since the operational sequence of the vehicle and the workstations is transmitted from the control station, there is no need for the vehicle and the workstations to memorise the sequence in advance. This makes the vehicle operation highly flexible.

In this network, a loop transmission system with SDLC protocol is employed to enable connection of up to 225 workstations, and the transmission speed is 1Mbyte/s. In addition, a work stocker and automatic battery changer are provided to complete a highly advanced unmanned transportation system.

Concluding remarks

This paper has presented a model system of unmanned transportation, with particular attention to the travelling function of robot vehicles. The robot vehicle can move in any direction without rotation of its chassis and can communicate with the terminals by the wireless method. C-MOS ICs are employed for the control circuit to save the battery power of the robot vehicle.

Mitsubishi is convinced that this transportation system is very effective for factory automation and intends to develop it further. It is also scheduled to develop the travelling function of the robot vehicle into more advanced applications such as state monitoring in factories. In this case it will be important to study further the environment recognition function and the more advanced inducing method.

References

[1] Hatschek, R. L. 1980. Guided carts link machines into 'system'. *American Machinist,* August 1980, 97-100.

[2] Fujiwara, K., Kawashima, Y., Kato, H., Watanabe, M. 1981. Development of guideless robot vehicle. In, *Proc. 11th Int. Symp. on Industrial Robots,* pp.203-207.

[3] Arai, T., Nakano, E., Hashino, S., and Yamaba, K. 1981. The control and application of 'Omni-Directional Vehicle (ODV)'. In, *Preprints of the 8th IPAC Congr.,* Vol. 14, pp.1-6.

A CONCEPT FOR AN AUTONOMOUS MOBILE ROBOT AND A PROTOTYPE MODEL

Y. Nakano, M. Fujie, T. Iwamoto, K. Kamejima, K. Sugiyama and M. Funabashi
Hitachi Ltd, Japan

A concept for an autonomous and self-reliant mobile robot is presented and a prototype model is described. The model can move about by its own decisions in an environment with corridors, stairs, doors and randomly located obstacles. It is equipped with a TV camera with a wide visual field angle for autonomous guidance. The arm is of the force-feedback multi-articulated type with six degrees of freedom. Carbon-fibre reinforced plastic is used for the mobile unit and the arm. Because of its light weight and high-speed guidance, the robot is able to drive at 2km/h.

Robots have established their status in the manufacturing industries with many successful results. As a result, robots will be widely applied in various new fields. The inspection and repair of large industrial plants and the maintenance of buildings are typical examples. In spite of strong demands for automation, most of this work is still carried out as a manual operation because of its diversified and irregular nature, and also because job sites are widely separated. The introduction of advanced robot technologies into these fields seems to be the most effective solution to the problem.

The necessary technologies for this type of robot are shown in Fig. 1 in comparison with conventional industrial robots. Viewing this diagram, one might say that autonomous mobile robots are the robots in demand for pioneering the new fields, and also the most effective means for developing advanced robot technologies for future industrial robots.

Concept

For such robots to be applied to the maintenance of say, nuclear power facilities, they will need to have the following capabilities:

- Accessibility – Ascent or descent of stairs, and passage through narrow and winding paths.
- Stability – Stable posture control while negotiating rough-roads.
- Independency – Power self-reliance and wireless communication.
- Operability – Close cooperation with operator.

Robots which have these capabilities must have the following functions:

- Mobile mechanism to enable rough-road negotiation.
- Compact and dexterous arm mounted on mobile mechanism.
- Multi-sensory system including vision for environment recognition.
- Wireless data transmission system.
- Autonomous navigation and guidance.
- On-board power supply system.

The authors intended to develop an autonomous mobile robot that fulfils all of these functions by integrating the most advanced robot technologies.

Prototype model and key technologies

Transformable crawler-type mobile mechanism

There are three major mobile mechanisms for robots: wheel, leg and crawler. Wheels are a very efficient method, especially on flat floors, but the height of the overstep is limited by the wheel diameter. Legs have the possibility of being the most nimble mobile mechanism, but they require a high degree of control to manage the complex multi-joint mechanism. They are expected to be a common method for future robots. Crawlers are simple mechanisms, and suitable for travelling on rough terrain if the crawler track is designed to be adaptive to different terrains. The authors decided on the crawler-type for the model.

Fig. 2 shows the newly developed transformable crawler-type mobile mechanism. The crawler unit consists of a pair of tracks, main wheels, planet wheels, main and sub-arms and side frames. This type of crawler is superior in compactness and mobility to multiple crawler-type mechanisms. By controlling a pair of planet wheels, the tracks around the wheels vary in shape to adapt to road undulations, thereby enhancing the travelling capability of the crawler. In order to ensure this track transformation, the following conditions must be satisfied. When all the wheels are of equal radius, the locus of the centre of a planet wheel has to be an elliptical orbit having two focal points at the centres of the front and rear main wheels in order to keep the track of constant tautness.

A unique idea is adopted to control the locus of the track so as to maintain constant track tautness. A planet wheel is mounted on the sub arm, which is connected to the main arm through a simple gear

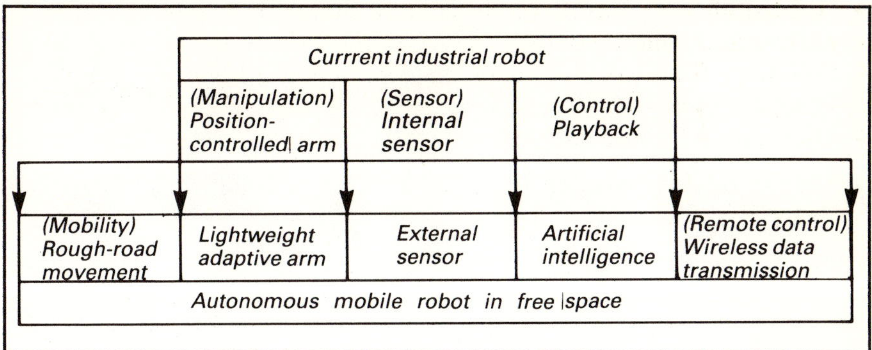

Fig. 1 Trend of robot technologies for autonomous mobile robots

mechanism (see Fig. 2). If the sub arm (length: r) is mounted on the top of the main arm (length: R) and rotates at the same angle (a) as that of the main arm but in the opposite direction, the centre $P(x,y)$ of the planet wheel is on the following ellipse:

$$x^2/(R+r)^2 + y^2/(R-r)^2 = 1$$

The ratio of RT to r is determined so as to satisfy the following relation:

$$F = 4\sqrt{Rr}$$

where F is half the distance between the two focal points, which determines the size of the crawler.

To attain a lightweight, CFRP (carbon-fibre reinforced plastic) was used for the mobile unit and arm.

As a result, a compact, lightweight and nimble mechanism has been developed. Fig. 3 shows the sequences of the crawler mechanism in overcoming obstacles.

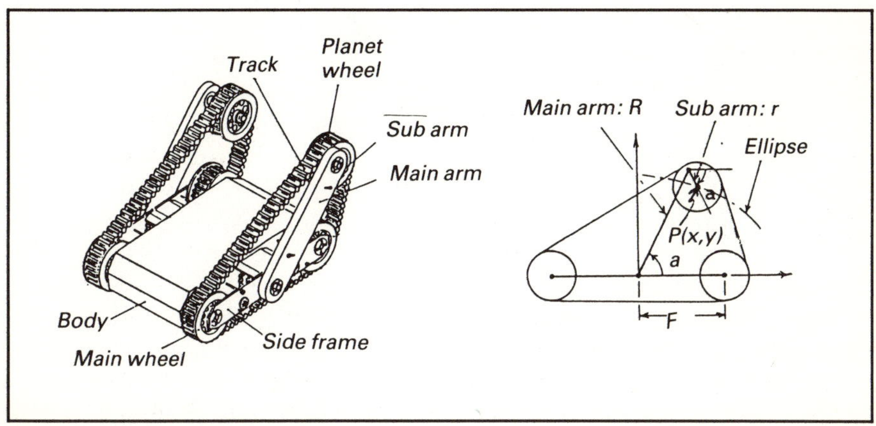

Fig. 2 Transformable crawler

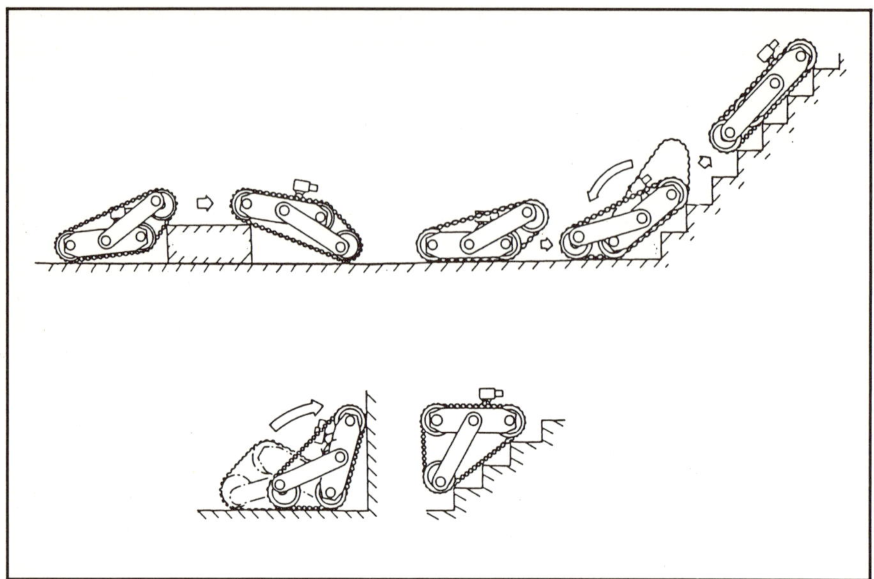

Fig. 3   Sequences of transformable crawler manipulation

Lightweight multi-articulated arm

The robot arm has been required to be lightweight and dexterous. This is especially important for an arm mounted on a mobile mechanism. The arm must perform a coordinated manipulation while moving.

The newly developed multi-articulated arm has been much improved in these areas. CFRP is employed not only for the beam but also for joint parts. A six-axis force-torque sensor, which is set on the wrist, feeds back control signals to each joint to enable highly sensitive and dexterous control of the arm. This allows coordination of arm manipulation with robot movement by its controlled elasticity function. Moreover, this control allows the arm to be of lightweight construction.

Perceptive guidance

An autonomous mobile robot for maintenance work is required to have a travelling speed around 1 to 2km/h, as it cooperates with human operators. Such 'high-speed' guidance has never before been attained. Moreover, the guidance should not be based on specially settled landmarks but the entire scene itself. The guidance system presented here is based on high-speed perception of the surroundings using information obtained from multiple sensors.

Fig. 4 shows the conceptual basis of the guidance system. The 'navigator' generates a predicted target pattern for guidance by using a map that is stored in the computer memory in the form of a three-dimensional wire-frame model, and the route is settled on the map. The 'image detector' receives an image through the vision sensor and extracts the edge pattern by a differential process. Autonomous guidance is then conducted by matching the two image patterns through a 'pattern potential guidance method' as follows. The disparity vector obtained by connecting corresponding points on the two image patterns denotes the gradient vector field caused by potentials sourced in the reference pattern. As the disparity occurs as the point of view of the moving robot changes, guidance can be conducted so as to reduce the value of the disparity vector. The guidance vector $U=[dx,dy]$ is obtained from the following equations:

$$dU/dt = W\int_s X[f_m]\, Dp\, ds, \quad W = \mathrm{diag}[w_1,w_2]$$

$$\delta p/\delta t = Lp - X[f_r]$$

where L is the Laplacian, D the gradient, X the characteristic function of set [.], w_1 and w_2 are positive constants, f_r is the predicted pattern, f_m the detected pattern, p the pattern potential, diag [.] the diagonal matrix and S the imaging screen.

Fig. 5 illustrates the mechanism which converts the sequence of detected images to the guidance signal. In this system, the differential operator L is implemented by the image processor I, while the gradient is implemented by image processor II.

Since prediction of the surroundings requires wide-range surveillance, an integrated perception system of proximity sensors and a TV

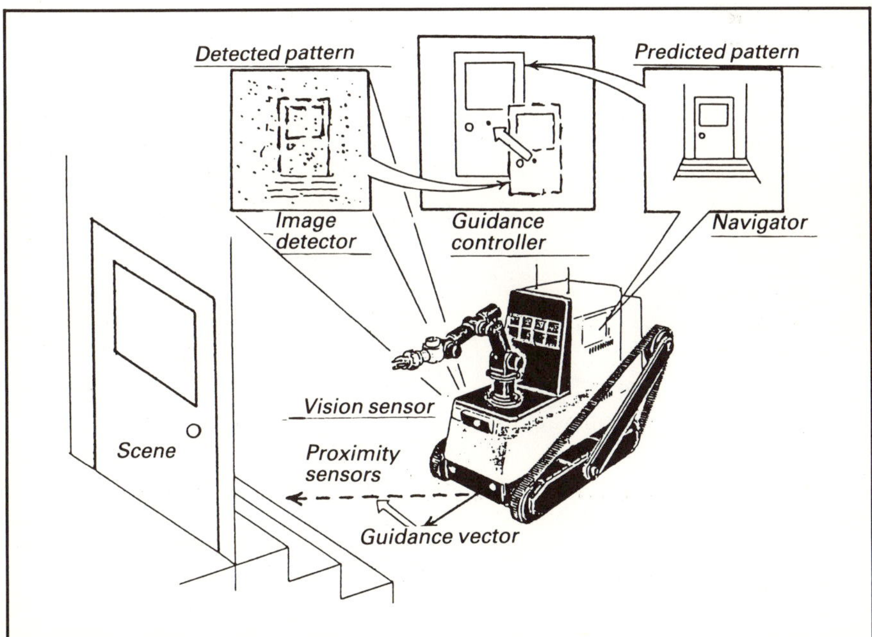

Fig. 4   Robot guidance by perception control

camera with a 162° visual field angle has been developed for the main observation channel.

Spread spectrum transmission

Wireless data transmission is essential for supervising the autonomous movement of the robot. As shown in Table 1, spread spectrum transmission is the most promising method for a mobile robot. However, present spread spectrum transmission systems require improvement.

Fig. 6 shows a newly developed spread spectrum transmission system. Blocks outlined by bold lines show areas of improvements, especially for increasing data transmission speed to the level of image transmission.

At the transmitter, the input signal is converted to another wider band by the spreading code. This code has a flat and wide band spectrum to realise high-quality transmission. The input signal is then modulated by carrier wave and the spectrum form is modified to compensate distortion in the high and low frequency regions of the wide band signal caused by characteristics of the transmission medium. At the receiver, an almost reverse process is executed by using modulation signals synchronised to the transmitter, and the original signal is regenerated.

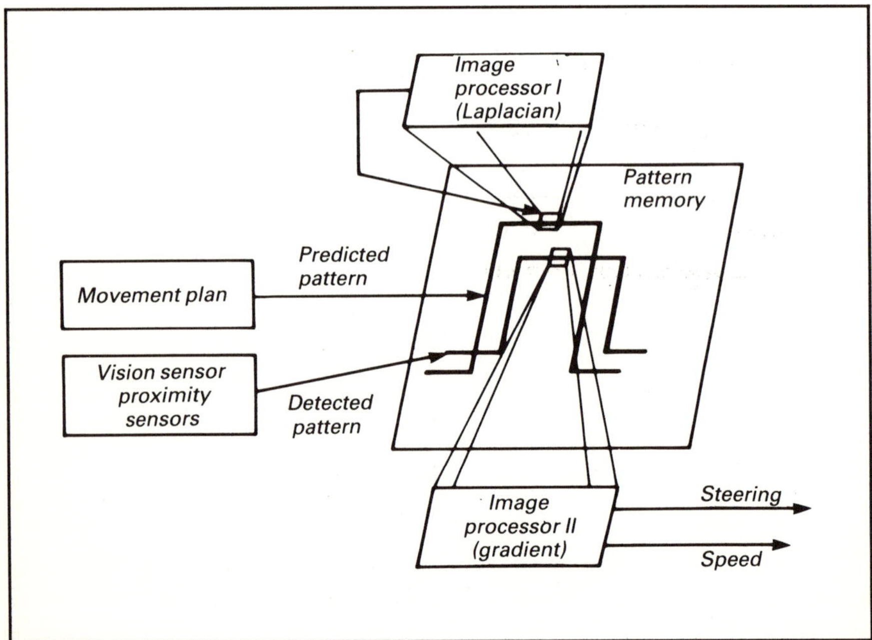

Fig. 5   *Pattern potential guidance controller*

Table 1 Comparison of wireless data transmission for robots (current methods)

	Optical beam	Inductive cable	Coaxial leakage cable	FM	Spread spectrum
All direction	N	M	M	M	G
Distance	G	N	N	G	G
Speed	G	N	G	G	M
Multi-station	N	G	G	G	G
Electro-magnetic hazard	G	M	M	N	G

The noise reduction effect during spread spectrum transmission is calculated from the following equation:

$$(S/N)_0 = (S/N)_1 \, (W_s/W_d)$$

where $(S/N)_1$ and $(S/N)_0$ are the signal-to-noise ratios of the received and output signals, respectively, and W_s and W_d the band-widths of the spreading code and transmitting data, respectively.

This last equation shows that a high S/N ratio can be attained with a large W_s/W_d ratio, which allows data transmission by low-power radio wave. On the other hand, stable and quick synchronisation of the spreading code and carrier demodulation is essential. In this system, this is attained by separating two synchronisations for receiving.

Prototype model performance

A prototype model was built that combined the newly developed technologies here described. The intended performance of each technology was confirmed and total system operation was demonstrated in a mock environment typical of the interior of buildings, with

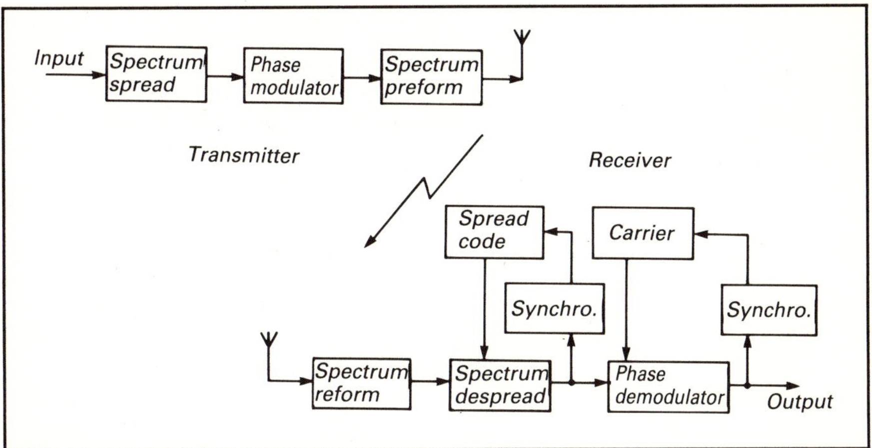

Fig. 6 High-speed spread spectrum transmission

corridors, corners, doors and obstacles in the passage way. The results are listed as follows:

- Autonomous running speed: 2km/h (flat), 0.6km/h (stairs).
- Size and weight:
     Mobile unit: weight 150kg, payload 150kg.
     Arm: weight 25kg, handling weight 3kg, length 850mm.
- Capacity of wireless transmission: speed 135kbyte/s, output 1mW (max.), distance 60m.

Concluding remarks

An autonomous mobile robot has been developed. This robot demonstrated the following new features which are essential for pioneering new robot application fields:

- Lightweight and compact transformable mobile mechanism.
- Viewpoint-controlled wide-range vision.
- Perception guidance using the pattern potential method.
- Lightweight force-feedback multi-articulated arm.
- High-speed spread spectrum wireless data transmission.

These newly developed features could greatly contribute to manufacturing autonomous mobile robots for use in areas such as:

- Automatic inspection and maintenance of nuclear power plants.
- Construction and material handling in space stations.
- Material handling and transportation for factory automation.

AN INTELLIGENT FACTORY TRANSPORT SYSTEM

J. J. Solberg and C. D. McGillem
Purdue University, USA

The value and feasibility of a novel material handling system intended to fulfil the needs of an intelligent factory are discussed. The goals of the system are to provide a high degree of flexibility while at the same time permitting a major reduction of work-in-process inventory. The system employs a fleet of small, fast robot vehicles able to travel freely throughout the factory. The vehicles would be in constant communication with a central computer that would dispatch them on demand. The vehicles would handle individual workpieces or in some cases two or three units at a time. Individual vehicles would have sufficient on-board intelligence to navigate throughout the work space, detect obstacles, avoid collisions, maintain speed and so forth. The central computer would optimise utilisation of the vehicles and the machines in a flexible manufacturing system. The various tasks involved in a research project designed to prove the feasibility of this system are described.

The work described herein relates to one of several major projects of the Center for Intelligent Manufacturing Systems, one of six Engineering Research Centers created in 1985 under funding from the National Science Foundation of the United States Government. The Center at Purdue has the goal of providing the engineering foundations for a new generation of discrete product factories and of developing the methodologies for designing and operating them. This new generation of technology goes under the heading 'intelligent manufacturing systems', a phrase intended to describe a higher order of automated manufacturing systems than any now known. This paper discusses the potential value and feasibility of a novel material handling system which might fulfil the transport needs of such an intelligent factory. The research project is still in its early stages and consequently the text makes more mention of problems than solutions. However, the project workers believe that there are strong motives for taking the direction chosen in this research.

There are two powerful forces acting upon the future development of manufacturing technology: the need for greater flexibility, and the need for reduction of work-in-process inventories. (Other frequently mentioned objectives – such as reducing flow times, or reducing batch sizes – can be shown to be equivalent to these two.) Many of the limiting factors in achieving improvements in flexibility and inventory reduction have to do with the time, cost and difficulty of set-ups. However, part of the problem lies purely within the realm of material handling.

The many types of factory material handling equipment available today can be grouped into two categories. The first category includes various kinds of conveyors (belt, roller, overhead, etc.). Conveyors are capable of transporting workpieces individually, which is essential for minimum in-process inventory, but their routing is inherently inflexible. They are suitable primarily for high-volume dedicated line production, a form of manufacturing which seems to be less and less justifiable as product design lifetimes shorten and diversity increases. The second category consists of various kinds of vehicles, such as fork-lift trucks, tractor/trailer carts, and the currently favoured automated guided vehicles (AGVs). Vehicles are considerably more flexible than conveyors, but those that are commercially available today are designed to carry large loads. The policy of accumulating loads for batch handling runs counter to the objective of keeping inventories low. Thus there appears to be no really suitable means of automated transport for the very flexible, very efficient factories we hope to see in the future.

Most of the larger flexible manufacturing systems (FMS) that are being installed incorporate AGV systems which use in-floor cables for guidance. This system concept seems to have evolved out of earlier transport methods employing trucks with human drivers. The driver was eliminated (at the expense of limiting somewhat the manoeuvrability of the vehicle), but the size and speed of the vehicles remained much the same. In particular, the vehicles are designed to carry heavy loads, they move slowly, and they are individually quite expensive. Of course, if a vehicle is slow and expensive, it is certainly desirable for it to be capable of carrying a large load. Conceptually, the present AGV systems are mass transit systems, analogous to a railroad.

Consider now, however, a very different concept. Suppose that one were able to purchase small and fast robot vehicles. Suppose that they were not confined to fixed guidepaths, but were instead able to travel freely anywhere throughout the factory. Imagine that they were in constant communication with a central computer, which was able to dispatch them on demand. This system would function more like a personal transit system, analogous to a fleet of taxi-cabs. The vehicles would handle individual workpieces, or perhaps in some cases two or three units at a time. The idea would be to advance the work from one workstation to the next as soon as it is ready to go, as opposed to waiting until a sufficient load has accumulated.

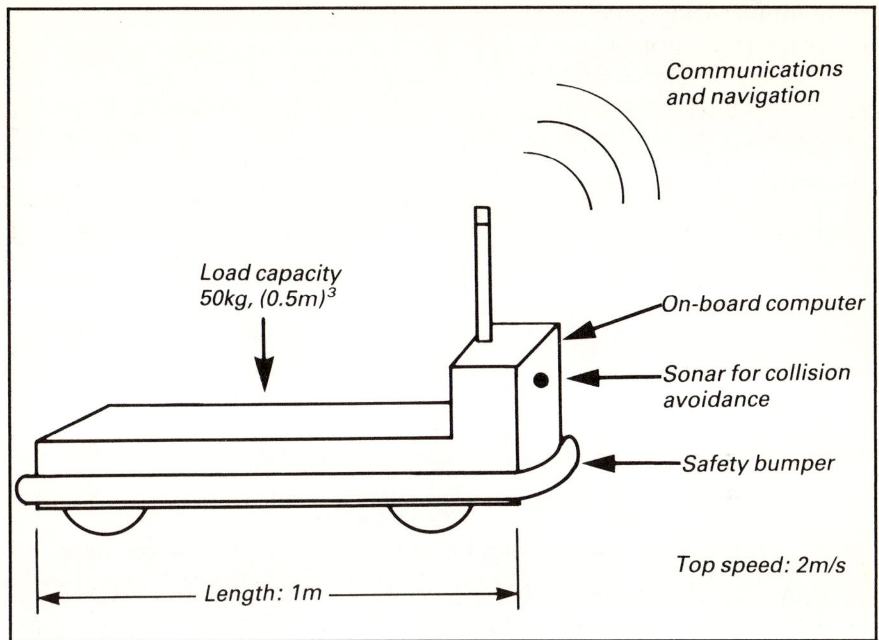

Fig. 1   Conceptual design of robot vehicle

The existence of such a system would open many opportunities for handling ancillary traffic. For example, routine tool replacements would be easy to implement. More significantly, non-routine errands could be accommodated without disrupting normal operations. An operator or repairman could send a cart to fetch a tool, a spare part, a fixture, or even his lunch.

All of the carts would communicate with a central 'traffic control' computer. This computer would receive requests from machines (or, in less automated facilities, human operators), dispatch a nearby available vehicle, identify unobstructed paths for the quickest available routes, and so forth. Individual vehicles would have sufficient on-board intelligence to detect obstacles, avoid collisions, maintain a heading, control speed, and so on.

Fig. 1 shows a preliminary conceptual design of one of these vehicles. A typical installation would use perhaps 20 to 30 carts. Because there are so many (relative to the number of vehicles accepted as normal in a conventional system), it would be essential that the cost of an individual vehicle be low. However, because the payload would be so much smaller, it should be possible to achieve low production costs (provided, of course, that the design of the vehicles is consciously orientated toward that goal).

Fig. 2 contrasts the (expected) characteristics of the new system concept with alternative approaches. The major penalty paid for achieving improved characteristics will be the much higher sophistication required to create the system. This is a one-time cost, however,

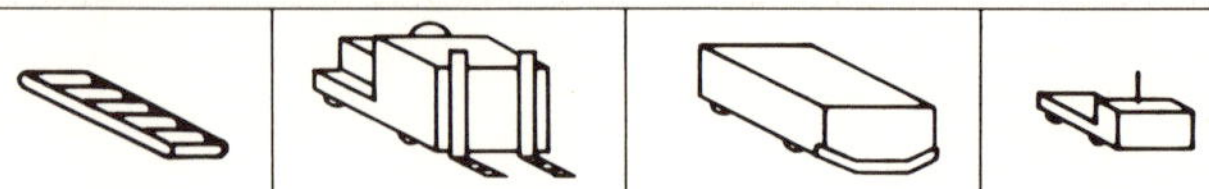

Characteristic / Type of system	Conveyor	Driven vehicle	AGV	New concept
Flexibility	poor	good	moderate	good
Responsiveness	good	poor	moderate	good
Inventory requirements	good	poor	moderate	good
Ease of installation	poor	good	moderate	good
Vulnerability to failures	poor	good	moderate	good
Simplicity	good	good	moderate	poor
Cost	poor	moderate	moderate	good

○ = poor ◉ = moderate ● = good

Fig. 2 *Comparison of new system characteristics with alternative approaches*

and once the system design is perfected it should be easy to employ – that is both easy to install and easy to operate. Vehicle repairs would most likely be beyond the capability of in-plant maintenance personnel, but a few back-up carts would take care of any problems relating to failures.

Origins of the concept

The possibility of achieving vast improvements in flow time and in-process inventory (without adversely affected production rates) was uncovered by mathematical modelling. The principal tool employed was CAN-Q[1,2], a model based on the theory of networks of queues. CAN-Q was developed at Purdue in the period 1975-80, and has proved its worth in hundreds of applications in both universities and industry[3,4].

As an indication of the potential benefits from reducing the transport load size (vs accumulating loads), consider the results of the CAN-Q analysis shown in Figs. 3, 4 and 5. These figures show the effects upon average flow time, average in-process inventory, and average production rate, respectively. Each graph contrasts the values obtained when the transport load size is 24 or 3, for systems consisting of from 8 to 20 machines. In each case, the same product mix with appropriate work requirements is assumed. The original data came from an actual factory, in which the material handling was by fork-lift trucks carrying an average load of 24 workpieces. The significant observation suggested by the graphs is that, regardless of the size of the system, both flow time and in-process inventory could be reduced without serious adverse impact upon production rates, simply by minimising the transport load sizes.

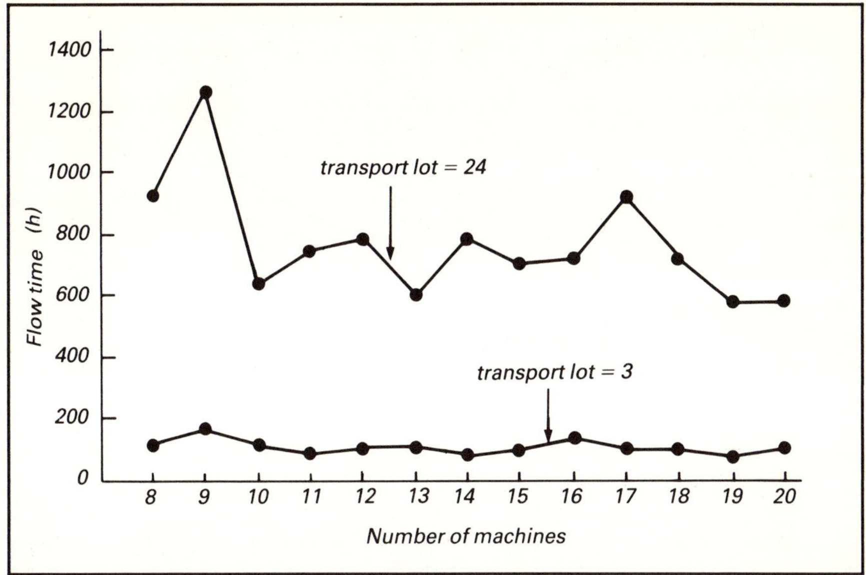

Fig. 3   Flow time vs number of machines for different lot sizes

At first, this analysis was viewed sceptically; the conclusion seemed too simple to be true. Upon reflection, however it is easy to understand how these gains are achieved. The policy of delivering to a machine a full load of workpieces to be processed one at a time and then also holding workpieces after an operation in order to build up the transport batch introduces both delays and in-process inventories

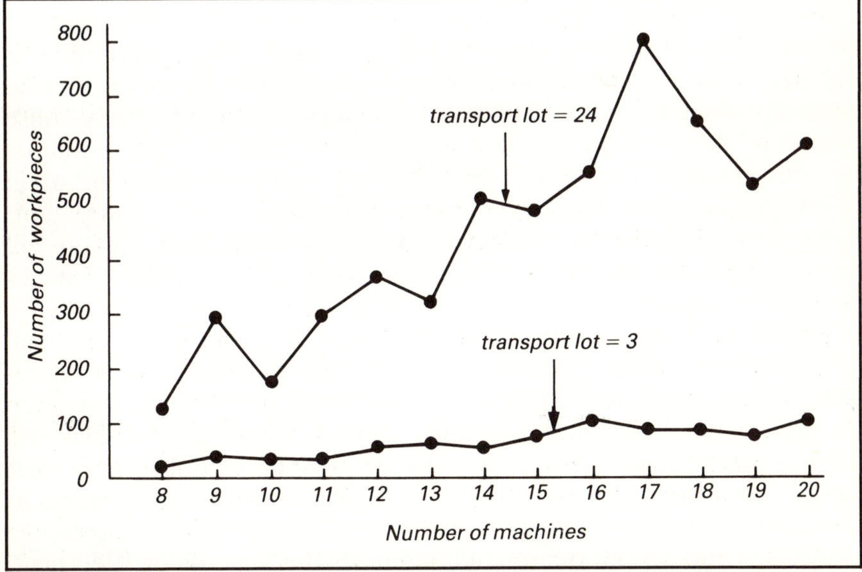

Fig. 4   Work-in-process inventory vs number of machines for different lot sizes

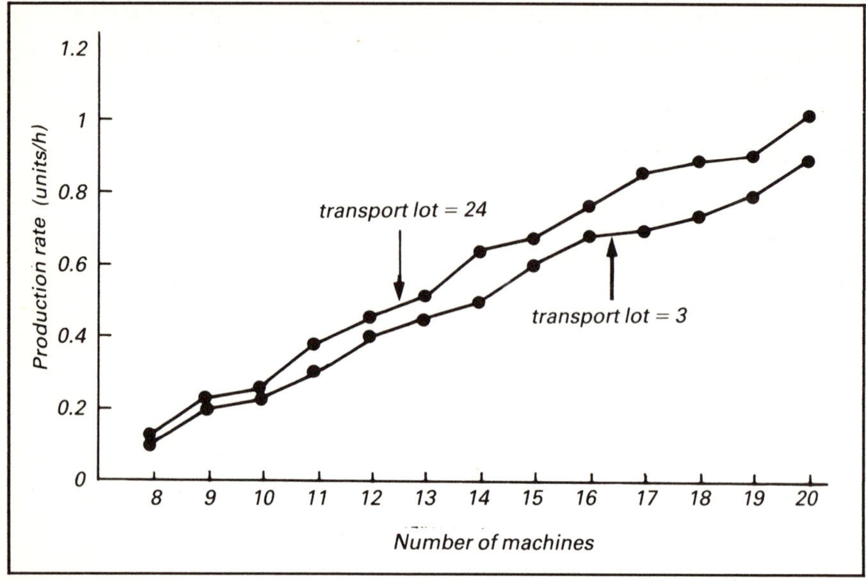

Fig. 5   *Production rate vs number of machines for different lot sizes*

which are theoretically unnecessary. The practice stems, of course, from the very real necessity of limiting the amount of transportation demands in a factory, which in turn is a direct consequence of the characteristics of the material handling systems available. Indeed, it is an established principle of facility layout and material handling system design that one should attempt to minimise trip demands, in order to avoid overloading the transport system.

The mathematical analysis cited above assumed that the number and speed of the vehicles were sufficient to handle whatever traffic was generated. This is an easy assumption to make, but the fact is that no currently available material handling equipment could satisfy those demands. In other words, it was discovered that sizeable potential benefits were to be found, but that these were dependent upon developing a whole new concept in material handling equipment. Listing the functional requirements of a hypothetical system led directly to the concept of small, fast, free-roaming robot vehicles described earlier.

Inasmuch as the concept is a sharp departure from existing material handling technology, considerable research in several areas is required. Basic feasibility studies on a small scale have been undertaken at Purdue to verify that the most serious technical obstacles are surmountable. The results have been encouraging. A full-scale project has begun within the Center for Intelligent Manufacturing Systems to develop a complete demonstration system. The project involves five professors and ten graduate students from the Schools of Electrical, Industrial and Mechanical Engineering.

The remainder of this report outlines the areas of research under investigation, with particular emphasis on the questions yet to be answered.

Research programme

Establishment of performance specifications

Since the transport system is intended to accomplish performance improvements in a new and untried manner there are not comparable systems available from which one can draw conclusions regarding design trade-offs. Consequently these choices must be based on simulation and on quantitative analysis. Just as computer models suggested the need for this new material handling system, similar models will guide its design. For example, the relation among the maximum speed of each vehicle, the number of vehicles required, and the total system cost will be mathematically expressed and used to guide the choice of motors, gears and wheel size for the vehicles. Similar cost/performance issues arise in almost every aspect of the design. The principal modelling technique that will be used to guide the early design decisions will be CAN-Q.

Vehicle design

This task is concerned with all of the issues related to selection of the vehicle configuration and the mechanical design of the vehicle itself. These include physical dimensions, weight suspension, shock absorption and propulsion. All of these considerations must be carefully balanced in view of the design speed requirement of 1–2m/s and the need to ensure the safety and health of the worker and machine alike in the factory environment where it is to be used. Another important design consideration will be the dynamic stability of the vehicle under normal manoeuvring and during collision conditions.

The propulsion system design will address the issues of the steering mechanism, the drive train and the energy source. The selection of the steering mechanism, the number of wheels and the driven wheel configuration will directly influence the vehicle navigation system. The vehicle should be operable on floors typically encountered in a factory environment.

Loading/unloading

Special mechanical handling systems will be investigated for loading and unloading of the vehicle. An important question to be answered is whether this should be done by a mechanism on the vehicle or off the vehicle. A closely related question is should the vehicle be guided into the final workstation position by an external mechanical system, or would this be accomplished by the vehicle itself? This will affect the way in which loading and unloading is carried out. If the loading mechanism is placed on-board, actuated with the energy from the battery pack, then it must be designed for maximum efficiency to ensure a frugal consumption of energy. Consideration will be given to expeditious loading and unloading techniques since the economy of time is at the heart of this project. The possibility of loading/unloading a moving vehicle will be considered. There are important considerations with respect to the prehension system for handling the parts, for example, whether the prehension system should be compliant for gripping the workpieces with varied boundary characteristics, and whether it should possess some intelligence to make its operation fast and efficient.

Location

Feasibility of the flexible materials handling system hinges upon development of an accurate, reliable and economical vehicle locating and navigating system. None of the systems described in the literature appear suitable for this application[5-11]. Preliminary studies at Purdue have shown that a system utilising infrared beacons mounted around the periphery of the work space can form the basis of a precision location system[12]. The angles between two or more pairs of beacons measured at the vehicle provide sufficient data to solve for the vehicle position. Alternatively, measurement of the differences in the radial distances from a master and two or more reference beacons also provides sufficient data for accurate location. Another technique that appears suitable is to employ an accurate dead reckoning system with special 'signposts' that provide precise heading and position information to the vehicle when it passes by. In theory, all of these systems can provide continuous vehicle location to accuracies in the order of 10cm in a typical factory location. The choice of the location system to be utilised with the mobile vehicles will be made on the basis of detailed design studies, simulation, empirically determined data characterising the factory environment, and experimental evaluation of breadboard models of the systems. A number of factors such as sensitivity to measurement errors, interfering radiation, contamination of sensor elements, and presence of slippery floors will be included in the analyses before a final selection is made.

Communication

An equally important problem that must be addressed in connection with the mobile vehicle development is that of communication to and from the vehicle. Because of the great cost reduction that has occurred in recent years in computer components it is feasible to include substantial computational capability on the vehicle itself. However, there are a number of advantages in keeping the majority of computations centralised. Part of the trade-off is between communications performance requirements and computer performance requirements. Another part of the trade-off relates to the advantage of having all pertinent information available at a central control location to allow more effective optimisation, easier reprogramming of routes, handling of emergency conditions, and similar items. The communication problem must be considered in terms of multiple vehicles in simultaneous operation. The number of vehicles may range from 20 to 100 or more. The amount of data that must be communicated depends on how much computation is carried out on the vehicle and how much is carried out at a central location. For example, the location measurement data could be continuously transmitted to a central computer and the vehicle control signals sent back. Or the location computations could be made on the vehicle and status reports sent to the central computer. In any case it will always be necessary for the control centre to know the location of all vehicles and their status regarding execution of their assigned tasks. The vehicles can have the capability of deciding their own route for task execution or the route can be specified by the control centre. Provisions will be necessary to detect abnormal vehicle behaviour and communicate it to the control centre. This would include such things as low battery charge, loss of speed, excessive vibration or listing, detection of obstacles in a normal path location, or occurrence of an accident. The data rates may be high both to and from the vehicles.

A fundamental problem in connection with the communication system is selection of the carrier to be used. The candidates are infrared (IR) or radio frequency (RF) radiation. If the location system employs IR beacons it may be possible to superimpose modulation on the location signals to handle communications. If IR is not used in the location system or is found to be inappropriate for other reasons, then an RF system will be considered. Whatever system is used, it appears highly desirable that it should have a multiple access/discrete address capability. The system must be able to operate in a hostile environment of both transient and steady-state interference. It is anticipated that error correction encoding of signals will be required and that some type of spread spectrum modulation will be necessary to combat the interference and provide the required reliability of communication. The requirements for communication data rate, nature of signals to be transmitted, quantisation levels, sampling frequencies, bandwidth and power level will depend heavily on the results of other aspects of the vehicle and system design and will be constantly changing as the system evolves.

Vehicle and system control

Control of the vehicles has two different connotations. One is the generation of the specific assignment for each vehicle and the monitoring and modification of that assignment from an overall system point of view. The other is control of the vehicle in its detailed execution of the task assigned to it. These two control functions are intimately connected with both the global and macroscopic characteristics of the system and their development will require continuous coordination across all tasks in the program. Overall system control will draw heavily on operations research techniques, queuing theory, optimisation procedures and modelling. Control of the vehicle during task execution will require development of steering signals to keep the vehicle on its desired path, and this in turn will require continuous calculation of heading and position signals from the location system. Coordinated turning will be required in rounding corners, in moving to and from loading stations, and for obstacle avoidance. Some type of obstacle or collision detection system will be required to prevent accidents from occurring. The control system will have to handle signals from such a system and cause the vehicle to take appropriate action.

It is anticipated that the mathematical models developed for performance specification will be useful for aggregate-type decisions for capacity planning, configuration selection, etc. Detailed facility design and operational control issues will be addressed separately. Optimisation and simulation methods will be applied to address issues related to the following:

- Material flow topology, including the determination of whether flow paths should be unidirectional (where vehicles travel along aisles in one direction only), or bidirectional (where vehicles are allowed to travel along some or all aisles in both directions).
- Vehicle routing to cover both static and dynamic scheduling rules.
- Zone assignment strategies whereby certain vehicles will be assigned jobs moved only within predefined areas.
- Traffic management strategies and traffic resolutions at conflict zones (e.g. vehicle priorities at intersections).

Previous studies[13] have shown that, for systems where vehicles follow an embedded-wire guidepath, extreme congestion and grid-locking can take place. Congestion results in poor system performance while gridlocking is not an acceptable condition in automated systems where down-time is usually expensive. When gridlocking takes place it is necessary to have a subordinate rule to release blocked queues. An intelligently planned and controlled system will be able to recognise conditions that lead to these problems and institute alternative actions to avoid blocking and gridlocking.

Bidirectional guidepaths for these systems also offer significant potential for improvement. In comparative studies between un-idirectional and bidirectional guidepaths[14], significant improvements in shop throughput are shown to be possible in systems where bidirectional flows are allowed. Additional benefits of reduced hardware for guidepaths and decreased requirements for aisle space are possible. However, for bidirectional flows the traffic control and scheduling problems are more difficult.

The use of simulation as a design tool for material handling systems is well documented[15,16]. However, simulation alone would not be adequate for the systems design problem for free-roaming robot vehicle systems. There are too many alternative configurations and simulating all of them would not be possible. The use of network-based optimisation procedures has also been proposed[17]. Previous work in this area, as far as is known, addressed issues related to an environment where vehicles follow each other along a single, fixed guidepath. The work proposed here will specifically address issues relating to free-roaming robot vehicles. These vehicles will have the ability to overtake one another along aisles, will have collision avoidance capability, and will be operated by an intelligent control system.

Acknowledgement

Work described in this article is being supported by the National Science Foundation under Contract CDR 8500022.

References

[1] Solberg, J. J. 1977. A mathematical model of computerized manufacturing systems. In, *Proc. 5th Int. Conf. on Production Research*, Tokyo, Japan.

[2] Solberg, J. J. 1980. *The CAN-Q user's guide.* School of Industrial Engineering, Purdue University.

[3] Solberg, J. J. and Nof, S. Y. 1980. Analysis of flow control in alternative manufacturing configurations. *Journal of Dynamic Systems, Measurement and Control,* 102: 151-157.

[4] Solberg, J. J. 1981. Capacity planning with stochastic workflow model. *AIIE Transactions,* 13: 116-122.

[5] Premi S. K. and Besant, C. B. 1983. A review of various vehicle guidance techniques that can be used by mobile robots or AGVS. In, *Proc. 2nd Int. Conf. on Automated Guided Vehicle Systems*, Vol.2, Stuttgart.

[6] Marce, L., Julliere, M. and Place, H. 1981. An autonomous computer-controlled vehicle. In, *Proc. 1st Int. Conf. on Automated Guided Vehicle Systems,* Stratford-upon-Avon, UK.

[7] Fujiwara, K., Kawashima, Y., Kato, H. and Watanabe, M. 1981. The development of guideless robot vehicles. In, *Proc. 11th Int. Symp. on Industrial Robots,* Tokyo, Japan, October 7-9, 1981.

[8] *Material Handling News,* 321 (September 1985). AGV that roams free.

[9] Kulweic, R. 1984. Guided vehicles controlled by LEDs, not wires. *Plant Engineering,* 38(12).

[10] Stevens, P., Rabins, M., Roberts, M. and Gek, V. K. 1983. Truck Location Using Retroreflectance Strips and Triangulation with Laser Equipment (TURTLE). In, *Proc. 2nd European Conf. on Automated Manufacturing,* Birmingham, UK.

[11] Moravec, H. 1983. The Stanford Cart and the CMU Rover. *Proc. IEEE,* 71(7).

[12] Rappaport, T. S. 1985. The design and development of a mobile robot location system. M.S.E.E. Thesis, Purdue University.

[13] Egbelu, P. J. and Tanchoco, J. M. A. 1985. Characterization of automatic guided vehicle dispatching rules. *International Journal of Production Research* 22(3).

[14] Egbelu, P. J. and Tanchoco, J. M. A. 1986. Potential for bidirectional guidepaths for automatic guided vehicle based systems. *International Journal of Production Research.*

[15] Egbelu, P. J. and Tanchoco, J. M. A. 1983. Design of automatic guided vehicle systems using AGVSim. In, *Proc. 2nd Int. Conf. on Automated Guided Vehicle Systems, Stuttgart.*

[16] Phillips, D.T. 1980. Simulation of material handling systems: When and which methodology. *Industrial Engineering,* 12(9).

[17] Maxwell, W. L. and Muckstadt, J. A. 1982. Design of automatic guided vehicle systems. *IIE Transactions,* 15(2)

NEW AGV WITH REVOLUTIONARY MOVEMENT

S. Jonsson
Mecanum AB, Sweden

The latest trends in flexible manufacturing systems (FMS) point to the use of grid systems and intelligent AGVs. The revolutionary Mecanum AGV presented in this paper makes this fully realistic. The resulting savings in time and space, together with the increase in flexibility, will usher in a new era for FMS.

The invention of the wheel is justly considered to be one of the most important inventions of mankind. However, the traditional wheel has one weakness – it only moves forwards/backwards. Mecanum Movement, however, is based on a unique steering principle that makes it possible to drive a vehicle in any direction from 0-360°, including sideways, at an angle of 90°. The key components in Mecanum Movement are the Mecanum wheels and the Mecanum steering systems. The Mecanum wheel is made up of a number of free-rolling rollers placed at an angle of 45° on a rim (Fig. 1). The Mecanum steering system controls a vehicle by altering the speed and/or direction of one or more wheels in relation to the other wheels.

Mecanum Movement – a unique and revolutionary innovation – is used today for a number of special vehicle functions: a wheelchair, a carrier for handling track-laying vehicle chassis, a cleaning vehicle, a working platform, a storage lift truck for X-ray film handling, a cable drum carrier, and a manipulator for precision assembly of heavy vehicle wheels and engines, etc.

The use of Mecanum Movement in an automated guided vehicle (AGV) raises a whole series of exciting possibilities. Within the field of flexible manufacturing systems (FMS) it is now possible to attain a far greater degree of flexibility, to save a considerable amount of workshop space, and to increase the speed and efficiency of AGV activities.

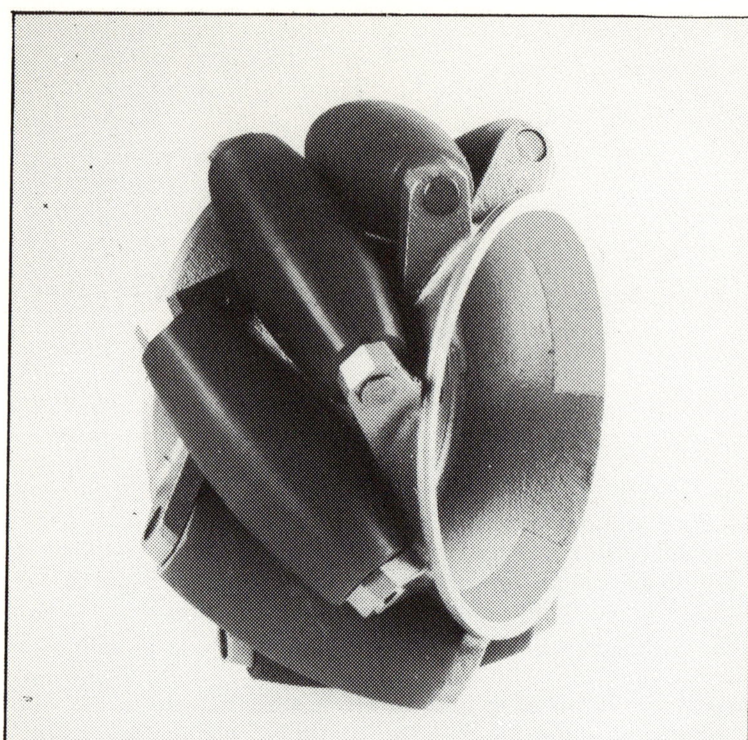

Fig. 1   The Mecanum wheel

The advantages of using an AGV based on Mecanum Movement

Flexible manufacturing systems are designed according to a host of different principles and varying transport needs. At the simplest level, FMS can be used to transport a pallet to or from a group of machines. This scheme can be augmented to include the transportation of tools, etc. But one problem that crops up is that the collection and delivery stations have to be situated so that the vehicle has access.

Traditional trucks have the important restriction that they are unable to collect or deposit materials unless the vehicle is directed towards the collection or delivery port or station. Using a truck based on Mecanum Movement, this restriction is removed. A Mecanum AGV can travel in any direction and is able to collect and deliver materials regardless of the direction of travel of the vehicle. This is an important breakthrough, since it means that machines can be situated without having to leave broad aisles or passageways for truck access, and the workshop planner or FMS designer does not need to concentrate on workshop layout based on the limitations of traditional vehicles.

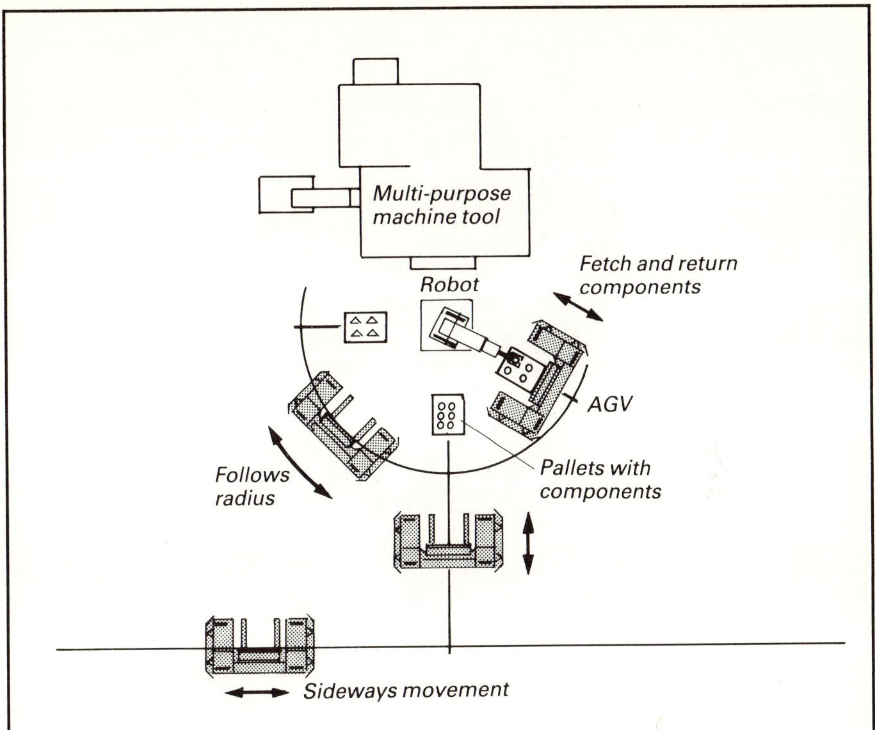

Fig. 2   Multi-purpose machine tool with robot arm served by Mecanum AGV

The following gives some examples of what this *new flexibility* can mean.

Fig. 2 shows a multi-purpose machine tool with a robot arm. The machine is served by an AGV which, because it is based on Mecanum Movement, is able to change its direction of travel without physically rotating. It thus requires no turning radius. The AGV can be programmed to travel in any direction, in an arc, or to rotate, depending on the specific needs of the station. An AGV based on Mecanum Movement is far better suited to this function than other, traditional truck types.

Another example of the way that a Mecanum Movement-based AGV can give increased flexibility and hence efficiency is in connection with lateral movement of pallets, as can be seen in Fig. 3. Having addressed the machine, the vehicle is able to travel directly in a lateral direction without the need to rotate the entire vehicle.

Thanks to Mecanum Movement, a Mecanum AGV has several advantages. It is capable of remarkable precision, speed and efficiency. The lateral movement that is possible using Mecanum Movement minimises transport times. Since a Mecanum AGV has no turning radius, considerable space saving is possible. Two examples of this last point are given here.

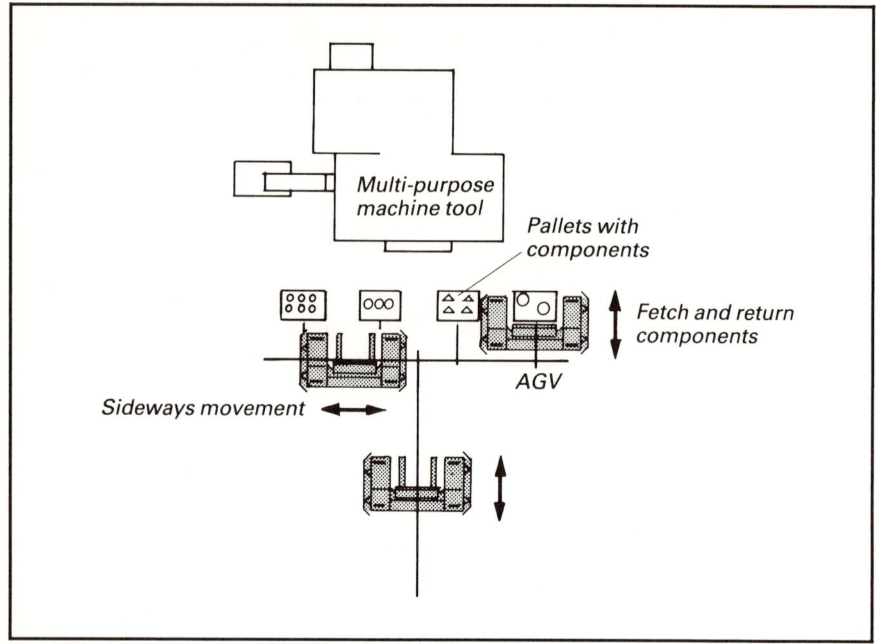

Fig. 3   Lateral movement of pallets by Mecanum AGV

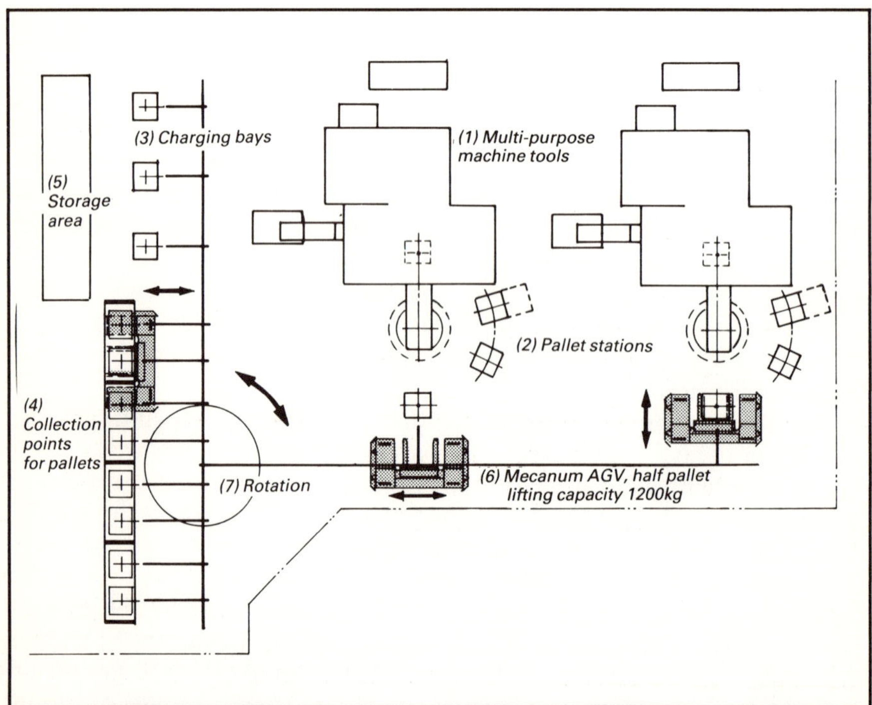

Fig. 4   FMS machining cell with Mecanum AGV half-pallet handling

Fig. 4 shows an FMS cell with AGV half-pallet handling. The Mecanum AGV, with a lifting capacity of 1200kg, here serves two multi-purpose machine tools. After collecting a pallet from the collection point (4), the AGV rotates 90° (7) and moves to the first station (2), where the half-pallet is deposited. From the pallet station, the pallet is moved into position for the machine tool (1). On completion of the operation the pallet is collected from the station. One AGV can serve two or more machines from one and the same storage area. Note the charging bays (3) and the storage area (5).

Fig. 5 shows another FMS cell, this time for heavier materials. The machines here are an NC-controlled carousel lathe equipped with a pallet changer, and a multi-purpose machine tool also equipped with a pallet changer (1). Here, a Mecanum AGV with a lifting capacity of 3500kg is used to collect pallets from the store and serve them to the respective machines. Notice how compact the cell is. This is a direct result of the fact that no turning radii are necessary. The fact that a Mecanum AGV is also capable of side-loading is both a space-saver and a time-saver. This side-loading capacity also provides the system with a great back-up potential for expansion, and reduces the number of vehicles needed – a considerable long-term investment saving.

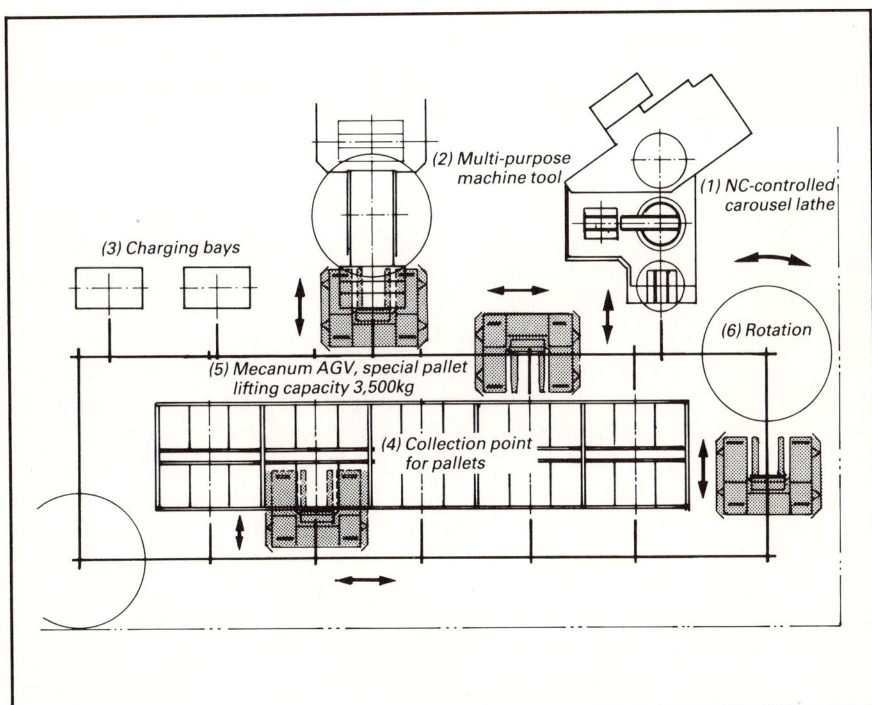

Fig. 5   Mecanum AGV system for handling heavy materials in an FMS machining cell

The Mecanum AGV

The *Mecanum wheel* is 300mm in diameter, and is made up of a number of free-rolling polyurethane rollers placed at an angle of 45° on a nodular iron rim. The roller axle is made of hard chrome steel, and the bearings are lubricant-free gliding bearings. The rollers are situated in such a way that a side view of the wheel forms a circle. The surface that is in contact with the ground travels from one end of the roller, along its length to its inner edge, and on to the next roller. This construction means that the only resistance occurs when the roller is pressed to the ground. This results in a low degree of wear to both the rollers and the ground surface.

Mecanum wheels are designed for both left and right mounting, and are set out in a specific pattern in relation to each other.

The *power system* consists of a direct-mounted transmission gear on each wheel. A dc permanent magnet motor is mounted in each transmission. The motors are together 4×1.2kW, and turn at 3000rpm at top speed.

Electromagnetic brakes are used for parking or for emergency stops. Braking is normally done using the motors. There are two batteries which can be charged automatically or manually. The battery effect is 207 Ah/5h, 80V.

Using the *Mecanum steering system*, the AGV can follow a painted line on the floor in any direction. To do this, two optical sensors are used which are coupled to the vehicle's automatic system for regulating the power to the wheels. The AGV has its own computer which registers the vehicle's position in relation to reference points in the floor. In addition to information concerning the vehicle's route, the computer also governs such information as collection and depositing points and heights, choice of alternative route, and – using a coding system on pallets – the nature of the material being carried. Altogether, this constitutes a simple, effective and safe system for use in FMS.

The AGV is capable of controlled acceleration and deceleration. Great precision is possible, thanks to the advanced nature of the steering system. Using three basic movements and three superimposed ones (see Fig. 6), an infinite number of manoeuvres is possible. In addition to its automatic function, each Mecanum AGV is provided with a hand-held override manoeuvring unit with a joystick and buttons for driving the AGV in any direction. A series of signal lamps indicate direction and malfunctions.

The chassis construction is designed to compensate for unevenness in the floor surface. Vehicle service is made easy by direct access hatches to all major components: computer, batteries, etc.

The hydraulic fork-lift is of a pillar type (Fig. 6). The forks can be set in two positions. The maximum lifting height is 3.2m.

Two-way radio communication is possible between the AGV's

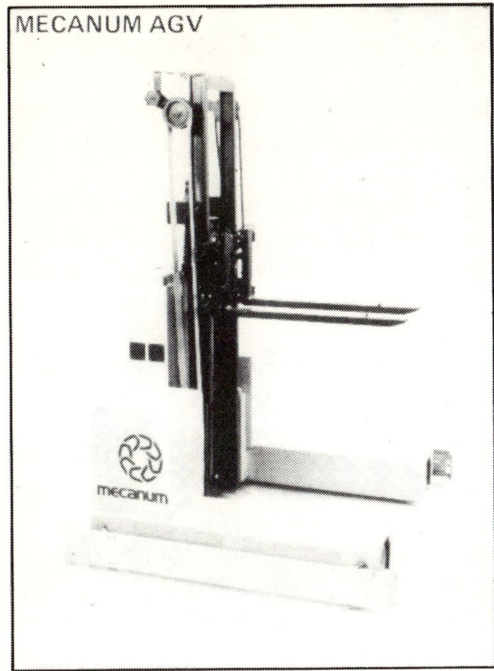

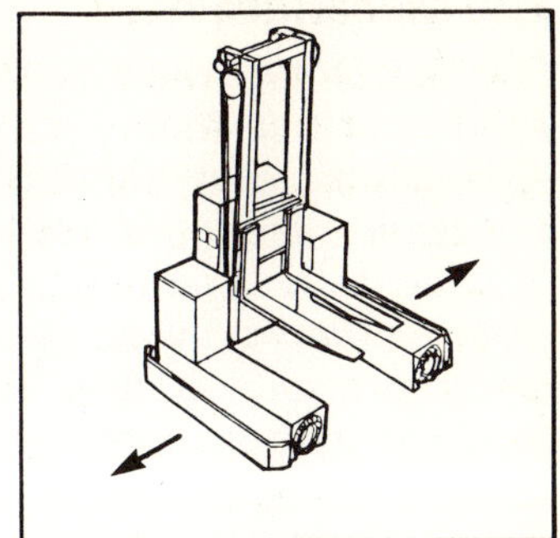

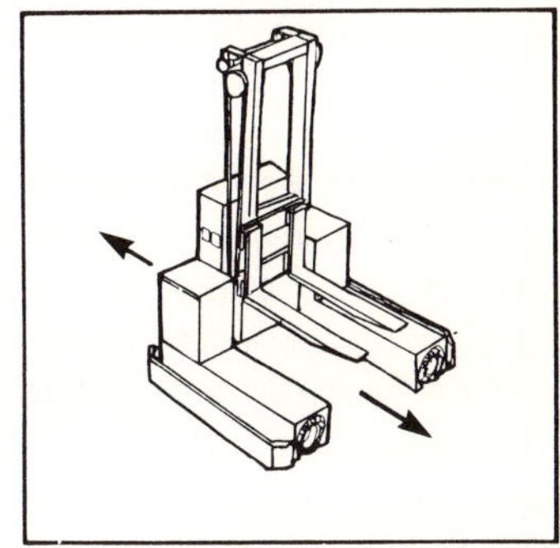

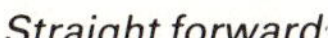

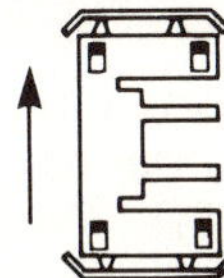

All wheels are driven at the same speed and in the same direction

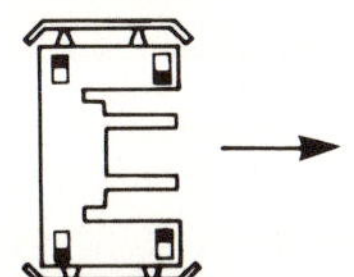

The wheels on the right side are driven towards each other at the same speed while the wheels on the left side are driven away from each other

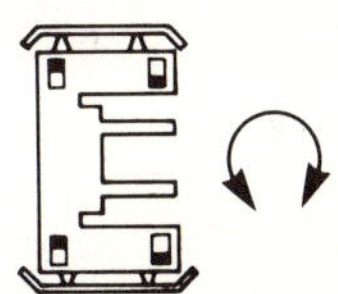

The wheels rotate at the same speed and in the same direction in which the vehicle is to rotate

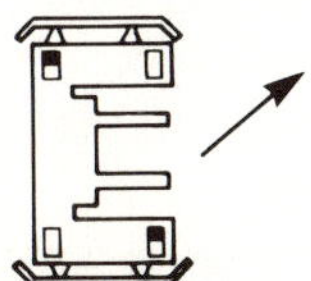

Same forward power to two diagonally opposed wheels, no power to the other wheels. The vehicle moves at an angle of 45°

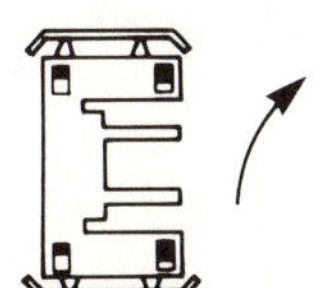

Two wheels on one side of the vehicle with full forward power, the other two with less power. The vehicle moves in a curve

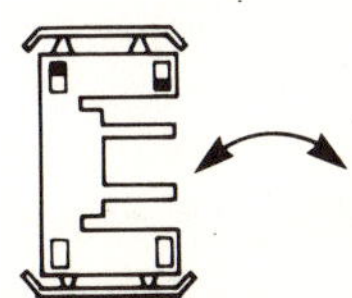

One front wheel forward power, the other reverse power, the rear wheels with no power. The vehicle moves in a lateral arc

Fig. 6   A Mecanum AGV (hydraulic fork lift) and its basic movements

internal computer and the main FMS system. This gives great flexibility for changing routes, in case of emergency, etc., while the vehicle is under way.

To ensure safety there is an anti-collision device which activates the emergency brakes if the vehicle should run into any object. The braking distance at full speed in any direction is less than the distance from the anti-collision sensor plates to the vehicle itself. In order to avoid such sudden stops, however, the AGV is equipped with sensors that register any object closer than 1.5–2m from the vehicle. A registration results in a reduction in speed, so that any following emergency stop will be less drastic.

If the AGV loses contact with the grid lines it stops automatically. The AGV's computer monitors the condition of the grid line, providing a warning if the lines are wearing out.

Other monitoring functions in the AGV include battery status, the travelling area for the AGV's lifting forks, correct positioning of loads on pallets on collection or depositing, temperature levels and levels in the hydraulic system, and the temperature in the computer housing.

The Mecanum AGV has a system for precision positioning which operates to a tolerance of ± 4mm for collection or depositing. This system operates using the Mecanum wheels in direct contact with optical sensors.

Principles for steering and navigating the AGV

The system is based on two computers, one located in the AGV itself and one central FMS computer. One of the main aims of an FMS system is to attain simplicity and flexibility. The simplest FMS layout consists of a grid with reference points. Lines should be painted on the floor which can be read by optical sensors in the AGV. Reference points – usually located at the intersections of grid lines – relay information from the vehicle to the FMS computer as to the AGV's position. Instructions are accordingly sent to the AGV's steering system.

A painted line grid is far more flexible than a loop wire or other system. An AGV with Mecanum Movement is well suited to working in a painted line grid since it is capable of side-loading, can travel in any direction, and requires no turning radius.

FMS software and the 'factory of the future'

Of course, the weakest link in the chain is the main FMS program, or rather the complexity of altering FMS functions. The Swedish Institute of Production Engineering Research (IVF) is fully engaged in researching this and other related problems. In its new 'factory of the future', the Institute has installed a multi-operating machine tool, a robot and a Mecanum AGV. The author is fully convinced that user-friendly software will soon be available. This will allow production supervisors to reprogram their FMS functions without recourse to outside expertise. This will be a great breakthrough, resulting in a maximising of the flexibility and efficiency of the FMS/AGV concept.

As one can imagine, the direct savings involved are quite remarkable. The factories of the future will be rationally designed and have the flexibility that modern production technology demands. One great step in this direction is the combination of new user-friendly FMS software and intelligent AGVs using Mecanum Movement.

Source of material

AGVS in Europe – Current techniques and future trends
First presented at the Executive Briefing on Automated Guided Vehicles, 6–7 November 1986, Stratford-upon-Avon, UK. Reprinted courtesy of the author and IFS (Conferences) Ltd.

Evolutionary AGVS – From concept to present reality
First presented at the Executive Briefing on Automated Guided Vehicles, 6–7 November 1986, Stratford-upon-Avon, UK. Reprinted courtesy of the author and IFS (Conferences) Ltd.

Automated guided vehicles
First presented at the 1st International Conference on Automated Materials Handling, 20–22 April 1983, London, UK. Reprinted courtesy of the author and IFS (Conferences) Ltd. Updated July 1986.

Economic and management considerations for AGV system purchase
First presented at the 3rd International Conference on Automated Guided Vehicle Systems, 15–17 October 1985, Stockholm, Sweden, under the title 'So you want to buy an AGV system'. Reprinted courtesy of the author and IFS (Conferences) Ltd.

Socio-technical considerations for AGV system implementation
First presented at the 1981 Spring Annual Conference and World Productivity Congress, USA. Reprinted courtesy of the author and the American Institute of Industrial Engineers, Inc.

AGV system safety – New developments to meet changing needs
Not previously published.

AGV designs
Not previously published.

A comparative evaluation of AGV navigation techniques
First presented at the 3rd International Conference on Automated Guided Vehicle Systems, 15–17 October 1985, Stockholm, Sweden.
Reprinted courtesy of the author and IFS (Conferences) Ltd.

Centralised and decentralised control of AGVs
First presented at the 3rd International Conference on Automated Guided Vehicle Systems, 15–17 October 1985, Stockholm, Sweden.
Reprinted courtesy of the author and IFS (Conferences) Ltd.

Design parameters and specification of an AGV system
First presented at the Executive Briefing on Automated Guided Vehicles, 6–7 November 1986, Stratford-upon-Avon, UK.
Reprinted courtesy of the author and IFS (Conferences) Ltd.

Capacity, availability and flexibility of AGV systems
First presented at the 3rd International Conference on Automated Guided Vehicle Systems, 15–17 October 1985, Stockholm, Sweden.
Reprinted courtesy of the author and IFS (Conferences) Ltd.

Operational characteristics of AGV systems
First presented at AIEE 1982 (NSF Report No. PB82-232646), under the title 'Design of automatic guided vehicle systems'.
Reprinted courtesy of the authors and National Science Foundation.

Characterisation of AGV dispatching rules
First published in the International Journal of Production Research. Vol. 22, No. 3, pp.359–374, 1984.
Reprinted courtesy of the authors and Taylor & Francis Ltd.

Simulation – An aid to the design of AGV networks
Not previously published.

Evaluating AGV circuits by simulation
First presented at the 3rd International Conference on Automated Guided Vehicle Systems, 15–17 October 1985, Stockholm, Sweden.
Reprinted courtesy of the authors and IFS (Conferences) Ltd.

Automated material handling in FMS and FAS
First presented at the 2nd International Conference on Automated Materials Handling, 15–17 May 1985, Birmingham, UK.
Reprinted courtesy of the author and IFS (Conferences) Ltd.

AGV systems in FMS
First presented at the 2nd International Conference on Flexible
Manufacturing Systems, 26–28 October 1983, London, UK,
under the title 'Flexible manufacturing systems with AGVs'.
Reprinted courtesy of the author and IFS (Conferences) Ltd.

Flexible manufacturing systems with automated material handling
First presented at the 2nd International Conference on Flexible
Manufacturing Systems, 26–28 October 1983, London, UK.
Reprinted courtesy of the author and IFS (Conferences) Ltd.
Updated November 1986.

Design and use of automated material flow systems
Not previously published.

**Mobile robots – Key elements for the integration of transport and
handling functions**
First presented at the 15th International Symposium on Industrial
Robots, 11–13 September 1985, Tokyo, Japan.
Reprinted courtesy of the authors and the Japan Industrial Robot
Association.

Flexible assembly through AGVs in the automotive industry
Not previously published.

Automated warehousing using AGVs and narrow-aisle trucks
First presented at the 3rd International Conference on Auto-
mated Materials Handling, 19–21 March 1986, Birmingham, UK.
Reprinted courtesy of the author and IFS (Conferences) Ltd.

Automated material handling in a distribution centre
First presented at the 1st International Conference on Auto-
mated Materials Handling, 20–22 April 1983, London, UK.
Reprinted courtesy of the author and IFS (Conferences) Ltd.

Load transfer techniques for AGV systems
Not previously published.

Automated guided fork-lift truck system
First presented at the 4th International Conference on Automa-
tion in Warehousing, 30 September – 2 October 1981, Tokyo,
Japan.
Reprinted courtesy of the author and the Japan Industrial Robot
Association.
Updated September 1986.

Free-ranging AGV and scheduling system
First presented at the 3rd International Conference on Automated Guided Vehicle Systems, 15–17 October 1985, Stockholm, Sweden.
Reprinted courtesy of the authors and IFS (Conferences) Ltd.

Advanced automated transportation system
First presented at the 3rd International Conference on Flexible Manufacturing Systems, 11–13 September 1984, Boeblingen, West Germany.
Reprinted courtesy of the authors and IFS (Conferences) Ltd.
Updated August 1986.

A concept for an autonomous mobile robot and a prototype model
First presented at the International Conference on Advanced Robotics, 9–10 September 1985, Tokyo, Japan.
Reprinted courtesy of the authors and the Japan Industrial Robot Association.

An intelligent factory transport system
First presented at the 3rd International Conference on Automated Materials Handling, 19–21 March 1986, Birmingham, UK.
Reprinted courtesy of the authors and IFS (Conferences) Ltd.

New AGV with revolutionary movement
First presented at the 3rd International Conference on Automated Guided Vehicle Systems, 15–17 October 1985, Stockholm, Sweden.
Reprinted courtesy of the author and IFS (Conferences) Ltd.

Authors' organisations and addresses

M. Annborn
AB Bygg-och Transport-
  ekonomi (BT)
Box 331
S-595 00 Mjölby
Sweden

P. Boegli
Schindler-Digitron AG
Erlenstrasse 32
CH-2555 Brugg-Biel
Switzerland

J. Burton
Ingersoll Engineers Inc.
Bourton Hall
Rugby
Warwicks CV23 9SD
England

D. A. Clayton
Jungheinrich (GB) Ltd
Southmoor Road
Manchester M23 9DU
England

B. Duffau
Seri Renault Automation
2 Avenue du Vieil-Etang
Montigny-le-Bretonneux
  (Yvelines)
BP 19
F-78391 Bois-D'Arcy Cedex
France

P. J. Egbelu
College of Engineering
Department of Industrial &
  Management Systems
  Engineering
The Pennsylvania State
  University
207 Hammond Building
University Park, PA 16802
USA

G. C. Hammond
GMI Engineering and
  Management Institute
1700 West Third Avenue
Flint, MI 48502–2276
USA

G. Handke
Automation Division
Industrial Products Group
Messerschmitt-Bölkow-Blohm
  GmbH (MBB)
PO Box 80 11 09
8000 München 80
West Germany

R. Hollingworth
Istel Ltd
Highfield House
Headless Cross Drive
Headless Cross
Redditch
Worcs B97 5EU
England

H. Imai
Manufacturing Development
 Laboratory
Mitsubishi Electric Corp.
1-1 Tsukaguchi Honmachi
 8 chome
Amagasaki
Hyogo 661
Japan

U. Jansson
Schindler-Digitron
Highfield House
Headless Cross Drive
Headless Cross
Redditch
Worcs B97 5EU
England

S. Jonsson
Mecanum AB
Domarevaegen 5
S-902 52 Umeau
Sweden

S. Karlsson
AB Bygg-och Transport-
 ekonomi (BT)
Box 331
S-595 00 Mjölby
Sweden

H. Lindgren
SattControl AB
Box 9034
20039 Malmo
Sweden

W. L. Maxwell
College of Engineering
School of Operations Research &
 Industrial Engineering
Cornell University
Upson Hall
Ithaca, NY 14853–7501
USA

P. McEllin
Materials Handling & Factory
 Automation
SKF Engineering Products Ltd
North Crawley Road
Newport Pagnell
Bucks MK16 9HB
England

T. Müller
EB 01
Fachgebeit Fordertechnik
Fachhochschule Hamburg
Berliner Tor 21
D-2000 Hamburg 1
West Germany

J.-O. Myhrman
MTS-Systems Norden AB
Box 1118
S-43699 Askim
Sweden

Y. Nakano
Mechanical Engineering
 Research Laboratory
Hitachi Ltd
502 Kandatu-machi
Tsuchiura-shi 300
Japan

I. A. Noble
Automated Systems Division
Jungheinrich (GB) Ltd
Southmoor Road
Manchester M23 9DU
England

I. Persson
ACS Autocarrier System AB
Argongatan 30
S-431 33 Mölndal
Sweden

S. G. Sanden
ACS Autocarrier System AB
Argongatan 30
S-431 33 Mölndal
Sweden

F. Schneider
Schindler-Digitron AG
Erlenstrasse 32
CH-2555 Brugg-Biel
Switzerland

R. D. Schraft
Fraunhofer-Institut für
    Produktionstechnik und
    Automatisierung (IPA)
Nobelstrasse 12
D-7000 Stuttgart 80
West Germany

J. J. Solberg
Engineering Research Center for
    Intelligent Manufacturing
    Systems
Purdue University
A. A. Potter Engineering Center
West Lafayette, IN 47907
USA

H. Takanami
Automatic Material Handling
    System Department
Komatsu Forklift Co. Ltd
No. 3-4
Akasaka 2-chome
Minato-ku
Tokyo
Japan

P. M. Tracey
Herga Electric Ltd
Northern Way
Bury St. Edmunds
Suffolk IP32 6NN
England

U. Wältken
VVA-Vereinigte
    Verlagsauslieferung GmbH
Postfach 7777
4830 Gütersloh 100
West Germany

S. P. Walker
Imperial College of Science and
    Technology
Department of Mechanical
    Engineering
Exhibition Road
London SW7 2BX
England